Metal-Based Composite Materials: Preparation, Structure, Properties, and Applications

Metal-Based Composite Materials: Preparation, Structure, Properties, and Applications

Editors

Andrey Suzdaltsev
Oksana Rakhmanova

Basel • Beijing • Wuhan • Barcelona • Belgrade • Novi Sad • Cluj • Manchester

Editors
Andrey Suzdaltsev
Scientific Laboratory of the
Electrochemical Devices and
Materials
Ural Federal University
Ekaterinburg
Russia

Oksana Rakhmanova
Laboratory of Electrode
Processes
Institute of High-Temperature
Electrochemistry UB RAS
Ekaterinburg
Russia

Editorial Office
MDPI
St. Alban-Anlage 66
4052 Basel, Switzerland

This is a reprint of articles from the Special Issue published online in the open access journal *Applied Sciences* (ISSN 2076-3417) (available at: www.mdpi.com/journal/applsci/special_issues/Metal_Based_Materials).

For citation purposes, cite each article independently as indicated on the article page online and as indicated below:

Lastname, A.A.; Lastname, B.B. Article Title. *Journal Name* **Year**, *Volume Number*, Page Range.

ISBN 978-3-0365-8953-4 (Hbk)
ISBN 978-3-0365-8952-7 (PDF)
doi.org/10.3390/books978-3-0365-8952-7

© 2023 by the authors. Articles in this book are Open Access and distributed under the Creative Commons Attribution (CC BY) license. The book as a whole is distributed by MDPI under the terms and conditions of the Creative Commons Attribution-NonCommercial-NoDerivs (CC BY-NC-ND) license.

Contents

About the Editors vii

Andrey Suzdaltsev and Oksana Rakhmanova
Special Issue on Metal-Based Composite Materials: Preparation, Structure, Properties and Applications
Reprinted from: *Appl. Sci.* **2023**, *13*, 4799, doi:10.3390/app13084799 1

Julian Popp, David Römisch, Marion Merklein and Dietmar Drummer
Joining of CFRT/Steel Hybrid Parts via Direct Pressing of Cold Formed Non-Rotational Symmetric Pin Structures
Reprinted from: *Appl. Sci.* **2022**, *12*, 4962, doi:10.3390/app12104962 5

Polina Viktorovna Polyakova, Julia Alexandrovna Pukhacheva, Stepan Aleksandrovich Shcherbinin, Julia Aidarovna Baimova and Radik Rafikovich Mulyukov
Fabrication of Magnesium-Aluminum Composites under High-Pressure Torsion: Atomistic Simulation
Reprinted from: *Appl. Sci.* **2021**, *11*, 6801, doi: 23

Roberto Ademar Rodríguez Díaz, Sergio Rubén Gonzaga Segura, José Luis Reyes Barragán, Víctor Ravelero Vázquez, Arturo Molina Ocampo and Jesús Porcayo Calderón et al.
The Synthesis of Aluminum Matrix Composites Reinforced with Fe-Al Intermetallic Compounds by Ball Milling and Consolidation
Reprinted from: *Appl. Sci.* **2021**, *11*, 8877, doi:10.3390/app11198877 34

Alexander Smirnov, Evgeniya Smirnova, Anatoly Konovalov and Vladislav Kanakin
Using the Instrumented Indentation Technique to Determine Damage in Sintered Metal Matrix Composites after High-Temperature Deformation
Reprinted from: *Appl. Sci.* **2021**, *11*, 10590, doi:10.3390/app112210590 53

Timofey Gevel, Sergey Zhuk, Natalia Leonova, Anastasia Leonova, Alexey Trofimov and Andrey Suzdaltsev et al.
Electrochemical Synthesis of Nano-Sized Silicon from $KCl–K_2SiF_6$ Melts for Powerful Lithium-Ion Batteries
Reprinted from: *Appl. Sci.* **2021**, *11*, 10927, doi:10.3390/app112210927 67

Aleksander A. Chernyshev and Evgenia V. Nikitina
A Porous Tungsten Substrate for Catalytic Reduction of Hydrogen by Dealloying of a Tungsten–Rhenium Alloy in an Aqueous Solution of Hydrochloric Acid
Reprinted from: *Appl. Sci.* **2022**, *12*, 1029, doi:10.3390/app12031029 79

Alexander Y. Galashev, Dmitriy S. Pavlov, Yuri P. Zaikov and Oksana R. Rakhmanova
Ab Initio Study of the Mechanism of Proton Migration in Perovskite $LaScO_3$
Reprinted from: *Appl. Sci.* **2022**, *12*, 5302, doi:10.3390/app12115302 89

Liudmila A. Yolshina, Aleksander G. Kvashnichev, Dmitrii I. Vichuzhanin and Evgeniya O. Smirnova
Mechanical and Thermal Properties of Aluminum Matrix Composites Reinforced by In Situ Al_2O_3 Nanoparticles Fabricated via Direct Chemical Reaction in Molten Salts
Reprinted from: *Appl. Sci.* **2022**, *12*, 8907, doi:10.3390/app12178907 107

Xiang Chen, Xinjie Peng, Xinjian Yuan, Ziliu Xiong, Yue Lu and Shenghai Lu et al.
Effect of Zinc Aluminum Magnesium Coating on Spot-Welding Joint Properties of HC340LAD + ZM Steel
Reprinted from: *Appl. Sci.* **2022**, *12*, 9072, doi:10.3390/app12189072 . . . . **125**

Denis Petrovich Borisenko, Alexander Sergeevich Gusev, Nikolay Ivanovich Kargin, Petr Leonidovich Dobrokhotov, Alexey Afanasievich Timofeev and Vladimir Arkhipovich Labunov et al.
Few-Layer Graphene as an Efficient Buffer for GaN/AlN Epitaxy on a SiO_2/Si Substrate: A Joint Experimental and Theoretical Study
Reprinted from: *Appl. Sci.* **2022**, *12*, 11516, doi:10.3390/app122211516 . . . . **138**

Anastasia M. Leonova, Oleg A. Bashirov, Natalia M. Leonova, Alexey S. Lebedev, Alexey A. Trofimov and Andrey V. Suzdaltsev
Synthesis of C/SiC Mixtures for Composite Anodes of Lithium-Ion Power Sources
Reprinted from: *Appl. Sci.* **2023**, *13*, 901, doi:10.3390/app13020901 . . . . **155**

Vladimir B. Vykhodets, Tatiana E. Kurennykh and Evgenia V. Vykhodets
Disks of Oxygen Vacancies on the Surface of TiO_2 Nanoparticles
Reprinted from: *Appl. Sci.* **2022**, *12*, 11963, doi:10.3390/app122311963 . . . . **166**

Alexander Smirnov, Vladislav Kanakin and Anatoly Konovalov
Neural Network Modeling of Microstructure Formation in an AlMg6/10% SiC Metal Matrix Composite and Identification of Its Softening Mechanisms under High-Temperature Deformation
Reprinted from: *Appl. Sci.* **2023**, *13*, 939, doi:10.3390/app13020939 . . . . **176**

About the Editors

Andrey Suzdaltsev

Andrey Suzdaltsev received his education from Ural Federal University (high school) and the Institute of High-Temperature Electrochemistry (post-graduate school). Andrey Suzdaltsev completed his PhD thesis on "Anodic processes on carbon in $KF-AlF_3-Al_2O_3$ melts" (2011), and his Dr. Habil. thesis on "Electrode processes at production of aluminum and its alloys in $KF-AlF_3-Al_2O_3$ melts" (2022). His scientific interests are related to the study of the kinetics and mechanisms of electrode and chemical processes during the production of pure metals, alloys, and composite materials in molten salts. He has solved various scientific and practical tasks: establishing the laws of physical and chemical processes in molten salts in order to develop the scientific foundations for new energy-efficient methods and technologies for the synthesis of popular elements, alloys, and materials. In particular, with his direct participation, the fundamentals of technologies for the electrolytic production of aluminum alloys and master alloys from oxide raw materials were developed. At present, the scientific laboratory under the leadership of Suzdaltsev Andrey deals with the establishment of laws governing the electrochemical synthesis of silicon and materials based on it for microelectronics and devices for converting and storing energy.

Oksana Rakhmanova

Dr. Oksana Rakhmanova graduated from the Ural State University (Yekaterinburg, Russia) in 2004 and obtained her degree in physics. During 2004-2008, she was a PhD student at the Institute of industrial ecology of the UB RAS. In 2009, she defended her dissertation and obtained her PhD degree in Physics. Her scientific work was devoted to the investigation of the absorption mechanisms of water clusters in the atmosphere interacting with different greenhouse (CO_2, N_2O, CH_4, O_3) and atmospheric (N_2, O_2, C_2H_2, C_2H_6) gases using the the method of molecular dynamics (MDs) simulation. During 2009-2015, she investigated the properties of two-dimensional structures, such as graphene and mechanisms of its cleaning from heavy metals (Pb, Hg) by bombarding the surface with Ar stream with different energies. In 2015, she started working for the Institute of High-Temperature Electrochemistry of the UB RAS (IHTE UB RAS) as a senior researcher. Since then, one of her scientific interests has been connected with the MD investigation of silicene as a promising anode material for lithium-ion batteries (LIBs) of a new generation. Her other scientific theme is devoted to MD searching the imitators for actinides (Pu, Am, Cm) in FLiNaK molten salt in the application to molten salt reactors. The dynamic, kinetic, and structural properties of molten FLiNaK with different lanthanides fluorides (Ce, Nd, La) are simulated. The model is verified according to the experimental data (density, viscosity, thermal conductivity). And the next step of the MD model is the prediction of the properties of molten salts containing actinides fluorides in different concentrations. Dr. Rakhmanova is the co-author of 72 WoS publications, her h-index is 12, and she is the co-Editor-in-Chief of the scientific journal *Electrochemical Materials and Technologies* published at the Ural Federal University.

Editorial

Special Issue on Metal-Based Composite Materials: Preparation, Structure, Properties and Applications

Andrey Suzdaltsev [1,2,*] and Oksana Rakhmanova [1,2]

1. Laboratory of the Electrochemical Devices and Materials, Ural Federal University, Mira St. 28, 620002 Yekaterinburg, Russia; oksana_rahmanova@mail.ru
2. Institute of High Temperature Electrochemistry, Ural Branch of the Russian Academy of Sciences, Akademicheskaya St. 20, 620990 Yekaterinburg, Russia
* Correspondence: suzdaltsev_av@mail.ru

The Special Issue is aimed at analyzing modern trends and recent advances in the synthesis of new metal-based composite materials. Such composites are increasingly used in civil, automotive, and aerospace engineering, shipbuilding, robotics, nuclear power, portable energy devices, biomedicine, electronic devices, and portable aircrafts.

Non-ferrous metals are often used as the matrix of composites, aluminum, magnesium, nickel, titanium and their alloys, and can act as modifiers with boron, carbon structures, borides, carbides, nitrides, and oxides of refractory metals and high-strength steel. For high-temperature composites, tungsten or molybdenum fibers are used. Despite the large number of scientific works in this area, new methods for the synthesis of such composites in order to improve and optimize their structure and properties are still needed. In this regard, experimental and theoretical works aimed at developing and optimizing methods for the synthesis of composite materials, as well as the search for new materials, have been successfully published in this Special Issue.

Thirteen papers devoted to the development of composite materials for different applications are published in this Special Issue [1–13]. Among them, the main scientific topics include the following:

- Light metals (Al)-based materials and alloys (Al–Mg, Al–Mg–Zn) [1–3,8,9,13] with modifying and reinforced additions (Fe [2], SiC [3,13], Al_2O_3 [8]) used as structural materials and coatings [9], as well as the steel-filament-reinforced carbon fiber for general use [6];
- Silicon-containing materials for lithium-ion batteries with increased specific capacity [4,12];
- Oxide materials for modern energetic devices [7,11];
- Multi-layer structures based on graphene for the synthesis of semiconductive materials [10];
- Materials for catalysis [5].

The applicability of these developed materials goes far beyond the declared topics of the Special Issue. Here, we provide a brief description of the published papers. The authors of work [1], using a molecular dynamics method, aim to investigate the details of the fabrication techniques of an Al–Mg composite with improved mechanical properties. It is shown that shear strain has a crucial role in the mixture process. According to the tensile tests, a fracture occurred in the Mg part of the final composite sample, which means that the interlayer region, where the mixing of Mg and Al atoms is observed, is much stronger than the pure Mg part.

The aim of work [2] is to investigate the microstructural and mechanical properties of a nanostructured Al-based matrix reinforced with Fe40Al intermetallic particles. The results indicate that the hardness varies linearly with the increase in the concentration

Citation: Suzdaltsev, A.; Rakhmanova, O. Special Issue on Metal-Based Composite Materials: Preparation, Structure, Properties and Applications. *Appl. Sci.* **2023**, *13*, 4799. https://doi.org/10.3390/app13084799

Received: 28 March 2023
Accepted: 7 April 2023
Published: 11 April 2023

Copyright: © 2023 by the authors. Licensee MDPI, Basel, Switzerland. This article is an open access article distributed under the terms and conditions of the Creative Commons Attribution (CC BY) license (https://creativecommons.org/licenses/by/4.0/).

of the Fe40Al intermetallic phase present in the composite material, and it grows almost linearly with the rising dislocation density and with the reduction in the grain size.

Paper [3] shows the applicability of data on the evolution of the elastic modulus measured by the instrumented microindentation technique to determine the accumulated damage in metal matrix composites under high temperature deformation. A composite material with the 7075 aluminum alloy matrix and SiC reinforcing particles is studied here. The results show that at the deformation temperature of 500 °C, the plastic properties of the material are significantly lower than at 300 and 400 °C.

The work [4] takes into consideration the questions regarding the electrolytic production of nanosized silicon from low-fluoride molten salts. The obtained silicon deposits are used to fabricate a composite Si/C anode for a lithium-ion battery. The galvanostatic cycling of the anode half-cells reveals a better capacity retention and higher coulombic efficiency of the Si/C composite based on silicon synthesized from a $KCl-K_2SiF_6-SiO_2$ melt. The capacity of the obtained material is estimated in the paper.

In work [5], a selective dissolution of a tungsten (85 wt.%)–rhenium (15 wt.%) alloy, with rhenium in hydrochloric acid at the temperature of 298 K, and anodic polarization modes are carried out. A thermodynamic description of the processes occurring during the anodic selective dissolution of a binary alloy is proposed. The existence of a bimodal structure on the tungsten surface after dealloying is proven.

The work [6] focuses the reader's attention to the investigation of the quasi-unidirectional continuous fiber reinforced thermoplastics joined with metal sheets. The unique novelty of the method is the use of non-rotational symmetric pin structures for material production. The created samples are consequently mechanically tested, and the failure behavior of the single lap shear samples is investigated.

Manuscript [7] provides results on the ab initio molecular dynamics studies of the mechanism of proton motion in a $LaScO_3$ perovskite crystal. It is shown that initial location and interaction between the proton and its nearest environment are of great importance to the character of the proton movement, while the magnitude and direction of the initial velocity and electric field strength are secondary factors characterizing its movement through the $LaScO_3$ crystal.

Paper [8] presents a new method for the industrial production of metal-matrix composites with improved properties, and an aluminum matrix reinforced by "in situ" $\alpha\text{-}Al_2O_3$ nanoparticles is fabricated and tested. Under static uniaxial tension, the cast aluminum composites containing aluminum oxide nanoparticles demonstrate the increased tensile strength, yield strength, and ductility. The microhardness and tensile strength of the composite material are 20–30% higher than those of the metallic aluminum.

The goal of article [9] is the consideration of the Zn–Al–Mg-coated steel HC340LAD + ZM due to its excellent corrosion resistance, self-healing properties, and good surface hardness. It is shown that the presence of the Zn–Al–Mg coating slightly affects the mechanical properties of welding joints. The corrosion current of the body material containing Zn–Al–Mg plating is 7.17 times that of the uncoated plate.

In the paper [10], single- and multi-layer graphene sheets are obtained on a highly textured Cu substrate using the chemical vapor deposition method. Plasma-assisted molecular beam epitaxy is applied to carry out the GaN graphene-assisted growth. The effect of graphene defectiveness and thickness on the quality of the GaN epilayers is studied. The density functional theory is used to calculate the energy of interaction between graphene and its substrates.

The work [11] investigates oxygen-deficient defects in TiO_2 nanoparticles synthesized by the sol–gel method and laser evaporation of ceramic targets. The nanopowders are subjected to vacuum annealings to modify the defective structure in nanoparticles. The behavior of the defects in TiO_2 nanoparticles is under consideration. According to the results, the concentration of the defects can vary in wide limits via vacuum annealings of nanopowders, which can lead to the formation of a solid film of titanium atoms 1–2 monolayers in thickness on the surface of oxide nanoparticles.

In work [12], a new and promising approach for the carbothermal synthesis of C/SiC composite mixtures with SiC particles of fibrous morphology, with a fiber diameter of 0.1–2.0 μm and with an increased energy density for lithium-ion batteries is proposed. As a result, the energy characteristics of the mixtures are determined, and the potential use of the obtained mixtures as anode materials for a lithium-ion battery is presented. The authors declare the Coulombic efficiency of the samples during cycling to be over 99%.

Paper [13] investigates the rheological behavior and microstructuring of an AlMg6/10% SiC metal matrix composite. The paper proposes a new method of adding data to a training sample, which allows neural networks to correctly predict the behavior of microstructure parameters, such as the average grain diameter, and the fraction and density of low-angle boundaries with little initial experimental data. It is shown that at strain rates ranging from 0.1 to 4 s^{-1} and temperatures ranging from 300 to 500 °C, the main softening processes in the AlMg6/10% SiC composite are dynamic recovery and continuous dynamic recrystallization, accompanied by geometric recrystallization.

In conclusion, some statistics pertaining to the papers published in this Special Issue should be mentioned (see Table 1). Of the 15 papers submitted, 13 of them were successfully accepted for publication and published. The average length of the publication period is 33 days, while the review process lasts from 3 up to 12 days. The rest of the time spent preparing this Special Issue involves the authors' processing of the article. Thanks to this Special Issue, MDPI journals have received more than 50 new citations and over 12,000 views. By the time this Editorial is published, some of the papers will already have significant view and citation metrics.

Table 1. Statistics and geography of published papers (27 March 2023).

Ref	Days before Publication	CrossRef	WoS	Scopus	Google Scholar	Views 16 March 2023	MDPI Refs
1	59	2	1	2	9	1128	0/42
2	33	1	1	1	1	1233	0/36
3	22	1	1	1	1	791	1/42
4	25	5	7	6	10	1476	0/54
5	52	2	2	2	2	1023	1/33
6	15	-	-	-	2	876	0/21
7	32	-	-	-	-	637	1/31
8	25	-	1	1	2	876	1/43
9	38	-	-	-	-	605	3/30
10	30	-	-	-	-	534	2/66
11	27	1	1	1	1	524	0/31
12	25	-	-	-	-	571	25/54
13	43	-	-	-	-	444	5/68
	Average 33	12	13	14	28	>12,000	All 39

Author Contributions: Conceptualization, A.S. and O.R.; writing—original draft preparation, A.S. and O.R.; writing—review and editing, A.S. and O.R.; project administration, A.S. All authors have read and agreed to the published version of the manuscript.

Acknowledgments: We are very grateful to the authors who agreed to take part in this Special Issue, and we wish the authors more interesting and significant scientific discoveries.

Conflicts of Interest: The authors declare no conflict of interest.

References

1. Polyakova, P.V.; Pukhacheva, J.A.; Shcherbinin, S.A.; Baimova, J.A.; Mulyukov, R.R. Fabrication of Magnesium-Aluminum Composites under High-Pressure Torsion: Atomistic Simulation. *Appl. Sci.* **2021**, *11*, 6801. [CrossRef]
2. Díaz, R.A.R.; Segura, S.R.G.; Barragán, J.L.R.; Vázquez, V.R.; Ocampo, A.M.; Calderón, J.P.; Mejía, H.C.; Rodríguez, C.A.G.; Fierro, J.I.B. The Synthesis of Aluminum Matrix Composites Reinforced with Fe–Al Intermetallic Compounds by Ball Milling and Consolidation. *Appl. Sci.* **2021**, *11*, 8877. [CrossRef]

3. Smirnov, A.; Smirnova, E.; Konovalov, A.; Kanakin, V. Using the Instrumented Indentation Technique to Determine Damage in Sintered Metal Matrix Composites after High-Temperature Deformation. *Appl. Sci.* **2021**, *11*, 10590. [CrossRef]
4. Gevel, T.; Zhuk, S.; Leonova, N.; Leonova, A.; Trofimov, A.; Suzdaltsev, A.; Zaikov, Y. Electrochemical Synthesis of Nano-Sized Silicon from KCl–K_2SiF_6 Melts for Powerful Lithium-Ion Batteries. *Appl. Sci.* **2021**, *11*, 10927. [CrossRef]
5. Chernyshev, A.A.; Nikitina, E.V. A Porous Tungsten Substrate for Catalytic Reduction of Hydrogen by Dealloying of a Tungsten–Rhenium Alloy in an Aqueous Solution of Hydrochloric Acid. *Appl. Sci.* **2022**, *12*, 1029. [CrossRef]
6. Popp, J.; Römisch, D.; Merklein, M.; Drummer, D. Joining of CFRT/Steel Hybrid Parts via Direct Pressing of Cold Formed Non-Rotational Symmetric Pin Structures. *Appl. Sci.* **2022**, *12*, 4962. [CrossRef]
7. Galashev, A.Y.; Pavlov, D.S.; Zaikov, Y.P.; Rakhmanova, O.R. Ab Initio Study of the Mechanism of Proton Migration in Perovskite $LaScO_3$. *Appl. Sci.* **2022**, *12*, 5302. [CrossRef]
8. Yolshina, L.A.; Kvashnichev, A.G.; Vichuzhanin, D.I.; Smirnova, E.O. Mechanical and Thermal Properties of Aluminum Matrix Composites Reinforced by In Situ Al_2O_3 Nanoparticles Fabricated via Direct Chemical Reaction in Molten Salts. *Appl. Sci.* **2022**, *12*, 8907. [CrossRef]
9. Chen, X.; Peng, X.; Yuan, X.; Xiong, Z.; Lu, Y.; Lu, S.; Peng, J. Effect of Zinc Aluminum Magnesium Coating on Spot-Welding Joint Properties of HC340LAD + ZM Steel. *Appl. Sci.* **2022**, *12*, 9072. [CrossRef]
10. Borisenko, D.P.; Gusev, A.S.; Kargin, N.I.; Dobrokhotov, P.L.; Timofeev, A.A.; Labunov, V.A.; Mikhalik, M.M.; Katin, K.P.; Maslov, M.M.; Dzhumaev, P.S.; et al. Few-Layer Graphene as an Efficient Buffer for GaN/AlN Epitaxy on a SiO_2/Si Substrate: A Joint Experimental and Theoretical Study. *Appl. Sci.* **2022**, *12*, 11516. [CrossRef]
11. Vykhodets, V.B.; Kurennykh, T.E.; Vykhodets, E.V. Disks of Oxygen Vacancies on the Surface of TiO_2 Nanoparticles. *Appl. Sci.* **2022**, *12*, 11963. [CrossRef]
12. Leonova, A.M.; Bashirov, O.A.; Leonova, N.M.; Lebedev, A.S.; Trofimov, A.A.; Suzdaltsev, A.V. Synthesis of C/SiC Mixtures for Composite Anodes of Lithium-Ion Power Sources. *Appl. Sci.* **2023**, *13*, 901. [CrossRef]
13. Smirnov, A.; Kanakin, V.; Konovalov, A. Neural Network Modeling of Microstructure Formation in an AlMg6/10% SiC Metal Matrix Composite and Identification of Its Softening Mechanisms under High-Temperature Deformation. *Appl. Sci.* **2023**, *13*, 939. [CrossRef]

Disclaimer/Publisher's Note: The statements, opinions and data contained in all publications are solely those of the individual author(s) and contributor(s) and not of MDPI and/or the editor(s). MDPI and/or the editor(s) disclaim responsibility for any injury to people or property resulting from any ideas, methods, instructions or products referred to in the content.

Article

Joining of CFRT/Steel Hybrid Parts via Direct Pressing of Cold Formed Non-Rotational Symmetric Pin Structures

Julian Popp [1,*], David Römisch [2], Marion Merklein [2] and Dietmar Drummer [1]

1 Institute of Polymer Technology, Friedrich-Alexander-Universität Erlangen-Nuremberg, 91058 Erlangen, Germany; dietmar.drummer@fau.de
2 Institute of Manufacturing Technology, Friedrich-Alexander-Universität Erlangen-Nuremberg, 91058 Erlangen, Germany; david.roemisch@fau.de (D.R.); marion.merklein@fau.de (M.M.)
* Correspondence: julian.georg.popp@fau.de

Citation: Popp, J.; Römisch, D.; Merklein, M.; Drummer, D. Joining of CFRT/Steel Hybrid Parts via Direct Pressing of Cold Formed Non-Rotational Symmetric Pin Structures. *Appl. Sci.* 2022, *12*, 4962. https://doi.org/10.3390/app12104962

Academic Editors: Oksana Rakhmanova and Suzdaltsev Andrey

Received: 28 April 2022
Accepted: 11 May 2022
Published: 13 May 2022

Publisher's Note: MDPI stays neutral with regard to jurisdictional claims in published maps and institutional affiliations.

Copyright: © 2022 by the authors. Licensee MDPI, Basel, Switzerland. This article is an open access article distributed under the terms and conditions of the Creative Commons Attribution (CC BY) license (https://creativecommons.org/licenses/by/4.0/).

Abstract: In this study, quasi-unidirectional continuous fiber reinforced thermoplastics (CFRTs) are joined with metal sheets via cold formed cylindrical, elliptical and polygonal pin structures which are directly pressed into the CFRT component after local infrared heating. In comparison to already available studies, the unique novelty is the use of non-rotational symmetric pin structures for the CFRT/metal hybrid joining. Thus, a variation in the fiber orientation in the CFRT component as well as a variation in the non-rotational symmetric pins' orientation in relation to the sample orientation is conducted. The created samples are consequently mechanically tested via single lap shear experiments in a quasi-static state. Finally, the failure behavior of the single lap shear samples is investigated with the help of microscopic images and detailed photographs. In the single lap shear tests, it could be shown that non-rotational symmetric pin structures lead to an increase in maximum testing forces of up to 74% when compared to cylindrical pins. However, when normalized to the pin foot print related joint strength, only one polygonal pin variation showed increased joint strength in comparison to cylindrical pin structures. The investigation of the failure behavior showed two distinct failure modes. The first failure mode was failure of the CFRT component due to an exceedance of the maximum bearing strength of the pin-hole leading to significant damage in the CFRT component. The second failure mode was pin-deflection due to the applied testing load and a subsequent pin extraction from the CFRT component resulting in significantly less visible damage in the CFRT component. Generally, CFRT failure is more likely with a fiber orientation of 0° in relation to the load direction while pin extraction typically occurs with a fiber orientation of 90°. It is assumed that for future investigations, pin structures with an undercutting shape that creates an interlocking joint could counteract the tendency for pin-extraction and consequently lead to increased maximum joint strengths.

Keywords: joining of CFRT/metal; hybrid technology; mechanical properties; pin geometry; continuous fiber reinforced thermoplastics

1. Introduction

Hybrid parts consisting of continuous fiber reinforced thermoplastics (CFRTs) and high strength steel components offer interesting possibilities in use-cases with challenges that cannot be met with mono-material parts. One possible example is the combination of a demand for lightweight construction, where CFRTs can offer superior weight-related properties [1], with locally high thermal or abrasive stress, where CFRTs typically meet limitations [2]. This could be resolved with the combination of metal components to form a hybrid part. In this approach, the joining operation is the key challenge due to strongly diverging physical and chemical properties such as stiffness and thermal expansion coefficients. In the state of the art, bolted and riveted joints are commonly used; as for these joints, a comprehensive knowledge is available in the literature and it is possible to create strong and durable joints [3]. However, despite their widespread use, these form-fitting joining techniques come with relevant disadvantages. First, the need for a precise bolt hole [4] requires an additional process step, which increases processing costs and typically destroys load-bearing fibers. This can require additional reinforcements such as locally embedded titanium sheets or additional layers in the laminate. Furthermore, the need for additional auxiliary elements such as bolts, nuts and rivets leads to increased weight of the structure, which contradicts the idea of lightweight construction [5]. A fiber-friendly joining technology is the use of adhesives. While adhesive joining leads to an even introduction of force into the laminate, which reduce concentration of tension and typically does not require additional reinforcements [6], it also comes with distinct challenges, including the demand for extensive process quality control to ensure a reliable joint and especially with thermoplastic composites, joinability via adhesives is oftentimes limited. Polyolephinic polymers especially require surface treatment to be joinable via adhesives that add process steps and complexity to the joining process [7].

A comparably new joining process is the creation of form fitting joints via cold formed pin structures, which can embedded into the locally heated CFRT component [8]. In the field of thermoset-based composite (CFRP)/metal joining, the use of pin-reinforced adhesive joints is already widely described. However, in CFRP/metal joining, the pin needs to be embedded during the manufacturing process of the part [9]. This can either be into the dry fabric prior to the infusion process via vacuum assisted resin infusion [10], vacuum assisted resin transfer molding [11] or into pre-impregnated sheet during manufacturing before the curing stage of the matrix [12]. Thus, a separate, subsequent joining process is not possible. In the field of CFRT/metal joining, the direct pin pressing process into a locally heated CFRT component has been shown to be an uncomplicated joining process with promising mechanical properties [13]. In the current state of the art, rotational-symmetric pins have been primarily investigated [14], although in Römisch et al. [15], the uses of non-rotational-symmetric pins have shown potential to increase the maximum load of the joint in the field of metal to aluminum hybrid joining. Due to the good performance in metal/metal joining, the use of non-rotational-symmetric pin structures also appears promising in the use of CFRT/metal joining. However, the underlying mechanisms in the direct pin pressing of CFRT/metal parts are fundamentally diverse for steel/aluminum joining and consequently, a direct transfer of the knowledge from Römisch et al. [15] is not possible. In particular, the joints' dependency on the pin- and fiber orientation of the anisotropic CFRT material in relation to the load direction needs to be investigated.

Considering the existing literature, the present study aims to create an understanding of the CFRT/metal joining process with non-rotational-symmetric pin structures. The focus thereby lies on the mechanical behavior in dependency of the pin- and fiber orientation in relation to the load direction and the investigation of occurring failure modes.

2. Materials and Methods

2.1. Used Material

In this study, a custom-fabricated quasi-unidirectional glass fiber reinforced polypropylene (GF/PP) material with a thickness of approximately 2 mm was used. Thus, glass fiber non-crimp fabrics from Saertex GmbH & Co. KG (Saerbeck, Germany) and a polypropylene (PP) type BJ100HP from Borealis AG (Vienna, Austria) were impregnated and consolidated on an interval hot press at the Neue Materialien Fürth GmbH (Fürth, Germany). The residual after incineration was measured to 71.1 wt.-%, which translates to a fiber volume content under the assumption of ideal consolidation of 47 vol.-%. The melt peak temperature of the composite was measured to 164.4 °C and the crystallization peak temperature was measured to 121.4 °C both as an average of nine differential scanning calorimetry (DSC) measurements.

As a steel component, a galvanized dual-phase steel type HCT590X (DP600) from Salzgitter AG (Salzgitter, Germany) was used. This cold-rolled steel variant was frequently used in the automotive sector for a crash relevant component of the car body. The thickness of the used steel sheets was 1.5 mm.

2.2. Definition of Pin Geometry

In the scope of this study, three different pin geometries with a height of 1.8 mm were investigated. As a baseline geometry, which aims to create a comparability between the different non-rotational symmetric pins as well as a comparability to sources in the literature, which primarily use cylindrical pin structures, a cylindrical pin with a diameter of 1 mm was chosen. As a pin with a strong directional dependency in means of fiber displacement and behavior under load, an elliptical pin with a small diameter of 1 mm and a large diameter of 2 mm was chosen. As a third geometry with an anisotropic behavior under compressive and tensile loading, a polygonal pin structure was chosen, which is based on a triangular geometry. It has an in-circle diameter of 1 mm; the sides are curved with a radius of 1.18 while the edges are rounded with a radius of 0.1 mm. Figure 1 displays the three investigated geometries.

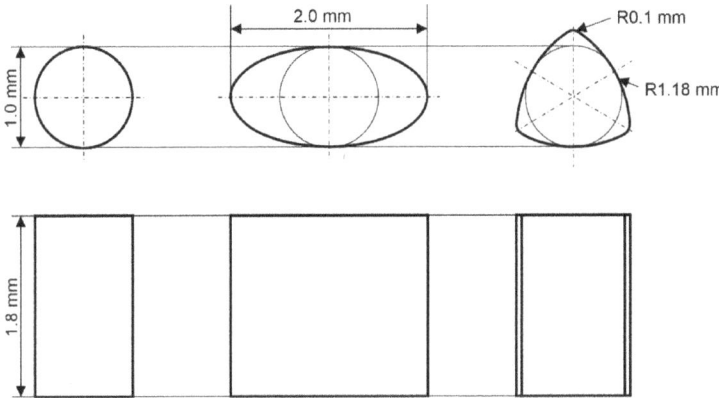

Figure 1. Used pin geometries: cylindrical (**left**), elliptical (**middle**), polygonal (**right**).

The investigated geometries feature different cross sections, volumes and projection areas (for elliptical pins in dependency of the viewing angle) which are summarized in Table 1. These values allow for assessing the efficiency of a pin-geometry in means of used foot-print on the sample and mechanical performance of the created joints.

Table 1. Geometric properties of pin geometries.

	Cylindrical	*Elliptical*	*Polygonal*
Cross section	0.79 mm^2	1.57 mm^2	0.98 mm^2
Volume	1.41 mm^3	2.83 mm^3	1.77 mm^3
Projection area	1.8 mm^2	3.8 mm^2 (90°) 1.8 mm^2 (0°)	2.13 mm^2

2.3. Pin Manufacturing via Cold Forming Process

For the production of the pin geometries described above, forward extrusion from the sheet metal plane was used. For this purpose, a multiple-acting tool is required to ensure independent control of the blank holder as well as the punch. The process sequence of the pin extrusion as well as the essential tool components for the extrusion are shown in Figure 2. First, the blank holder moves axially downwards onto the test specimen and initially applies a blank holder pressure (σBH) of 250 MPa. This is necessary to prevent the sheet from bulging due to local deformation and the resulting stresses. Furthermore, the applied pressure and the resulting friction between the blank and the blank holder reduces the radial flow of material into the sheet metal plane, thus increasing the material flow into the die during extrusion. Once the pressure is applied, the forming punch made of carbide with a diameter (dP) of 3 mm, which is guided in the blank holder, moves axially downwards at a constant speed of 5 mm/min and penetrates the steel sheet made of DP600 with an initial sheet thickness (t0) of 1.5 mm. In the process, the material underneath the punch is displaced both axially into the die and laterally outward into the sheet plane and laterally inward into the die. Three different dies made of carbide with a shrink ring and the geometries described above were used for the investigations performed in this work. Each die had an additional entry radius of 0.1 mm to reduce the stress on the die, especially on the edges, and to promote the axial material flow into the die and decrease the required punch penetration depth (s). To achieve the required pin height, the punch penetration depth is controlled in order to adjust the amount of displaced material. For this purpose, mechanical stops with varying heights are used, with which the forming punch comes into contact, initiating the end of the forming process. Subsequently, first the blank holder and the punch and then the ejector punch move axially upwards to eject the formed pin structure with a height (h), from the die. For the investigations carried out in this work, a pin height of 1.8 mm was targeted for the pin extrusion.

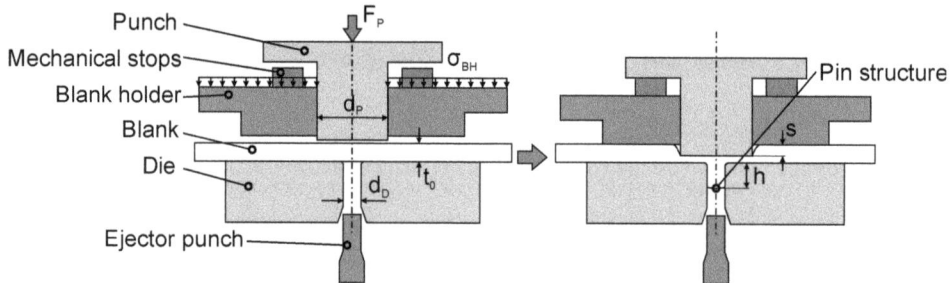

Figure 2. Schematic illustration of the pin extrusion process and the essential tool components.

2.4. Joining Process

For the joining process, a custom joining device was used (see Figure 3). This device is equipped with an infrared spot type 600.5100.1 (Optron Infrared Systems GmbH, Garbsen, Germany) and a pneumatic piston type ADN-80–25-A-P-A (Festo SE & Co. KG, Esslingen, Germany). The infrared spot has a maximum power output of 150 watts and is focused to a nominal spot diameter of 10 mm at a focus distance of 50 mm, and is used to locally heat the CFRT component above the melting point of the matrix. Despite the nominal focus diameter, the actual irradiated area is not sharply limited and exceeds the nominal spot diameter of 10 mm. Consequently, a mask with a circular opening of 12 mm is used to sharply define the irradiated area. The pneumatic piston is used to press the pin structure into the CFRT sheet and to reconsolidate the CFRT sheet during the cooling stage after the joining process. The piston diameter is 80 mm resulting in a maximum force of 3000 N at an operation pressure of 0.6 MPa. To control the joining speed, a choke valve is equipped, which allows to adjust the piston velocity between approx. 0.01 m/s and 0.1 m/s. A polyetherehterketone (PEEK) insert is utilized to thermally insulate the CFRT sample from the steel positioning device that would otherwise lead to excess heat loss at the bottom of the sample.

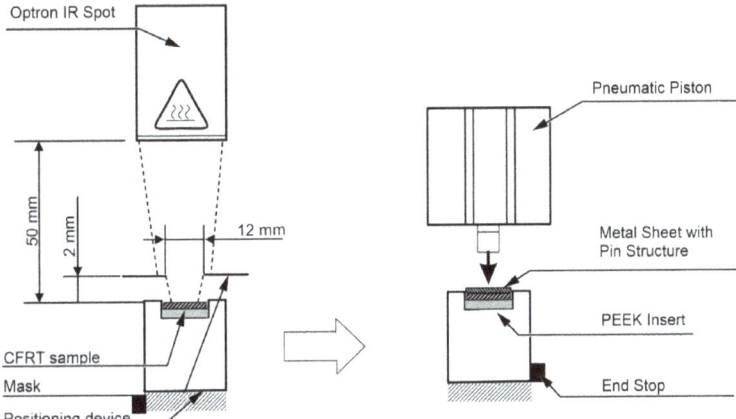

Figure 3. Schematic illustration of the joining device.

The joining process is conducted in the following three steps:

First, the sample is placed in the positioning device and is heated in such way, that the matrix is molten through the entire thickness of the sample. The heating parameters have been identified in previous investigations [16] and are directly taken over.

Second, the positioning tool is placed under the pneumatic piston and the metal component is placed on top of the CFRT sample while a form-fit in the positioning device allows proper concentrical alignment of the pin structure with the heated zone in the CFRT component. Detailed measurements of the samples used can be seen in Figure 4.

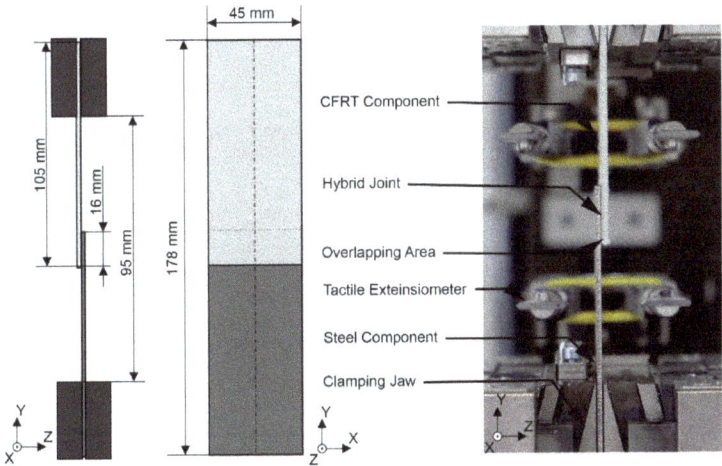

Figure 4. Sample dimensions and test setup for single lap shear tests.

Third, the pin is pressed into the CFRT component via the pneumatic piston. During the cooling stage, the pressure remains applied onto sample, to allow a complete reconsolidation of the laminate. After the cooling stage, which was defined to approximately 60 s in this study, the pressure is removed and the joined sample is removed from the positioning device.

The heating as well as joining parameters in this study were identical with the settings in [16], where a good reconsolidation quality without matrix damage due to overheating could be achieved. The settings are summarized in Table 2.

Table 2. Used heating and joining parameters.

Mask Diameter	IR Spot Power	Heating Time	Joining Force	Piston Speed
12 mm	39 Watt	210 s	2500 N	0.1 m/s

2.5. Mechanical Characterization—Single Lap Shear Test

The joined specimen were mechanically tested using the single lap shear test according to the technical bulletin DVF EFB 3480-1 [17]. Both components of the hybrid sample had a length of 105 mm and a width of 45 mm; the joining position was placed 8 mm from the short side of the samples, resulting in an overlap of 16 mm and a total length of the hybrid sample of 178 mm. The clamping free length measured 95 mm. The elongation of the sample was measured via tactile extensometers with a measuring distance of 50 mm. The test setup and sample dimensions are summarized in Figure 4.

Before testing, the samples were stored in normalized climate (23/50) according to DIN EN ISO 291 [18]. The test was performed on a universal testing machine type Z1465 from ZwickRoell AG (Ulm, Germany) in quasi-static condition with a testing speed of 1 mm/s and a sampling rate of 100 Hz.

In the scope of this study, 10 different variations were investigated consisting of a variation in the non-rotational-symmetric pin geometries' orientation in relation to the load direction and a variation in the fiber orientation in relation to the load direction. The sample size of each parameter setting was n = 3 resulting in 30 tested samples. Thereby, the two investigated orientations of the elliptical pin were named "0°" and "90°", which correspond to the orientation of the long axis of the pin in relation to the load direction. The investigated pin orientations of the polygonal pin are described as "sharp" and "blunt", which distinguishes whether the edge or the side of the triangular geometry transmit the force between both joining partners in the single lap shear test. A variation in the orientation of the cylindrical pin was not possible due to the rotational symmetric shape. Table 3 summarizes the conducted variations. The fiber orientation is described to 0° and 90°, which corresponds to the fiber orientation in relation to the load direction.

Table 3. Summary of the conducted variations for the mechanical characterization.

Pin Type	Pin Orientation	Fiber Orientation	Nomenclature
Cylindrical	n/a	0°	Cylindrical 0°
Cylindrical	n/a	90°	Cylindrical 90°
Elliptical	0°	0°	Elliptical 0°/0°
Elliptical	0°	90°	Elliptical 0°/90°
Elliptical	90°	0°	Elliptical 90°/0°
Elliptical	90°	90°	Elliptical 90°/90°
Polygonal	Sharp	0°	Polygonal Sharp/0°
Polygonal	Sharp	90°	Polygonal Sharp/90°
Polygonal	Blunt	0°	Polygonal Blunt/0°
Polygonal	Blunt	90°	Polygonal Blunt/90°

2.6. Microscopic Investigation

In order to create an understanding about the failure behavior of the joint, microscopic investigations of the CFRT and metal component after failure were conducted. For this, a stereo microscope type Discovery V12 and a microscopy camera of the type Axio Cam MRc5 (both Carl Zeiss Microscopy GmbH, Jena, Germany) were used. The images were created so that the X/Y plane could be analyzed in detail (compare Figure 4). In order to create a sufficient depth of focus and to be able to analyze sections of the samples with different elevations, multiple single images with different focus distances were stacked to one image. Before investigation, the metal and CFRT components were separated and were investigated individually. Of each variation, one CFRT component and one metal component were investigated. Thus, the samples, which showed the highest maximum forces of each variation, were chosen for the microscopic investigation. To create a better understanding about the initial failure behavior under shear load, two cylindrical samples with fiber orientations in 0° and 90° were tested in the above-described procedure until a first drop in reaction force could be seen. After this initial drop in the reaction force, the testing procedure was manually terminated and the samples were examined in the same manner as described above.

3. Results
3.1. Pin Manufacturing Process

To gain a deeper understanding of the pin extrusion process, especially for non-rotationally symmetrical geometries, to verify the repeatability during extrusion and to investigate the extrusion process in more detail, the force–displacement data of the pin extrusion process were analyzed. Figure 5 shows the punch force–displacement curves averaged from four experiments for each of the pin geometries investigated. Furthermore, the standard deviation of the respective geometries is shown using an error band. Here, it can be seen that the standard deviation with a maximum value of ±0.13 kN across all the investigated geometries is very low compared to the maximum forces between 18.23 kN for the elliptical 0° pins and 21.74 kN for the cylindrical pins.

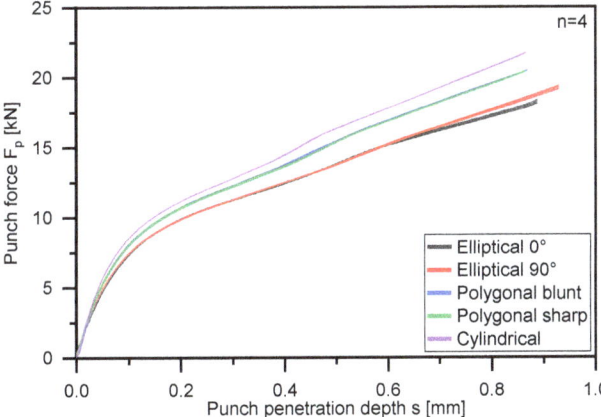

Figure 5. Punch force F_p to punch penetration depth s for the different pin geometries and different orientations. The curves are the average from four tests and the standard deviation is shown using error bands.

Comparing the maximum standard deviation of the different pin geometries, the cylindrical and polygonal blunt pins show the smallest deviations with ±0.08 kN. In contrast, the deviation of the other geometries with ±0.12 kN (polygonal sharp), ±0.13 kN (elliptical 0°) and ±0.13 kN (elliptical 90°) are on a slightly higher level. If we compare the force–displacement curves of the geometries formed in different orientations, i.e., the polygonal sharp and blunt as well as the elliptical 0° and 90°, we see that the rotation of the polygonal geometry has only a very slight influence on the forming and thus the force–displacement curve, and thus the curves are almost identically superimposed. This can be explained by the fact that the general orientation of the geometry in relation to the rolling direction of the sheet does not change and thus similar forming conditions are present. Conversely, a comparison of the elliptical pins at 0° and 90° shows a deviation in the force starting at a punch penetration depth of around 0.6 mm. Here, the slope of the force–displacement curve is greater for the elliptical 90° pins than for the elliptical 0° pins. In this case, the rolling direction during cold rolling and thus the orientation of the grains of the sheets used for the extrusion of the pin structures is along the long axis of the ellipse in 0° and therefore along the short axis for the ellipse in 90°. Consequently, it is possible that for the 90° ellipses along the short side, a greater number of linear dislocations are necessary for the pin extrusion. However, this effect can only be detected as the punch penetration depth increases, as can be seen from Figure 5, since at the beginning of extrusion, the radial flow of material outward into the sheet plane is initially more dominant, and this is directionally independent [19]. The smaller the residual sheet thickness, the more dominant the extrusion into the die, and here the directional dependence due to the pin geometry can be detected.

In Table 4, in addition to the force–displacement curves shown in Figures 4 and 5, the corresponding pin height, punch penetration depth, sheet thickness, relative punch penetration depth and residual sheet thickness averaged from a sample of n = 4 pins are listed for the different investigated pin geometries. Looking at the force–displacement curves shown in Figure 5 and the associated standard deviations, as well as the parameters relevant for pin extrusion shown in Table 4 and the respective standard deviations, it can be seen that both the process and the resulting pin structures can be assumed to be repeatable.

Table 4. Pin height h, punch penetration depth s, sheet thickness, relative punch penetration depth in relation to initial sheet thickness and residual sheet thickness for the investigated and extruded pin geometries (n = 4).

Pin Type	Cylindrical	Elliptical 0°	Elliptical 90°	Polygonal Sharp	Polygonal Blunt
Pin-height h [mm]	1.80 ± 0.01	1.78 ± 0.02	1.82 ± 0.01	1.79 ± 0.02	1.81 ± 0.01
Punch penetration depth s [mm]	0.87 ± 0.00	0.90 ± 0.00	0.93 ± 0.00	0.87 ± 0.01	0.87 ± 0.00
Sheet thickness [mm]	1.50 ± 0.00	1.50 ± 0.00	1.49 ± 0.01	1.50 ± 0.00	1.50 ± 0.00
Rel. punch penetration depth [%]	57.91 ± 0.15	59.58 ± 0.20	62.58 ± 0.31	58.11 ± 0.39	58.10 ± 0.18
Residual sheet thickness [mm]	0.63 ± 0.00	0.61 ± 0.00	0.56 ± 0.01	0.63 ± 0.01	0.63 ± 0.00

When considering the pin heights and the required punch penetration depths, it can be seen that, compared to polygonal and cylindrical pins, the elliptical pin structures require a greater penetration depth. This effect has already been demonstrated in [15] and it could be shown that with an increasing pin cross-sectional area compared to the punch cross-sectional area, it leads to a greater material utilization of the material displaced by the punch, since the reduction in the flow resistance into the die reduces the radial material flow into the sheet plane and increases the axial material flow into the die. However, due to the larger cross-sectional area, a larger volume of material is required to achieve the same pin height as in the other pin geometries with smaller cross-sectional areas. This larger volume of material is not entirely compensated for by the increase in material utilization, which is why a greater punch penetration depth is necessary for the elliptical pins [15]. A comparison of the punch penetration depths of the elliptical 0° and 90° pins shows it is noticeable that the elliptical 90° pins have a greater punch penetration depth than the elliptical 0° pins. However, these also have a greater pin height of 1.82 mm on average compared to 1.78 mm. Thus, less material was displaced from the punch with the elliptical 0° pins, which can result in the lower average pin height. However, the orientation of the geometry to the rolling direction may also have a certain influence on the punch penetration depth and the corresponding pin height cannot be conclusively clarified based on the present results, since the penetration depths are not identical. However, an influence can be suspected, since due to the higher force requirement, which results from the orientation of the geometry and thus represents a greater flow resistance, the amount of radial material flow increases compared to the elliptical 0° pins. This would result in a greater punch penetration depth being required for the elliptical 90° pins to achieve the same pin height as the elliptical 0° pins.

3.2. Mechanical Characterization

As it can be expected, the results from the single lap shear tests show a strong dependency on the pin type and pin-orientation as well as the fiber orientation in relation to the load direction. For all pin types and pin orientations, it can be seen that a significant increase in the maximum force comes with a fiber orientation of 90°. Table 5 and Figure 6 give a summary of the measured maximum forces in the single lap shear test.

Table 5. Summary of maximum shear force results.

Pin Type	Cylindrical		Elliptical 0°		Elliptical 90°		Poly. Sharp		Poly. Blunt	
Fiber Orientation	0°	90°	0°	90°	0°	90°	0°	90°	0°	90°
Max. force [N]	179.3 ± 8.7	239.0 ± 8.0	187.5 ± 15.5	291.3 ± 16.0	319.7 ± 19.7	415.0 ± 13.6	185.3 ± 23.8	281.7 ± 16.7	232.0 ± 7.0	346.7 ± 7.5

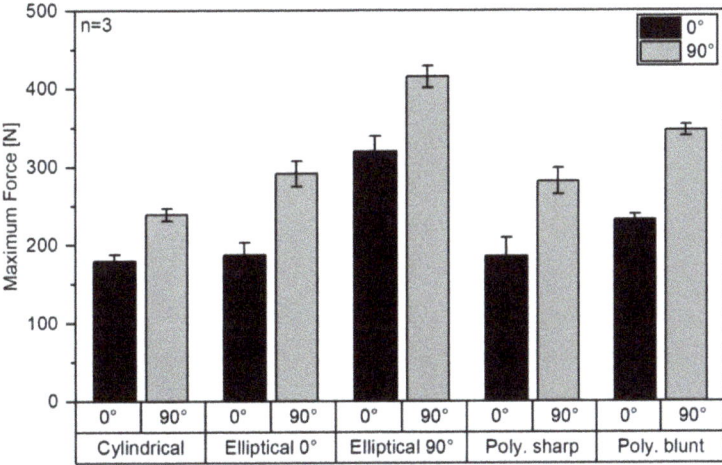

Figure 6. Maximum force in single lap shear tests.

Therefore, the increase in force between fibers in 0° and 90° lies between 33.3% for the cylindrical pins and 55.4% for elliptical pins with an orientation of 0° in relation to the load direction. It is noticeable that the non-rotational symmetric pins lead to increased maximum forces in comparison to cylindrical pins independently on the pin- and fiber orientation. However, the increase in maximum force for an elliptical pin 0°/0° and the polygonal sharp/0° pin compared to a cylindrical 0° pin is relatively low with only 4.6% and 3.3%, respectively, in comparison to the corresponding cylindrical pins tested with the same fiber orientation.

When comparing elliptical pins with different orientation, it shows that a 90° pin orientation leads to an increase in maximum force of 70.5% (0° fiber orientation) and 73.6% (90° fiber orientation), which leads to the conclusion that for elliptical pins in CFRT/metal joining, an orientation of the pin in 0° to the load direction is strongly inferior in comparison to a pin orientation of 90°. When comparing this result with Römisch et al. [15], where a material combination of a DP600 and aluminum EN AW-6014 was investigated, this finding is contrary. Römisch et al. found an increased joint strength under shear load, when the pin was oriented in 0° of 878 ± 7 N in comparison to 811 ± 9 N with a pin oriented in 90° while in both cases the failure occurred due to a broken-off pin structure. This is explained with a higher area moment of resistance of the pin of 0.39 mm^3 in 0° in comparison to 0.2 mm^3 in 90°, which leads to pin failure at higher loads when the pin is oriented in 0°. In the present study, however, the joint did not fail due to pin breakage and consequently other mechanisms were found (compare Section 3.3). For polygonal pins, it shows that a "blunt" orientation in relation to the load direction is superior in means of maximum transmitted force with an increase of 25.2% (0° fiber orientation) and 23.1% (90° fiber orientation) in comparison to the "sharp" orientation.

When analyzing the foot print related joint strength, which can be seen as an indication for the efficiency of a certain pin geometry, especially with the perspective of multi-pin joints, which will be required to achieve higher total joint stabilities, a different situation shows. The baseline joint strength of the cylindrical pins is 227.0 ± 11.1 MPa and 302.5 ± 10.2 MPa, respectively, for fiber orientations in 0° and 90°. Table 6 and Figure 7 summarize the resulting foot print related joint strength.

Table 6. Summary of joint strength results.

Pin Type	Cylindrical		Elliptical 0°		Elliptical 90°		Poly. Sharp		Poly. Blunt	
Fiber Orientation	0°	90°	0°	90°	0°	90°	0°	90°	0°	90°
Joint strength [MPa]	227.0 ± 11.1	302.5 ± 10.2	119.4 ± 9.9	185.6 ± 10.2	203.6 ± 12.5	264.3 ± 8.7	189.1 ± 24.3	287.4 ± 17.1	236.7 ± 7.1	353.7 ± 7.7

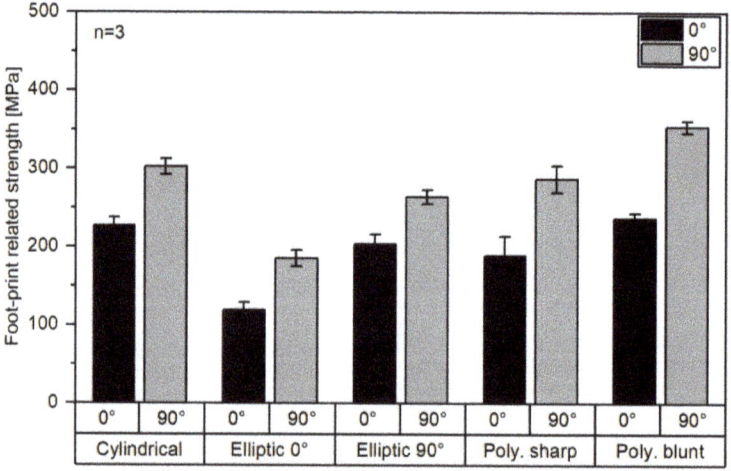

Figure 7. Foot print related strength in single lap shear tests.

Elliptical pins show reduced joint strength when compared to cylindrical pins. This phenomenon is most pronounced for elliptical 0°/0° samples where a joint strength of only 119.4 ± 9.9 MPa is measured, which is a decrease of 107.6 MPa or 47.4% when compared to cylindrical 0° samples. For elliptical 0°/90° pins, the drop in joint strength is in a similar range with 116.9 MPa or 38.6%. Elliptical pins with an orientation of 90°/0° and 90°/90° perform better but still lead to a reduction of 23.4 MPa (0° fiber orientation) and 38.2 MPa (90° fiber orientation) respectively.

In contrast to elliptical pins, polygonal pins lead to either increased or decreased foot print related joint strength, depending on the pin-orientation. While pins with "sharp" orientation lead to a reduction in joint strength of 37.9 MPa/16.7% (0° fiber orientation) and 15.1 MPa/5.0% (90° fiber orientation), respectively, an increase by 9.7 MPa/4.3% (0° fiber orientation) and 51.2 MPa/16.9% can be seen for polygonal pins with a "blunt" pin orientation in relation to the load direction. Generally, it can be summarized that from a joint strength perspective, an elliptical pin is disadvantageous, especially when oriented in the load direction while a polygonal pin can be beneficial when oriented so that the blunt side introduces the load into the CFRT component but is disadvantageous when the sharp edge is oriented towards the interfacing surface.

When comparing the joint strength in relation of the projection area of the pin geometries in the direction of the testing load, it shows that, especially with fiber orientations of 0°, the measured values are in a comparable range between 87.0–103 MPa (compare Figure 8 and Table 7). This can be explained with the dominant failure behavior of these samples (compare Section 3.3 and Table 8), which for fiber orientations of 0°, typically is a failure of the CFRT component, which occurs at similar compressive stresses induced into the CFRT laminate. The lowest projection related strength is measured for elliptic 90°/0° pins, which show a combination of CFRT failure and subsequent pin extraction, which can be assumed as the reason for the reduced strength.

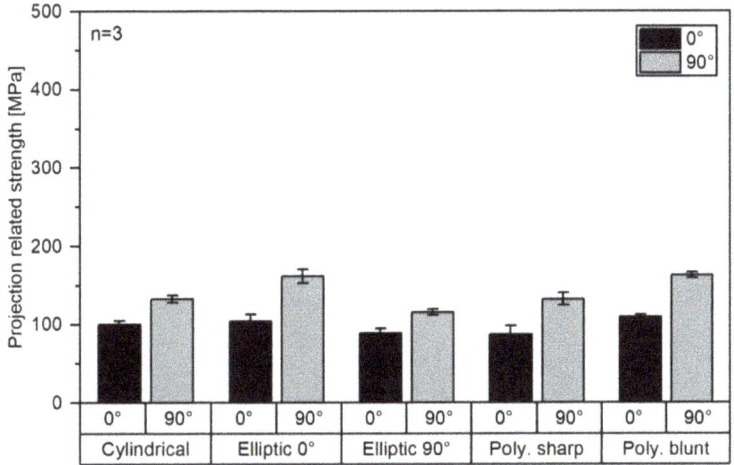

Figure 8. Projection area related strength in single lap shear test.

Table 7. Summary of projection area related joint strength.

Pin Type	Cylindrical		Elliptical 0°		Elliptical 90°		Poly. Sharp		Poly. Blunt	
Fiber Orientation	0°	90°	0°	90°	0°	90°	0°	90°	0°	90°
Joint strength [MPA]	99.6 ± 4.9	132.8 ± 4.5	104.2 ± 8.6	161.9 ± 8.9	88.8 ± 5.5	115.3 ± 3.8	87 ± 11.2	132.3 ± 7.9	103 ± 3.3	162.8 ± 3.5

Table 8. Summary of failure modes.

Pin Type	Cylindrical		Elliptical 0°		Elliptical 90°		Poly. Sharp		Poly. Blunt	
Fiber Orientation	0°	90°	0°	90°	0°	90°	0°	90°	0°	90°
Failure mode	CFRT failure	extraction	CFRT failure	CFRT failure	CFRT failure/extraction	extraction	CFRT failure	CFRT failure	CFRT failure	extraction

With 90° fiber orientation, the values vary between 115.3 and 162.8 MPa, which is a significantly higher range. This higher fluctuation can be explained with two approaches: first, more diverse failure modes (both pin extraction and CFRT failure occur); second, for samples with a failure due to pin extraction, the area moment of resistance of the pin structure strongly varies leading to pin bending and subsequent pin extraction at different loads. So consequently, elliptical 0°/0° pins have a significantly increased strength in comparison to cylindrical 0° pins as it has a higher area moment of resistance with the same projection area as the cylindrical pin avoiding the failure due to pin extraction, and consequently, failure at higher loads but with the same projection area leading to higher projection are related joint strength.

Figure 9 summarizes the force–displacement curves of all tested samples in dependency of pin type, pin orientation and fiber orientation. Therefore, curves with dotted lines represent samples with a 90° fiber orientation while solid lines represent 0° fiber orientation. Generally, it can be seen that a fiber orientation of 90° leads to higher maximum forces and typically to a more abrupt drop in reaction force after the maximum force is reached. The exceptions to this phenomenon are elliptical 0°/90° and polygonal sharp/90°

pins, where the reduction in reaction force is relatively smooth as it is the case for their 0° fiber orientation counter parts but with the distinction of higher force levels. A possible explanation for this behavior is that these pin samples have a comparably high resistance against pin bending, which avoids pin deflection and consequent pin extraction but are shaped in such way that the force is introduced into the CFRT sample over a relatively small and sharp edge. This leads to concentrations of tension and following gradual failure of the fibers in the CFRT component, leading to a less abrupt failure (compare Section 3.3). Another interesting phenomenon is that the force–displacement curves of cylindrical 0° and elliptical 0°/0° samples are very similar and that the measured average maximum forces are identical in the range of the standard deviation. This can be explained with the failure behavior that is for both samples a failure of the CFRT sample and no deformation of the pin structure (compare Section 3.3). As the projected area of both pin types in the direction of force application is identical, similar pressure is introduced into the CFRT component by the pin and consequently, the maximum force before failure of the CFRT in both sample types is almost identical.

For all samples, with the exception of elliptical 90°/90° samples, it can be seen that the initial stiffness of sample pairs with 0° and 90° fiber orientation are very similar, but that the force–displacement curves of samples with 0° fiber orientation flattens, indicating a reduction in stiffness, before reaching the maximum force level while a 90° fiber orientation typically leads to a steeper curve until the maximum force is reached. In combination with the typically more abrupt drop in force with 90° degree fiber orientation, it is concluded that in the course of this study with non-undercutting pin geometries, 0°-degree fiber orientation leads to a less critical failure behavior than a fiber orientation of 90°, but at lower absolute force levels.

The comparison of the mechanical performance in the present study with studies in the literature which deal with pin based CFRT/metal joining is not easily achievable. Reasons for this are the use of different CFRT [14] or metal [20] components as well as divergent pin manufacturing methods [21]. When comparing with Kraus et al. [14], where a maximum force of 118 N was reached with a pin diameter of 1.32 mm and a pin height of 1.48 mm, both the maximum force as well as the foot print related joint strength in the present study are higher. This could possibly be explained with the lower pin height of the samples, which leads to early pin extraction as well as a different sample geometry. In [20], a pin-array with 16 pins was tested with a maximum reaction force under shear load of 5570 N and consequently a maximum force per pin of 348 N. This is higher than the values in the present study, but only partially comparable. First, the thickness of the CFRT component is significantly higher with 4 mm, potentially leading to higher reaction forces. Second, the sample geometry is fundamentally different and avoids pin extraction due to bending of the single lap shear sample, which could also lead to increased reaction forces. Finally, the exact pin geometry was not presented in [20], which makes it difficult to assess the performance based on the foot print.

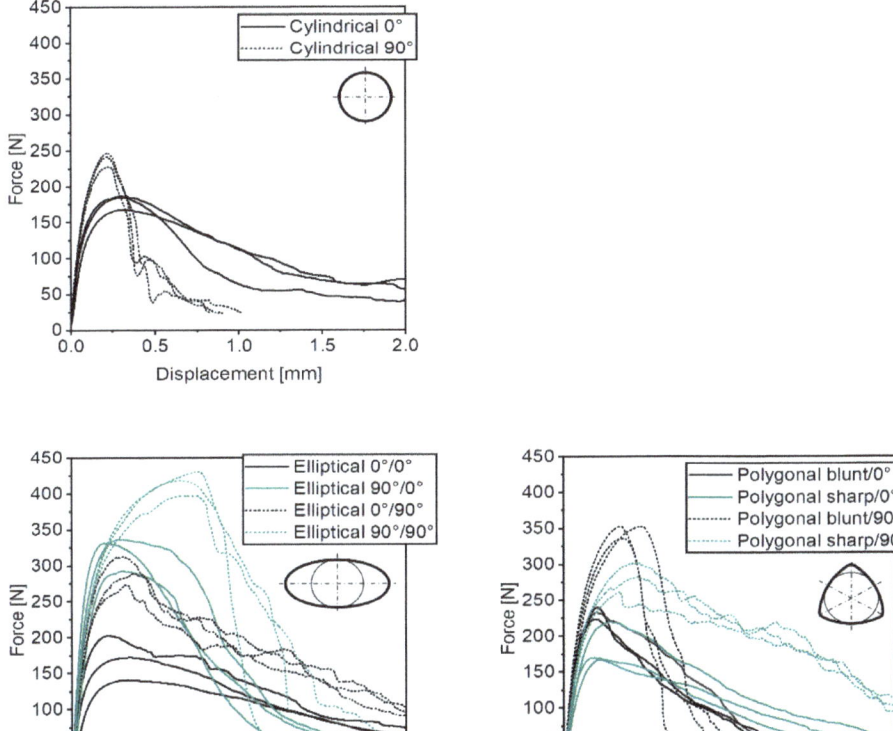

Figure 9. Force–displacement curves for cylindrical (**top**), elliptical (**bottom left**) and polygonal pins (**bottom right**).

Generally, the presented mechanical performance was relatively good, when compared to other studies, but it has to be acknowledged, that the comparability is limited [14,20,21]. Therefore, in future studies it is required to investigate pin-arrays which on one hand increase the general load capacity of the joint but also allow to relate the maximum forces with the actual joint surface to obtain comparable joint strength values, which also allows to relate the results with established joining technologies such as adhesive joining or arrays of bolts and rivets.

3.3. Failure Analyzation

In the following section, the observed failure behavior of the samples is described and first explanations for the underlying failure mechanisms are given. In the course of this study, two distinct failure modes occurred depending on the fiber orientation in the CFRT component and the type and orientation of the pin geometry in relation to the load direction. The study shows that with a fiber orientation of 0°, the samples tend to fail due to failure of the CFRT component while with fibers oriented in 90°, the samples more often fail due to an extraction of the pin structure at higher loads, as was also found in [13].

Figure 10 shows images of samples created after manually stopping the single lap shear test after the initial drop of force was noticed. On the left, an example of a sample with 90° fiber orientation and a cylindrical pin is shown with a clearly visible separation between the CFRT and metal component which is a sign for beginning pin extraction and also a significant bending of the CFRT sample, which is a result of the samples' low bending stiffness perpendicular to the fiber orientation and the resulting rotary moment induced by the testing forces. In the middle, a microscopic image of an exemplary CFRT sample with a 90° fiber orientation after the extraction of the steel component is shown, which shows no significant deformation of the pin hole. Despite the lack of hole-deformation, dark areas can be seen below the pin hole that is the area where the testing forces are introduced into the laminate, which can be interpreted as matrix cracks due to the introduced compressive stress perpendicular to the fiber orientation. These cracks are a result of the unidirectional composite's low strength under compressive loads perpendicular to the load direction. However, despite the low compressive strength perpendicular to the load direction, samples loaded in 90° to the fiber orientation still show higher maximum forces before failure of the joint, which needs further explanation. This can be explained by the more efficient introductions of the testing load into the fibers in comparison to 0° fiber orientation. In 90°, the fibers take up the testing load similar to an arresting cable used at aircraft carriers to rapidly decelerate landing airplanes and translate the testing load into a tensile load in the fibers.

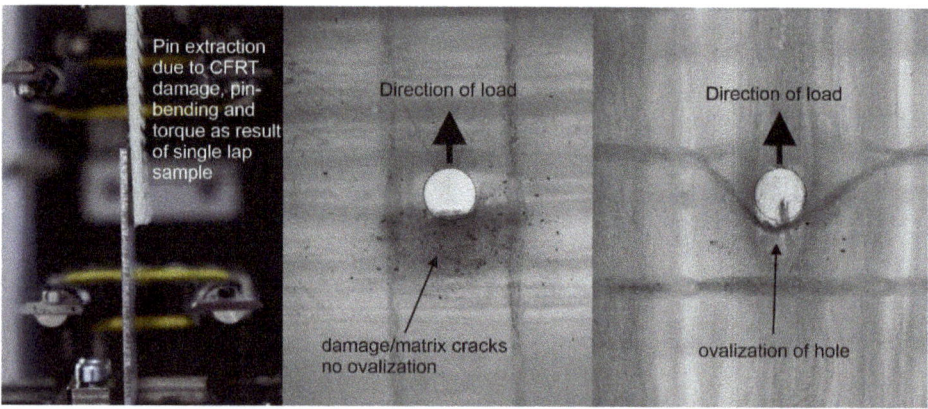

Figure 10. Summary sample behavior directly after maximum force: single lap shear sample with fiber orientation in 90° to load direction (**left**), CFRT component with removed cylindrical pin (fiber orientation: 90°, (**middle**)) and CFRT component with ovalized pin hole (fiber orientation: 0°, (**right**)).

On the right, a sample with fibers in 0° orientation is shown. This sample clearly shows an ovalization of the pin hole as a result of the introduced compressive stresses. An explanation for this deformation can be found in the distinct fiber morphology, which occurs, in the direct pin pressing process (compare [8]) which leads to practically fiber free, matrix rich zones, which are located next to the pin hole in the direction of fibers. When samples are loaded in 0° to the fiber orientation, the testing force is directly introduced into these matrix rich zones, which have no reinforcing fibers and consequently are less stiff leading to the shown ovalization of the pin-hole and to a reduction in force in the mechanical tests.

Table 8 summarizes the observed failure modes. A tendency for failure of the CFRT component with 0° fiber orientation and pin extraction failure with 90° fiber orientation can be seen, which also corresponds with the findings in the previous section.

Exceptions to this are elliptical 0°/90° and polynomial blunt/90° samples, which also showed divergent force–displacement curves in Figure 9. This can be explained by the comparably sharp edge of the pin structure at the point of load transmission in combination with a pin geometry, which has an increased resistance against pin bending and consequently less likely pin extraction. This leads to a failure of the CFRT component where the fibers are cut perpendicular to the fiber orientation. In Figure 11, a summary of microscopic images after failure of the sample is shown. It can be clearly seen that samples that failed with pin extraction are significantly less damaged than samples that failed due to CFRT failure. Furthermore, the elliptical 0°/90° and polygonal sharp/90° example clearly show a CFRT failure with cut fibers in the direction of load.

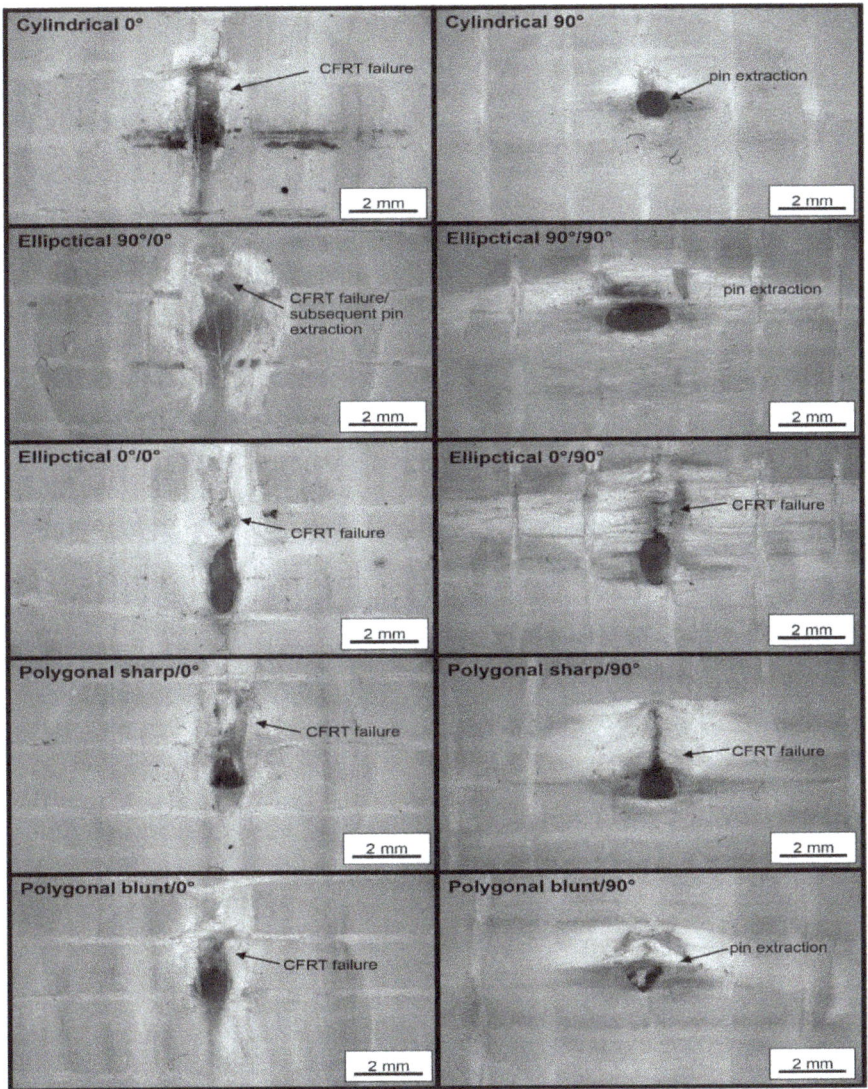

Figure 11. Summary of CFRT samples after single lap shear test.

One factor that amplifies the tendency for pin extraction is that pins tested with 90° fiber orientation bend due to the higher testing forces in comparison to 0° fiber orientation. This force in combination with the comparably soft CFRT component, which does not provide significant support against bending, leads to the pin plastically deflecting under load. Figure 12 shows cylindrical and polygonal pins after failure. It can be clearly seen that pins with a 90° fiber orientation strongly deflect in the direction of testing while pins tested with a 0° fiber orientation are not or significantly less deformed. In comparison to the shown images, elliptical 90°/90° pins are similarly deformed while polygonal sharp/90° pins and elliptical 90°/0° pins show a strongly reduced deformation. Elliptical 0°/90° as well as 0°/0° and polygonal sharp/0° pins do not show a significant deformation. Pins tested with fiber orientation in 0° generally did not significantly deform.

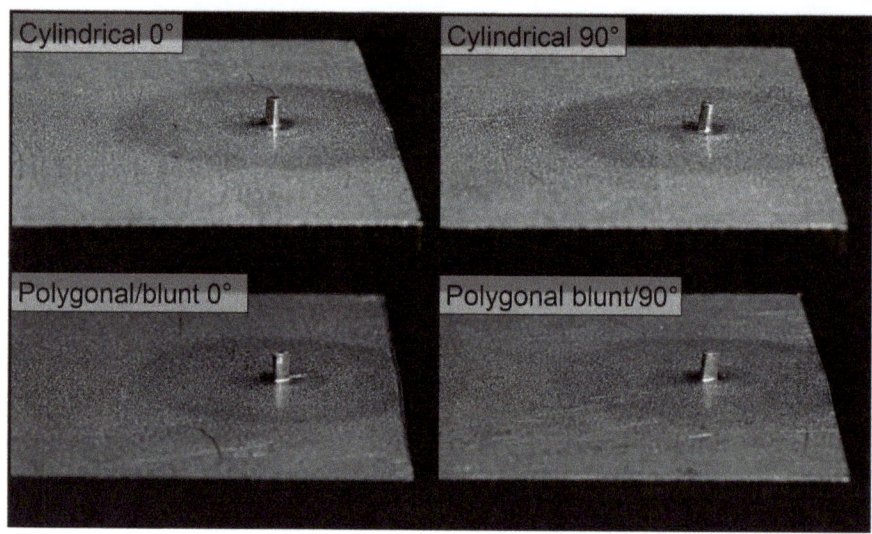

Figure 12. Cylindrical and polygonal blunt pins with fiber orientations of 0° and 90° after testing.

The shown deformation of the pin structure corresponds with the earlier described failure phenomena. Pins of samples with failure due to pin extraction are strongly deformed while samples with CFRT failure typically show only slightly deformed pins. The shown deflection leads to a changed flux of force during experimental testing with a resulting force component that facilitates the pin extraction and ultimately leads to a slipping of the pin from the joint and consequent joint failure. A possible way to counteract this tendency for pin extraction is the use of undercutting pin geometries, which interlock in the CFRT component and consequently counteract a pin extraction. This was already shown by Ucsnik et al. in [20] where in the field of metal/epoxy based composite joining, interlocking pin geometries with ball heads lead to 37.4% higher maximum forces in a double lap shear test.

4. Summary

In the scope of this study, it could be shown that in the field of direct pin pressing of cold formed pin structures into locally heated CFRTs, an increase in maximum force before failure can be achieved with non-rotational symmetric pins in comparison to the investigated cylindrical reference pin. The maximum transmittable force depends strongly on the factors of fiber orientation and pin orientation in relation to the load direction as well as the tested pin geometry. Therefore, elliptical pins and fibers both oriented in 90° to the load direction show the highest average maximum force with 415.0 ± 13.6 N. However, when relating the maximum force to the foot print of the pin structure as an indication for the efficiency of a pin structure, it shows that elliptical pins are inferior to cylindrical pins and that polygonal pin structures are, in dependency of the pin orientation, either slightly superior or slightly inferior to cylindrical pins.

Investigation of the failed samples shows that two distinct failure modes occur:

Failure of the CFRT component, which typically occurs when a sample with a fiber orientation in 0° is tested or when pins with a sharp edge such as elliptical pins with a 0° orientation or polygonal pins in the "sharp" orientation are combined with a fiber orientation of 90° leading to cut fibers.

Pin bending and subsequent pin extraction, which occurs at higher loads than CFRT failure, which are sufficient to deform the pin structure. This failure typically occurs with fiber orientations of 90°, which leads to a more efficient force introduction into the reinforcing fibers as the fibers are oriented perpendicular to the introduced load and support the introduced load similar to an arresting cable in naval aviation.

5. Conclusions

Based on the findings in this study, the selection of a certain pin structure must be made based on the expected loads in the use case of the hybrid parts. If an isotropic behavior of the joint is required, cylindrical pin structures seem favorable; if a main direction of load is expected, a polygonal pin with specifically chosen orientation could be beneficial. Thus, the findings of this work enable the design of customized joints adapted to the application with the aid of adapted pin geometries and orientations, which increases the versatility of the joining process.

Furthermore, a more detailed investigation of the pin extrusion of non-rotationally symmetrical geometries based on the different orientation of the elliptical pin structures has shown that the orientation in relation to the rolling direction of the blank has an influence on the required forming force, which can consequently affect the work hardening of the pin structure.

In future studies, it is necessary to investigate multi-pin arrays in order to increase the maximum transmittable forces. Therefore, it is required to determine minimum pin spacing in dependency of the pin's dimensions, fiber orientations and restrictions from a pin manufacturing standpoint. In addition, the effect of the orientation of non-rotationally symmetric pin geometries on the strain hardening of the pin structures should be investigated in future work, since strain hardening of the pin structure, especially in the pin base, plays a crucial role in the subsequent shear strength of the joint. In addition, the production of undercutting pin geometries via cold extrusion requires further studies, as it is expected that undercuts can increase the joint strength under shear loads, especially with a fiber orientation of 90°, where pin extraction is the dominant failure mode, which could be counteracted with an undercutting pin. Furthermore, future studies are planned to investigate different laminate structures, such as bi-axial laminates.

Author Contributions: Conceptualization, J.P.; methodology, J.P. and D.R.; formal analysis, J.P. and D.R.; investigation, J.P. and D.R.; resources, D.D. and M.M.; writing—original draft preparation, J.P. and D.R.; writing—review and editing, J.P., D.R., D.D. and M.M.; visualization, J.P. and D.R.; supervision, D.D. and M.M.; project administration, D.D. and M.M.; funding acquisition, D.D. and M.M. All authors have read and agreed to the published version of the manuscript.

Funding: This work was funded by the Deutsche Forschungsgemeinschaft (DFG, German Research Foundation)-TRR 285 C01-Project-ID 418701707. We acknowledge financial support by Deutsche Forschungsgemeinschaft and Friedrich-Alexander-Universität Erlangen-Nürnberg within the funding programme "Open Access Publication Funding".

Institutional Review Board Statement: Not applicable.

Informed Consent Statement: Not applicable.

Data Availability Statement: All relevant data is represented in this publication. Data sharing is not applicable to this article.

Conflicts of Interest: The authors declare no conflict of interest. The funders had no role in the design of the study; in the collection, analyses, or interpretation of data; in the writing of the manuscript, or in the decision to publish the results.

References

1. Siebenpfeiffer, W. *Leichtbau-Technologien im Automobilbau: ATZ/MTZ-Fachbuch, 1*; Springer Vieweg: Wiesbaden, Germany, 2014.
2. Barkoula, N.-M.; Karger-Kocsis, J. Effects of fibre content and relative fibre-orientation on the solid particle erosion of GF/PP composites. *Wear* **2002**, *252*, 80–87. [CrossRef]
3. Martinsen, K.; Hu, S.J.; Carlson, B.E. Joining of dissimilar materials. *CIRP Ann.—Manuf. Technol.* **2015**, *64*, 696–699. [CrossRef]
4. Dawai, Z.; Qi, Z.; Xiaoguang, F.; Shengdung, Z. Review on Joining Process of Carbon Fiber Reinforced Polymer and Metal: Methods and Joining Process. *Rare Met. Mater. Eng.* **2018**, *47*, 3686–3696. [CrossRef]
5. Mitschang, P.; Velthuis, R.; Rudolf, R. Fügeverfahren für FKV. In *Handbuch Verbundwerkstoffe*, 2nd ed.; Neitzel, M., Mitschang, P., Eds.; Carl Hanser: München, Germany; Wien, Autria, 2014; pp. 469–482.
6. Schürmann, H. Klebeverbindungen. In *Konstruieren mit Faser-Kunststoff-Verbunden*; Springer: Berlin/Heidelberg, Germany, 2005; pp. 569–604.
7. Brockmann, W. Adhesive Bonding of Polypropylene. In *Polypropylene: An A–Z Reference*; Karger-Kocsis, J., Ed.; Kluwer: London, UK, 1999; pp. 1–6.
8. Popp, J.; Kleffel, T.; Römisch, D.; Papke, T.; Merklein, M.; Drummer, D. Fiber orientation Mechanism of Continuous Fiber Reinforced Thermoplastic Hybrid Parts Joined with Metallic Pins. *Appl. Compos. Mater.* **2021**, *28*, 951–972. [CrossRef]
9. Eberl, L.; Avila Gray, L.; Zaremba, S.; Drechsler, K. The effect of fiber undulation on the strain field for pinned composite/titanium joints under tension. *Compos. Part A* **2017**, *103*, 148–160. [CrossRef]
10. Ucsnik, S.; Scheerer, M.; Zaremba, S.; Pahr, D.H. Experimental investigation of a novel hybrid metal-composite joining technology. *Compos. Part A* **2010**, *41*, 369–374. [CrossRef]
11. Graham, D.P.; Rezai, A.; Baker, D.; Smith, P.A.; Watts, J.F. A Hybrid Joining Scheme for High Strength Multi-Material Joints. In Proceedings of the 18th International Conference on Composite Materials, Jeju, Korea, 21–26 August 2011.
12. Parkes, P.N.; Butler, R.; Meyer, J.; de Oliveria, A. Static strength of metal-composite joints with penetrative reinforcement. *Compos. Struct.* **2014**, *118*, 250–256. [CrossRef]
13. Popp, J.; Kleffel, T.; Drummer, D. Influence of pin geometry on the joint strength of CFRT-metal hybrid parts with metallic pins. *Join. Plast. Fügen Kunstst.* **2021**, *3*, 177–183.
14. Kraus, M.; Frey, P.; Kleffel, T.; Drummer, D.; Merklein, M. Mechanical joining without auxiliary element by cold formed pins for multi-material-systems. *AIP Conf. Proc.* **2019**, *2113*, 050006.
15. Römisch, D.; Kraus, M.; Merklein, M. Investigation of the influence of formed, non-rotationally symmetrical pin geometries and their effect on the joint quality of steel and aluminium sheets by direct pin pressing. *Proc. Inst. Mech. Eng. Part L J. Mater. Des. Appl.* **2022**. [CrossRef]
16. Popp, J.; Drummer, D. Joining of continuous fiber reinforced thermoplastics/steel hybrid parts via undercutting pin geometries and infrared heating. *J. Adv. Join. Process.* **2022**, *5*, 100084. [CrossRef]
17. Deutscher Verband für Schweißen und Verwandte Verfahren. *DVS/EFB 3480-1 Testing of Properties of Joints—Testing of Properties of Mechanical and Hybrid (Mechanical/Bonded) Joints*; DVS Media GmbH: Düsseldorf, Germany; Beuth Verlag GmbH: Berlin, Germany, 2007.
18. DIN e.V. *DIN EN ISO 219:2008*; Plastics-Standard Atmospheres for Conditioning and Testing. Deutsches Institute für Normung: Berlin, Germany, 2008.
19. Ghassemali, E.; Tan, M.J.; Jarfors, A.; Lim, S. Progressive microforming process: Towards the mass production of micro-parts using sheet metal. *Int. J. Adv. Manuf. Technol.* **2013**, *66*, 611–621. [CrossRef]
20. Thakkar, R.; Ucsnik, S. Cost Efficient Metal to Fibre Reinforced Composite Joining. In Proceedings of the ECCM16—16th European Conference on Composite Materials, Seville, Spain, 22–26 June 2014.
21. Feistauer, E.E.; Guimaraes, R.P.M.; Ebel, T.; dos Santos, J.F.; Amancio-Filho, S.T. Ultrasonic joining: A novel direct-assembly technique for metal-composite hybrid structures. *Mater. Lett.* **2016**, *170*, 1–4. [CrossRef]

Article

Fabrication of Magnesium-Aluminum Composites under High-Pressure Torsion: Atomistic Simulation

Polina Viktorovna Polyakova [1,*], Julia Alexandrovna Pukhacheva [2], Stepan Aleksandrovich Shcherbinin [3], Julia Aidarovna Baimova [1,2] and Radik Rafikovich Mulyukov [1,2]

1. Institute for Metals Superplasticity Problems of RAS, St. Khalturina, 39, 450001 Ufa, Russia; julia.a.baimova@gmail.com (J.A.B.); radik@imsp.ru (R.R.M.)
2. Department of Physics and Technology of Nanomaterials, Bashkir State University, St. Zaki Validi, 32, 450076 Ufa, Russia; coldday87@gmail.com
3. Peter the Great St. Petersburg Polytechnic University, Polytechnicheskaya, 29, 195251 St. Petersburg, Russia; stefanshcherbinin@gmail.com
* Correspondence: polina.polyakowa@yandex.ru

Citation: Polyakova, P.V.; Pukhacheva, J.A.; Shcherbinin, S.A.; Baimova, J.A.; Mulyukov, R.R. Fabrication of Magnesium-Aluminum Composites under High-Pressure Torsion: Atomistic Simulation. *Appl. Sci.* **2021**, *11*, 6801. https://doi.org/10.3390/app11156801

Academic Editors: Oksana Rakhmanova and Suzdaltsev Andrey

Received: 25 May 2021
Accepted: 9 July 2021
Published: 24 July 2021

Publisher's Note: MDPI stays neutral with regard to jurisdictional claims in published maps and institutional affiliations.

Copyright: © 2020 by the authors. Licensee MDPI, Basel, Switzerland. This article is an open access article distributed under the terms and conditions of the Creative Commons Attribution (CC BY) license (https://creativecommons.org/licenses/by/4.0/).

Abstract: The aluminum–magnesium (Al–Mg) composite materials possess a large potential value in practical application due to their excellent properties. Molecular dynamics with the embedded atom method potentials is applied to study Al–Mg interface bonding during deformation-temperature treatment. The study of fabrication techniques to obtain composites with improved mechanical properties, and dynamics and kinetics of atom mixture are of high importance. The loading scheme used in the present work is the simplification of the scenario, experimentally observed previously to obtain Al–Cu and Al–Nb composites. It is shown that shear strain has a crucial role in the mixture process. The results indicated that the symmetrical atomic movement occurred in the Mg–Al interface during deformation. Tensile tests showed that fracture occurred in the Mg part of the final composite sample, which means that the interlayer region where the mixing of Mg, and Al atoms observed is much stronger than the pure Mg part.

Keywords: composite; molecular dynamics; magnesium; aluminum; mechanical properties

1. Introduction

Magnesium (Mg) alloys are of considerable interest nowadays in the automotive and aerospace industry since they have lightweight properties[1,2]. Moreover, magnesium-based alloys can be biocompatible and biodegradable. However, magnesium has quite low strength, elastic modulus, creep resistance, and formability [3], which considerably limits its applications. Thus, the search for new possibilities to fabricate composites based on Mg and other metals that can demonstrate improved mechanical properties is of high importance. For example, aluminum (Al) can prevent the corrosion process of Mg alloy and possess better properties for both Mg and Al [4–6].

One of the possible ways to obtain metallic composites is high-pressure torsion (HPT), during which high compressive and shear stresses result in the formation of the composite structures. To date, several types of composite materials were obtained by HPT from metallic plates: Al–Cu [7–13], Al–Mg [14–16], Al–Nb [17], and Al–Ti [18]. Since severe plastic deformation can increase the diffusion in the materials [19,20], in situ composites can be obtained by HPT through the bonding of different metals, which can lead to the formation of new intermetallic phases. According to the phase diagram, several intermetallic Al–Mg phases can be obtained, which are AlMg, Al_3Mg_2-$β$, $Al_{30}Mg_{23}$-$ε$, and $Al_{12}Mg_{17}$-$γ$ [21]. The intermetallic $Al_{12}Mg_{17}$ can significantly influence the corrosion and mechanical properties of the Al–Mg structure.

Fundamental aspects of deformation behavior between Al and Mg and the formation of intermetallic compounds can be effectively studied by molecular dynamics (MD). Due to some limitations of experimental methods, MD is widely used to study phase transformations and to determine structural properties of different materials [22–25]. Moreover, it allows scientists to visualize the structure on the atomistic level, analyze the distribution of atoms during different processing stages, and calculate physical or mechanical properties. MD was used for studying of effect of grain boundary segregation on the deformation mechanisms and mechanical properties of Al–Mg alloys [26–28], reduction of tensile strength for Al matrix with Mg inclusion [29], strengthening effects of basal stacking faults in Mg [30], etc.

In this work, the interactions between Al and Mg on Al–Mg interface under compression combined with shear load using the MD simulation method is studied. This combination of compression and shear is the attempt to realize HPT which was successfully used to obtain in situ composites. The present work is the continuation of the previous work by authors of [31], in which preliminary results were published. To study the strength of obtained mixed structure on the Al–Mg interface, a tensile strain is applied to the sample.

2. Computational Methods

To study the process of atomic mixing under deformation treatment near the interface of different metals, MD modeling of Mg–Al sample, containing an interfacial boundary was used. The cubic sample contained 54,170

atoms and was a simulation box with a dimension of $10.0 \times 10.0 \times 10.0$ nm^3. The contact surfaces of Mg and Al were (0001) and (001) planes, respectively. The Al and Mg layers were 5.0 nm thick. The schematic of the initial structure is shown in Figure 1, where the face-centered cubic (FCC) Al is bordered by the hexagonal (HPC) Mg. The atomic radius of Al was 143 pm and for Mg was 160 pm; atomic masses were 26.98 for Al and 24.307 for Mg. The interlayer distance between two crystals was calculated as $(a_{Mg} + a_{Al})/2$ = 3.6 Å. Periodic boundary conditions were applied along x, y, and z. Melting temperatures of Al and Mg were close and equal to 660 °C and 650 °C, respectively.

The sample with the minimized interfacial boundary energy was used for modeling. To obtain the initial structure in an equilibrium state, the minimal energy of the system was achieved by the multiple corrections of the atomic positions with the help of the steepest descent method, terminated if the variation in the energy or force was less than a given value. The initial structure was relaxed until the local or global minimum of the potential energy was reached. The simulation run was terminated when one of the stopping criteria (energy or force) was satisfied. The main goal of the relaxation process, in this case, was to obtain equilibrium stresses on the interface. After several numerical experiments with different parameters of minimization, stopping tolerance energy 10^{-24} and stopping tolerance force 10^{-26} eV/Å were chosen. In the present case, the setting of tolerance force 10^{-26} means no x, y, and z component of the force on any atom will be larger than 10^{-26} eV/Å after minimization.

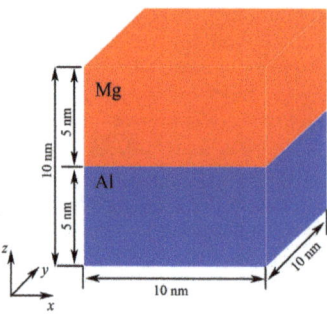

Figure 1. Schematic of the initial sample.

After the minimization process at 300 K, the simulation box was initially stress-free. The equilibrium state implies that normal stresses σ_{zz} in the two crystals are zero; the summation of stress σ_{xx} in the two crystals and the summation of stress σ_{yy} in the two crystals are zero. Moreover, the two dimensions of the sample (x and y) were not arbitrarily chosen because of the incommensurate nature of the FCC and HCP crystals but were determined such that the strains imposed on Al and Mg semi-infinite perfect crystals were minimized, ensuring periodic boundary conditions and equilibrium of the initial structure.

The simulations were carried out by MD using the large-scale atomic/molecular massively parallel simulator (LAMMPS) package. For temperature control, the Nose–Hoover thermostat was applied. Verlet algorithm to integrate the Newtonian equation of motion with an integration time-step of 2 fs was used. The well-approved Mg–Al embedded atom method (EAM) potential [32] was used in the molecular dynamics model. The EAM potentials for Al and Mg have been constructed on the basis of the experimental data and successfully used for a previous simulation in [33].

To obtain the composite structure and mixing of the atoms near the Al–Mg interface, uniaxial compression normal to the Al–Mg interface ($\varepsilon_{zz} < 0$), combined with shear in the interface plane (ε_{xy} in this case), was applied. It should be noted that in accordance with the deformation mechanism inserted in the LAMMPS simulation package, shear deformation is simply the change of the xy, xz, and yz tilt factors of the simulation box with the given strain rate [34]. To apply compression deformation, change of the specified dimension of the box via constant displacement (in this case, it was applied along z-dimension), which is effectively a constant engineering strain rate, occurred.

As previously shown [31,33], pure compression is not efficient enough to obtain composite. Thus, in this study, compression was combined with shear to reproduce analog to high-pressure torsion. Strain rates were $\dot{\varepsilon}_{zz} = 6.2 \times 10^{-8}$ ps^{-1} and $\dot{\varepsilon}_{xy} = 6.2 \times 10^{-7}$ ps^{-1}. Numerical experiments were conducted at room temperature 300 K, since it is the usual temperature for experiments.

In Figure 2, the stress–strain curve of the sample under compression is presented, along with the snapshots of the structure at three strain stages. There are several different possibilities to present stress–strain curves and calculate equivalent stress and strain in experiments when HPT is conducted [35–38]. However, the most useful equations for calculations of equivalent stress and strain proposed previously are not always suitable [38], especially when considering such a simple simulation, in which just two strain components are realized. For the present model, recalculation of compressive and shear stress and strain to the equivalent one can become physically unreasonable. Thus, stress–strain curves were presented for both components. Further, the shear strain is mentioned as characteristic for the description of the obtained results, since shear strain, in particular, results in better mixing of atoms.

The course of the curve for compression (see Figure 2a) before step II is almost linear, without considerable stress fluctuations, in comparison with the curve after step III. As it can be seen from the snapshots of the structure, even at $\varepsilon_{xy} = 0.5$ ($\varepsilon_{zz} = 0.05$), very good mixing of the atoms occurred, However, the strength of the interface and microstructure peculiarities should be estimated. In comparison with compressive components, shear stress is low (about 1 GPa), while the achieved strain is 10 times higher.

To check the strength of the obtained composite, tensile loading normal to the Al–Mg interface was applied ($\varepsilon_{zz} > 0$ in this case). Since this work is an attempt to reproduce the experimental work, all the parameters were chosen to be close to the experimental. Thus, tensile numerical tests were conducted at 300 K, which is usual for experiments. The tensile strain imposed in this simulation was performed by deforming the simulation box. During the dynamic loading, the stress was attained by the averaged stress, and the strain was derived from the positions of the periodic boundaries along the z-axis.

Tensile strain was applied after three steps of deformation and to the initial structure. The structure of the composite, obtained after compression to stage I, was considered as initial for the tensile test and was not additionally relaxed or changed. The same principle was applied for structures after compression to stage II and III. Thus, structures at points I, II, or III in Figure 2 are considerably stressed and at non-equilibrium.

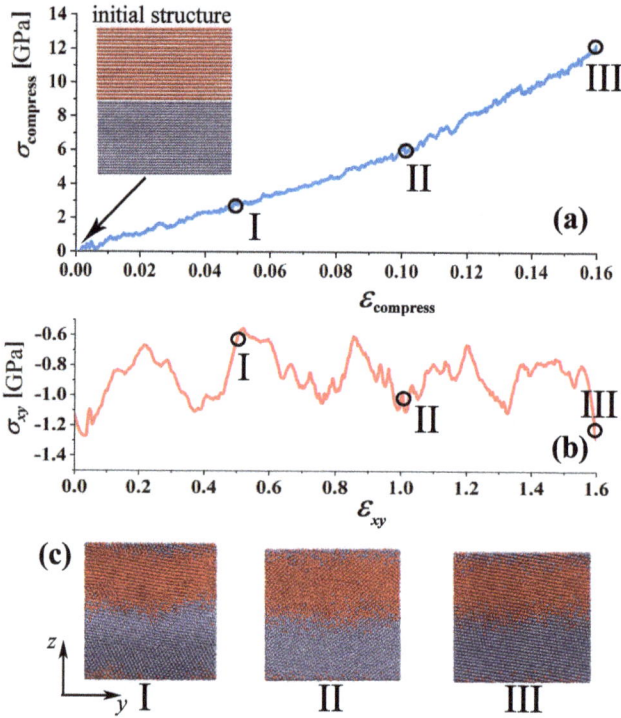

Figure 2. Stress–strain curves during uniaxial compression normal to the Al–Mg interface combined with shear: (a) compressive strain and stress; (b) shear strain and stress; (c) the snapshots of the structure at three stages of deformation. The red color is for Mg atoms, the blue color is for Al atoms.

Visualization of MD simulation data and structure analysis were carried out using the VMD [39] and OVITO [40] tools.

3. Results

3.1. Composite Fabrication

During deformation treatment, considerable mixing of Al and Mg atoms occurred through the interface. In Figure 3, changes of the atomic positions through Al–Mg interface for $0.0 < \varepsilon_{xy} < 1.6$ are presented. Mg block in the model is shifted to the right by about 100 Å for a clearance. As it can be seen, mostly, atomic mixing occurred during the first deformation stages $\varepsilon_{xy} < 0.4$. The first atomic movement occurred at $\varepsilon_{xy} = 0.017$ simultaneously with the appearance of base-centered cubic (BCC) lattice defects in the Mg part of the sample. Atomic positions between $\varepsilon_{xy} = 0.027$ and $\varepsilon_{xy} = 0.4$ are not presented since the continuous atomic movement occurred: Al atoms move towards the Mg part of the sample and vise versa.

The process of atomic migration can be better described by the average and maximum distances of an atomic displacement in comparison with the initial position of the boundary, which is presented in Figure 4. The value of Δz is calculated as the average movement of the atoms from the initial position of the interface. At $\varepsilon_{xy} = 0.017$, Al atoms moved from the atomic planes closest to the interface for $\Delta z_{max} = 2.2$ Å, and Mg atoms at the same strain moved for $\Delta z_{max} = 2.3$ Å. Considerable displacement of Al (Mg) atoms inside Mg (Al) part of the sample occurred before $\varepsilon_{xy}=0.4$, with the average displacement for 4 Å, while for $0.04 < \varepsilon_{xy} = 1.6$, the average atomic displacement is about 2 Å.

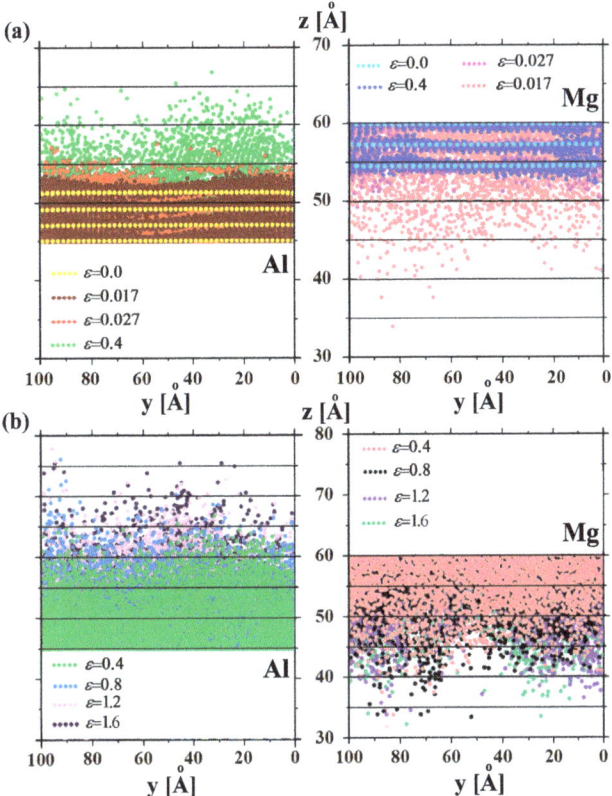

Figure 3. Changes of the atomic positions through Al–Mg interface: (**a**) for $0.0 < \varepsilon_{xy} < 0.4$ and (**b**) for $0.4 < \varepsilon_{xy} < 1.6$. Different colors correspond to different deformation stages. Only part of the sample is presented.

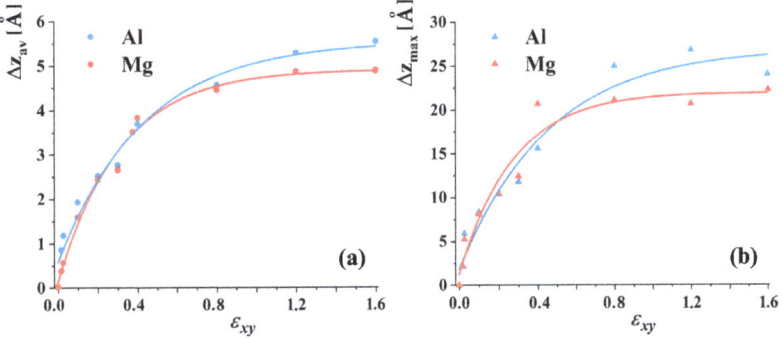

Figure 4. Changes of the atomic positions Δz as the function of compression strain. (**a**) Δz_{av}, (**b**) Δz_{max}.

As it can be seen from Figure 3, both average distances and changes of the maximum atomic displacements can be described by the power function for both metals. It is well seen that $\varepsilon = 0.4$ is enough to obtain good mixing of the atoms near the interface. After $\varepsilon = 1.6$, compression strain begins to prevent mixing, since the structure near the boundary is strongly compressed. As it is found, Al atoms move towards the Mg part of the sample slightly better, which is connected with the difference of the lattices: Mg HPC lattice is not as densely packed as FCC Al lattice. However, it should be mentioned that the number of Mg atoms moving into the Al part is almost equal to the number of Al atoms moving into the Mg matrix. This can be explained by the similarity of the atomic radii of Mg and Al. The lack of difference in the melting point for Mg (923 K) and Al (933.5 K) is worth considering, which means the bonds in both metals are similarly stronger to fracture. For both metals, a symmetrical binding movement occurred during the deformation of the initial sample.

In accordance with common-neighbor analysis (CNA), during the first stage of deformation (until $\varepsilon_{xy} = 0.4$), the Al part of the sample preserves FCC lattice, and the Mg part of the sample preserves HCP lattice with the appearance of BCC lattice. After $\varepsilon_{xy} = 0.0015$, a considerable number of dislocations appeared in the Al part, about two times more than in Mg. Dislocations in the Mg part rapidly move and disappear during shear, while in Al, they change type but never entirely disappear. Numerous dislocations appeared on the interface during a mixture of Al and Mg atoms.

For the first chosen stages I and II (before $\varepsilon = 1.0$), in the Mg part, mostly the HCP lattice can be seen from CNA, but further, the share of the BCC lattice increases, which is the main difference between structure at stage I and II, in comparison with stage III.

Although mixing dynamics is quite similar, structural transformations during compression to $\varepsilon_{xy} = 0.5$, $\varepsilon_{xy} = 1.0$, and $\varepsilon_{xy} = 1.6$ are different. The structure obtained after stages II and III are similar, in comparison with structure after stage I, which means that the strength of the final composite would be different.

In the experiment [9,10,12,13,17], annealing at 450 °C is applied after the initial deformation of the Al–Cu composite. It is observed, that after annealing, two more intermetallic phases have appeared in the structure, and microhardness increase two times. Although this work is also based on the experiment on Al–Mg composite fabrication, there are no published results for Al–Mg annealed after HPT.

In the present work, annealing at temperatures between 250 and 450° was conducted. However, no noticeable effect is observed for all the considered temperatures. Limitations of the MD model do not allow us to simulate annealing since considerable time is required for this process. Even four times bigger annealing time is not enough.

3.2. Al–Mg under Tension

In Figure 5, stress–strain curves with characteristic marks during tensile loading normal to the Al–Mg interface are presented for the tension of the initial sample (black curve, 1) and tension after compression at stages I (red curve, 2), II (green curve, 3), and III (blue curve, 4). Several pop-in events are observed on the stress–strain curves, which can be attributed to the release of strain energy accumulated during the deformation through defect activities. To analyze deformation mechanisms and strength of the composite, snapshots of the structure during tensile loading are presented for curves 1–3 in Figure 6, in accordance with CNA, and for curve 4 in Figure 7, in accordance with CNA and dislocation analysis.

The first structural changes of the initial sample occurred at $\varepsilon = 0.03$ in the Mg part of the sample (see Figure 6a). Fracture is observed in the Mg part of the sample at $\varepsilon = 0.045$ (shown by a circle in Figure 6a). Close to this limit, dislocations appeared in Mg together with other structural defects. This result is quite expected since Mg is well known for having low strength. However, this result is important for further comparison.

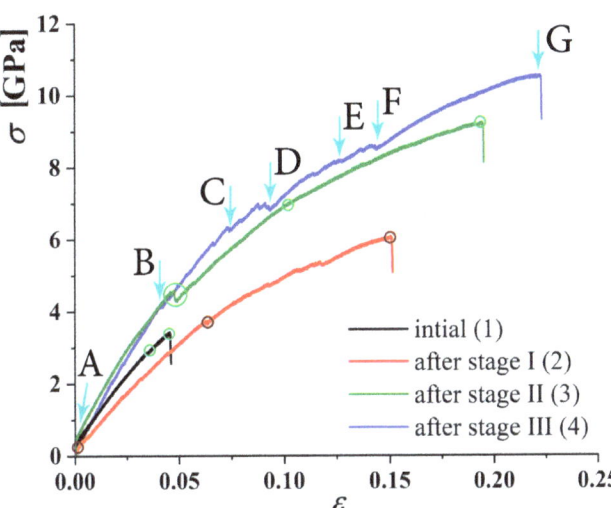

Figure 5. Stress–strain curves during tension normal to the interface region after different initial compression strain: after $\varepsilon_{xy} = 0.5$ (2); $\varepsilon_{xy} = 1.0$ (3), and $\varepsilon_{xy} = 1.6$ (4).

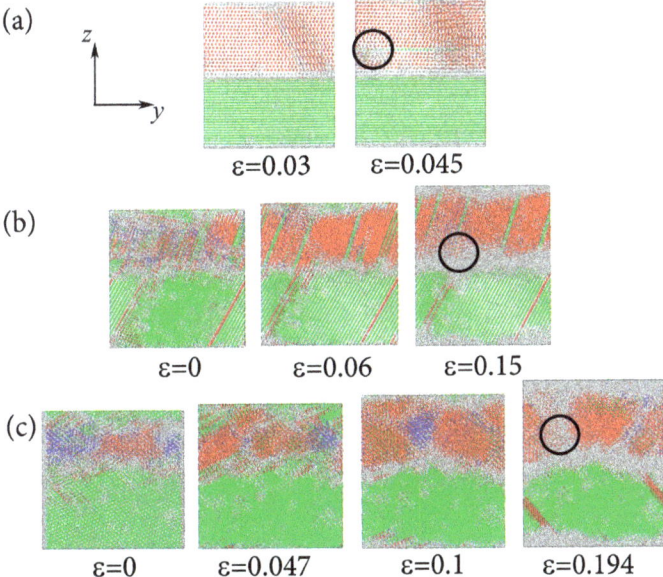

Figure 6. Snapshots of the structure in accordance with stress–strain curves shown in Figure 5: (**a**) tension of the initial sample; (**b**) tension of the sample after compression, stage I; (**c**) tension of the sample after compression, stage II. All atoms are colored by the CNA parameter, where HCP atoms are red, FCC atoms are green, BCC atoms are blue, and other atoms are gray.

From Figure 6b, it can be seen that fracture of the sample compressed to stage I occurred at $\varepsilon = 0.15$ on the boundary between Al and Mg, where the mixed structure is obtained. This means that the interlayer part is even weaker than the Mg part (see also Figure 5, curves 1 and 2). At tensile strain $\varepsilon = 0.0$, there are two lattices in the MG part of the sample—HPC (red atoms) and BCC (blue atoms); however, after $\varepsilon = 0.06$, BCC phase almost disappears. It should be noted that a single HCP atom layer represents a twin boundary, two adjacent HCP atom layers present an intrinsic stacking fault, and an FCC atom layer in the middle of two HCP atom layers stands for an intrinsic stacking fault. As in the experiments or other MD simulations, twinning is one of the effective mechanisms of deformation for Mg [41,42].

From Figure 6c, it can be seen that fracture of the sample compressed to stage II occurred at $\varepsilon = 0.194$ in the Mg part of the sample, between two boundary regions. At $\varepsilon = 0.047$, the share of the BCC phase considerably decreases; however, the Mg part of the sample contains the BCC phase during the whole deformation process.

Figure 7 presents a series of snapshots exhibiting the defect evolution during tension after stage III of compression corresponding to the characteristic points marked in the stress–strain curve 4 of Figure 5. This case is chosen for detailed analysis because the strength of the sample after stage III is the highest. The other atoms are colored grey, the HCP atoms are colored red, BCC atoms are colored blue, and the FCC atoms are colored green. In accordance with the OVITO dislocation extraction algorithm, the stair-rod dislocation lines are colored purple, the Hirth dislocation lines are colored yellow, Frank dislocation lines are colored light blue, the Perfect dislocation lines are colored blue, and the Shockley dislocation lines are colored green.

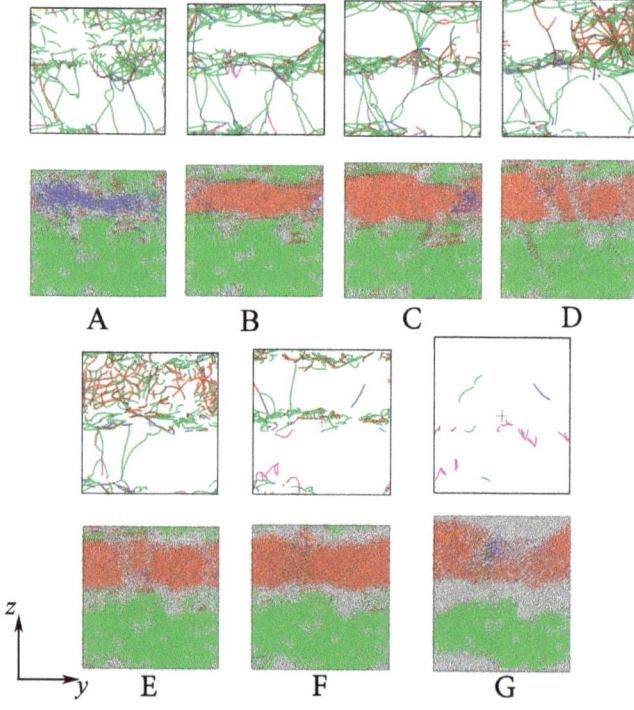

Figure 7. Snapshots of the structure in accordance with stress–strain curve shown in Figure 5 for tension after stage III. All atoms are colored by the CNA parameter, where HCP atoms are red, FCC atoms are green, BCC atoms are blue, and other atoms are gray.

After compression to $\varepsilon = 0.16$, the BCC phase is dominant in the Mg part of the sample. It should be mentioned that an even boundary region with mixed Al and Mg atoms is shown by green FCC lattice. At tensile strain $\varepsilon = 0.04$, the BCC phase almost disappears, and this region on the stress–strain curve is almost linear. Curves 3 and 4 from Figure 5 coincide before $\varepsilon = 0.045$, and both have similar structural transformations. Numerous dislocations appeared on the interface between Al and Mg.

At point C, new dislocations appeared in the Mg part, while dislocation distribution in Al remained almost the same from $\varepsilon = 0.05$ to $\varepsilon = 0.1$. After $\varepsilon = 0.1$, dislocation network is developed, mainly in the Mg part, until a tensile strain equal to 0.15 is achieved. After that, as it can be seen from Figure 5, on curve 4, no pop-in events are observed, which is connected with the total change of the dislocation structure (Figure 7F,G).

Fracture occurred at $\varepsilon = 0.22$ on the opposite side of the crystal, which cannot be seen in Figure 7G, but the crack appeared in the Mg part of the composite, where the BCC phase is localized. Close to the strength limit, almost all dislocations disappeared from the structure.

4. Conclusions

Molecular dynamics simulation was used to study and analyze atomic movement on the Al–Mg interface under high pressure combined with shear strain. The proposed model was based on the scenario, experimentally observed previously in [9,10,12,13,17] for Al–Nb and Al–Cu composites. It is found that compression combined with shear is an effective method to obtain composite structure on the Al–Mg interface.

Considerable mixing of Al and Mg atoms occurred after compressive strain 0.04, with simultaneously applied shear strain 0.4; however, it is found that at this stage, the mixed Al–Mg region is weak, and fracture occurred on the boundary region. Further deformation treatment is required to obtain the formation of a strong interface. From the tension tests, it is found that there is a critical compression level after which no considerable structural changes can be achieved. Moreover, the higher the applied shear strain is, the higher the strength of the composite is.

In the present work, no effect is observed after annealing of the compressed samples, which is in contradiction with previous experimental results on the Al–Cu interface [9,10,12,13,17]. This can be explained by the difference in the metals used for composite fabrication, the weaknesses of the methodology, or by the lack of understanding of how to find better annealing temperatures. The temperature of 450 °C chosen from the experiments for Al–Cu does not facilitate the mixing of the atoms.

Author Contributions: Conceptualization, J.A.B. and R.R.M.; methodology, P.V.P.; formal analysis, P.V.P.; resources, S.A.S.; investigation, J.A.P.; writing—original draft preparation, J.A.B. and R.R.M. All authors have read and agreed to the published version of the manuscript.

Funding: This research was funded by Russian Science Foundation Grant Number 18-12-00440. Work of R.M. supported by the program of fundamental researches of Government Academy of Sciences of IMSP RAS. The work of S.S. was supported by the Government of the Russian Federation (state assignment 0784-2020-0027).

Acknowledgments: The authors wish to acknowledge Peter the Great Saint-Petersburg Polytechnic University Supercomputer Center "Polytechnic" for computational resources.

Conflicts of Interest: The authors declare no conflict of interest.

References

1. Agnew, S.; Nie, J. Preface to the viewpoint set on: The current state of magnesium alloy science and technology. *Scr. Mater.* **2010**, *63*, 671–673. [CrossRef]
2. Suh, B.C.; Shim, M.S.; Shin, K.; Kim, N.J. Current issues in magnesium sheet alloys: Where do we go from here? *Scr. Mater.* **2014**, *84-85*, 1–6. [CrossRef]
3. Mordike, B.; Ebert, T. Magnesium. *Mater. Sci. Eng. A* **2001**, *302*, 37–45. [CrossRef]
4. Bamberger, M.; Dehm, G. Trends in the Development of New Mg Alloys. *Annu. Rev. Mater. Res.* **2008**, *38*, 505–533. [CrossRef]
5. Zhang, Z.; Couture, A.; Luo, A. An investigation of the properties of Mg-Zn-Al alloys. *Scr. Mater.* **1998**, *39*, 45–53. [CrossRef]
6. Yakubtsov, I.; Diak, B.; Sager, C.; Bhattacharya, B.; MacDonald, W.; Niewczas, M. Effects of heat treatment on microstructure and tensile deformation of Mg AZ80 alloy at room temperature. *Mater. Sci. Eng. A* **2008**, *496*, 247–255. [CrossRef]
7. Bazarnik, P.; Bartkowska, A.; Romelczyk-Baishya, B.; Adamczyk-Cieślak, B.; Dai, J.; Huang, Y.; Lewandowska, M.; Langdon, T.G. Superior strength of tri-layered Al–Cu–Al nano-composites processed by high-pressure torsion. *J. Alloys Compd.* **2020**, *846*, 156380. [CrossRef]

8. Kocich, R.; Kunčická, L. Development of structure and properties in bimetallic Al/Cu sandwich composite during cumulative severe plastic deformation. *J. Sandw. Struct. Mater.* **2021**, 109963622199388. [CrossRef]
9. Korznikova, G.; Nazarov, K.; Khisamov, R.; Sergeev, S.; Shayachmetov, R.; Khalikova, G.; Baimova, J.; Glezer, A.; Mulyukov, R. Intermetallic growth kinetics and microstructure evolution in Al-Cu-Al metal-matrix composite processed by high pressure torsion. *Mater. Lett.* **2019**, *253*, 412–415. [CrossRef]
10. Mulyukov, R.R.; Korznikova, G.F.; Nazarov, K.S.; Khisamov, R.K.; Sergeev, S.N.; Shayachmetov, R.U.; Khalikova, G.R.; Korznikova, E.A. Annealing-induced phase transformations and hardness evolution in Al–Cu–Al composites obtained by high-pressure torsion. *Acta Mech.* **2021**. [CrossRef]
11. Oh-ishi, K.; Edalati, K.; Kim, H.S.; Hono, K.; Horita, Z. High-pressure torsion for enhanced atomic diffusion and promoting solid-state reactions in the aluminum–copper system. *Acta Mater.* **2013**, *61*, 3482–3489. [CrossRef]
12. Khisamov, R.K.; Korznikova, G.F.; Khalikova, G.R.; Sergeev, S.N.; Nazarov, K.S.; Shayachmetov, R.U.; Mulyukov, R.R. Fabrication of an in situ Al-Nb Metal Matrix Composite by Constrained Shear Strain and Its Emission Efficiency in a Glow Discharge. *Tech. Phys. Lett.* **2020**, *46*, 1200–1202. [CrossRef]
13. Danilenko, V.; Sergeev, S.; Baimova, J.; Korznikova, G.; Nazarov, K.; Khisamov, R.K.; Glezer, A.; Mulyukov, R. An approach for fabrication of Al-Cu composite by high pressure torsion. *Mater. Lett.* **2019**, *236*, 51–55. [CrossRef]
14. Castro, M.; Pereira, P.H.; Figueiredo, R.; Langdon, T. Developing magnesium-based composites through high-pressure torsion. *Lett. Mater.* **2019**, *9*, 541–545. [CrossRef]
15. Castro, M.; Wolf, W.; Isaac, A.; Kawasaki, M.; Figueiredo, R. Consolidation of magnesium and magnesium-quasicrystal composites through high-pressure torsion. *Lett. Mater.* **2019**, *9*, 546–550. [CrossRef]
16. Kawasaki, M.; Ahn, B.; Lee, H.; Zhilyaev, A.P.; Langdon, T.G. Using high-pressure torsion to process an aluminum–magnesium nanocomposite through diffusion bonding. *J. Mater. Res.* **2015**, *31*, 88–99. [CrossRef]
17. Korznikova, G.; Korznikova, E.; Nazarov, K.; Shayakhmetov, R.; Khisamov, R.; Khalikova, G.; Mulyukov, R. Structure and Mechanical Behavior of Al–Nb Hybrids Obtained by High-Pressure-Torsion-Induced Diffusion Bonding and Subsequent Annealing. *Adv. Eng. Mater.* **2020**, *23*, 2000757. [CrossRef]
18. Bartkowska, A.; Bazarnik, P.; Huang, Y.; Lewandowska, M.; Langdon, T.G. Using high-pressure torsion to fabricate an Al–Ti hybrid system with exceptional mechanical properties. *Mater. Sci. Eng. A* **2021**, *799*, 140114. [CrossRef]
19. Goddard, W.A. III; Brenner, D.; Lyshevski, S.E.; Iafrate, G.J. *Handbook of Nanoscience, Engineering, and Technology*; CRC Press: Boca Raton, FL, USA, 2002; Chapter 22, pp. 22-1–22-41.
20. Divinski, S.V.; Reglitz, G.; Rösner, H.; Estrin, Y.; Wilde, G. Ultra-fast diffusion channels in pure Ni severely deformed by equal-channel angular pressing. *Acta Mater.* **2011**, *59*, 1974–1985. [CrossRef]
21. Murray, J.L. The Al-Mg (Aluminum-Magnesium) system. *J. Phase Equilib.* **1982**, *3*. [CrossRef]
22. Poletaev, G.M.; Zorya, I.V.; Starostenkov, M.D. Role of point defects in self-diffusion along low-angle twist boundaries in fcc metals: A molecular dynamics study. *J. Micromech. Mol. Phys.* **2018**, *03*, 1850001. [CrossRef]
23. Ghaffari, M.A.; Zhang, Y.; Xiao, S. Molecular dynamics modeling and simulation of lubricant between sliding solids. *J. Micromech. Mol. Phys.* **2017**, *2*, 1750009. [CrossRef]
24. Zorya, I.; Poletaev, G.; Rakitin, R.; Ilyina, M.; Starostenkov, M. Interaction of impurity atoms of light elements with self-interstitials in fcc metals. *Lett. Mater.* **2019**, *9*, 207–211. [CrossRef]
25. Zakharov, P.V.; Poletaev, G.; Starostenkov, M.; Cherednichenko, A. Simulation of the shock waves propagation through the interface of bipartite bimetallic Ni-Al particles. *Lett. Mater.* **2017**, *7*, 296–302. [CrossRef]
26. Babicheva, R.I.; Dmitriev, S.V.; Bai, L.; Zhang, Y.; Kok, S.W.; Kang, G.; Zhou, K. Effect of grain boundary segregation on the deformation mechanisms and mechanical properties of nanocrystalline binary aluminum alloys. *Comput. Mater. Sci.* **2016**, *117*, 445–454. [CrossRef]
27. Koju, R.; Mishin, Y. Atomistic study of grain-boundary segregation and grain-boundary diffusion in Al-Mg alloys. *Acta Mater.* **2020**, *201*, 596–603. [CrossRef]
28. Kazemi, A.; Yang, S. Effects of magnesium dopants on grain boundary migration in aluminum-magnesium alloys. *Comput. Mater. Sci.* **2021**, *188*, 110130. [CrossRef]
29. Pogorelko, V.V.; Mayer, A.E. Influence of titanium and magnesium nanoinclusions on the strength of aluminum at high-rate tension: Molecular dynamics simulations. *Mater. Sci. Eng. A* **2016**, *662*, 227–240. [CrossRef]
30. Zhang, J.; Dou, Y.; Wan, X.; Chen, J. The strengthening effects of basal stacking faults on {10–12} twin in magnesium: A molecular dynamics study. *Comput. Condens. Matter* **2020**, *23*, e00466. [CrossRef]
31. Polyakova, P.V.; Baimova, J.A. Molecular dynamics simulation of diffusion in Mg-Al system under pressure. *IOP Conf. Ser. Mater. Sci. Eng.* **2021**, *1008*, 012052. [CrossRef]
32. Liu, X.Y.; Adams, J. Grain-boundary segregation in Al–10%Mg alloys at hot working temperatures. *Acta Mater.* **1998**, *46*, 3467–3476. [CrossRef]
33. Polyakova, P.V.; Nazarov, K.S.; Khisamov, R.K.; Baimova, J.A. Molecular dynamics simulation of structural transformations in Cu-Al system under pressure. *J. Phys. Conf. Ser.* **2020**, *1435*, 012065. [CrossRef]
34. Available online: https://docs.lammps.org/Manual.html (accessed on 6 July 2021).
35. Zhilyaev, A.P.; McNelley, T.R.; Langdon, T.G. Evolution of microstructure and microtexture in fcc metals during high-pressure torsion. *J. Mater. Sci.* **2006**, *42*, 1517–1528. [CrossRef]

36. Kuznetsov, R.I.; Bykov, V.I.; Chernyshov, P.; Pilyugin, V.; Yefremov, N.; Posheye, V. *IFM UNTS AN USSR*; Preprint 4/85; USSR: Sverdlovsk, Russia, 1985. (In Russian)
37. Eichinger, A. *Handbuch der Werkstoffprüfung*; Springer: Berlin, Germany, 1955.
38. Jonas, J.J.; Ghosh, C.; Toth, L.S. The equivalent strain in high pressure torsion. *Mater. Sci. Eng. A* **2014**, *607*, 530–535. [CrossRef]
39. Available online: https://www.ks.uiuc.edu/ (accessed on 5 July 2021).
40. Available online: https://www.ovito.org (accessed on 6 July 2021).
41. Li, B.; Ma, E. Zonal dislocations mediating 10-11<10-1-2> twinning in magnesium. *Acta Mater.* **2009**, *57*, 1734–1743. [CrossRef]
42. Serra, A.; Pond, R.; Bacon, D. Computer simulation of the structure and mobility of twinning disclocations in H.C.P. Metals. *Acta Metall. Mater.* **1991**, *39*, 1469–1480. [CrossRef]

Article

The Synthesis of Aluminum Matrix Composites Reinforced with Fe-Al Intermetallic Compounds by Ball Milling and Consolidation

Roberto Ademar Rodríguez Díaz [1,*], Sergio Rubén Gonzaga Segura [2], José Luis Reyes Barragán [3], Víctor Ravelero Vázquez [3], Arturo Molina Ocampo [2], Jesús Porcayo Calderón [4], Héctor Cruz Mejía [5], Carlos Alberto González Rodríguez [5] and Jesús Israel Barraza Fierro [6]

1. Departamento de Ingeniería de Materiales, Tecnológico Nacional de México, Tecnológico de Estudios Superiores de Coacalco, Coacalco de Berriozábal 55700, Mexico
2. Centro de Investigación en Ingeniería y Ciencias Aplicadas–(IICBA), Universidad Autónoma del Estado de Morelos, Cuernavaca 62209, Mexico; sergio_65_2005@hotmail.com (S.R.G.S.); arturo_molina@uaem.mx (A.M.O.)
3. Departamento de Ingeniería en Diseño, Universidad Politécnica de la Zona Metropolitana de Guadalajara, Tlajomulco de Zúñiga 45640, Mexico; jlbecario@yahoo.com (J.L.R.B.); victor.ravelero@gmail.com (V.R.V.)
4. Departamento de Ingeniería Química y Metalurgia, Universidad de Sonora, Hermosillo 83000, Mexico; jporcayoc@gmail.com
5. División de Ingeniería en Nanotecnología, Universidad Politécnica del Valle de México, Tultitlan 54910, Mexico; hcruzmejia@gmail.com (H.C.M.); carlos.gonzalez@upvm.edu.mx (C.A.G.R.)
6. Escuela de Preparatoria, Universidad La Salle Nezahualcóyotl, Ciudad de México 57300, Mexico; jbarraza@ulsaneza.edu.mx
* Correspondence: ademar@tesco.edu.mx

Citation: Díaz, R.A.R.; Segura, S.R.G.; Barragán, J.L.R.; Vázquez, V.R.; Ocampo, A.M.; Calderón, J.P.; Mejía, H.C.; Rodríguez, C.A.G.; Fierro, J.I.B. The Synthesis of Aluminum Matrix Composites Reinforced with Fe-Al Intermetallic Compounds by Ball Milling and Consolidation. Appl. Sci. 2021, 11, 8877. https://doi.org/10.3390/app11198877

Academic Editor: Suzdaltsev Andrey

Received: 21 August 2021
Accepted: 16 September 2021
Published: 24 September 2021

Publisher's Note: MDPI stays neutral with regard to jurisdictional claims in published maps and institutional affiliations.

Copyright: © 2021 by the authors. Licensee MDPI, Basel, Switzerland. This article is an open access article distributed under the terms and conditions of the Creative Commons Attribution (CC BY) license (https://creativecommons.org/licenses/by/4.0/).

Abstract: In this study, a nano-composite material of a nanostructured Al-based matrix reinforced with Fe40Al intermetallic particles was produced by ball milling. During the non-equilibria processing, the powder mixtures with the compositions of Al-XFe40Al (X = 5, 10, and 15 vol. %) were mechanically milled under a low energy regime. The processed Al-XFe40Al powder mixtures were subjected to uniaxial pressing at room temperature. Afterward, the specimens were subjected to a sintering process under an inert atmosphere. In this thermal treatment, the specimens were annealed at 500 °C for 2 h. The sintering process was performed under an argon atmosphere. The crystallite size of the Al decreased as the milling time advanced. This behavior was observed in the three specimens. During the ball milling stage, the powder mixtures composed of Al-XFe40Al did not experience a mechanochemical reaction that could lead to the generation of secondary phases. The crystallite size of the Al displayed a predominant tendency to decrease during the ball milling process. The microstructure of the consolidated specimens indicated a uniform dispersion of the intermetallic reinforcement phases in the Al matrix. Moreover, according to the Vickers microhardness tests, the hardness varied linearly with the increase in the concentration of the Fe40Al intermetallic phase present in the composite material. The presented graphs indicate that the hardness increased almost linearly with the increasing dislocation density and with the reduction in grain sizes (both occurring during the non-equilibria processing). The microstructural and mechanical properties reported in this paper provide the aluminum matrix composite materials with the ideal conditions to be considered candidates for applications in the automotive and aeronautical industries.

Keywords: Al composite; Fe40Al intermetallic; mechanical milling

1. Introduction

Al-based alloys are widely used as parts and components in the automotive and aerospace industry sectors due to their high ductility, specific strength, formability, and stiffness [1]. However, both pure Al and Al alloys possess low hardness and poor wear resistance. A well-known strategy to improve the wear resistance of Al and Al alloys is to produce Al composites. In this sense, it is well known that Al composites provide better wear resistance and improved bulk mechanical properties [2]. In particulate metal matrix composites (MMCs), the function of reinforcement particles is to strengthen the metal matrix. The mechanical behavior of the MMCs is greatly influenced by the properties of the matrix and reinforcement, volume fraction, their distribution, and size of reinforcement, as well as the interfacial strength between the matrix and reinforcement [1]. Adequate distribution of the particulate reinforcement in the metal matrix can be attained by the ball milling technique [3–8].

Aluminum matrix composites have been produced using two methods—liquid state processing (ex-stir casting) and solid-state processing (powder metallurgy, mechanical alloying). In the powder metallurgy processing method, the Al matrix and reinforcement are mixed in order to be compacted with a subsequent sintering process to gain strength. Aluminum MMCs are typically reinforced by oxides or ceramic particles. Considering that in ex situ processing, the reinforcement is added externally to the liquefied metal, problems of wettability and particle agglomeration needed to be resolved [9]. In this regard, the powder metallurgy method emerged with the aim of avoiding wettability and agglomeration problems and promoting a uniform dispersion of reinforcement in the Al matrix. Another strategy to counter the disadvantages of utilizing ex situ processing is to use phases other than ceramics as reinforcements. For example, intermetallic compounds could be considered candidate materials to reinforce Al matrix composites since ordered intermetallic compounds based on the aluminides of transition metals, such as Fe, Ni, Nb, Ti, and Co, have been assessed and studied for their potential application as structural materials at medium and high temperatures [10]. Moreover, the Al contents of these compounds induce the formation of a passive layer of Al oxide, which is responsible for very good corrosion and oxidation resistance at elevated temperatures [11]. Iron aluminides of the type Fe-Al and Fe3Al exhibit excellent mechanical resistance at high temperatures, low densities, high melting points, and good structural stability; however, efforts were applied to enhance the ductility and impact resistance of these compounds [12].

In terms of producing an MMC with improved mechanical properties along with low density, intermetallic compounds are very suited to these purposes. For example, the intermetallic phases of the Al-Ca binary alloy system are among the materials that can produce very light metal matrix composites, as these phases are extremely low in density [13]. The mechanical resistance of MMCs is significantly influenced by the interfacial strength between the matrix and reinforcement, which depends on the wettability between the metal and reinforcing particles. In this sense, it has been reported that intermetallics can be wetted by molten metal without difficulty. Additionally, they exhibit comparable mechanical behavior and a coefficient of thermal expansion closer to that of Al when compared to hard ceramic phases [14]. Therefore, in recent times, intermetallics based on Ni-Al and Ti-Al systems have been investigated as a new type of reinforcement for MMCs [15–18]. It is worth noting that the stiffness and wear behavior of the MMCs can be enhanced by the addition of intermetallics.

Thus, the purpose of this investigation was to take into account the advantages provided by the powder metallurgy route and the excellent properties that intermetallic compounds possess, in order to fabricate Al matrix composites reinforced with intermetallics of the Fe-Al system. Another goal was to analyze and study the influence of the processing parameters on the microstructure and hardness of the composites in order to achieve desired properties for automotive or aeronautic applications.

2. Materials and Methods

Processing route. The stages that were carried out in the processing route developed in this work are described below.

First, cast ingots of binary Fe40Al intermetallic compound were prepared using a high-frequency induction furnace at around 1500 °C, as shown in Figure 1a. High purity (99.9%) Fe and Al were placed inside a silicon carbide crucible for induction melting.

The molten Fe–40 wt.% Al (labeled Fe40Al hereafter) alloy was poured into a rectangular parallelepiped steel mold and subsequently solidified as it cooled until it reached room temperature; these cooling conditions resulted in a coarse-grained microstructure. Then, in order to pulverize the ingots of the intermetallic compound, small pieces of the ingot with approximate measurements of 1 cm in height × 1 cm in length × 0.5 cm in width were cut. Then, these small ingot pieces were mechanically struck with a hammer, taking advantage of the fragility of the Fe40Al intermetallic compound in order to obtain a powder of this material with a particle size of less than 1 mm. Subsequently, the intermetallic composition powders were subjected to mechanical milling, and thus, their final size was obtained at this stage (see a schematic representation in Figure 1b). After, the powder mixture of Al and Fe40Al was subjected to planetary ball milling for various periods in order to reduce the particle sizes of both phases, obtain a uniform distribution of the Fe40Al intermetallic compound in Al, and induce the homogenization of the chemical elements in the powder mixture (see the illustrative image in Figure 1c). Then, the powder mixture of the Al and Fe was uniaxially pressed at room temperature in a cylindrical mold, as shown in Figure 1d. Subsequently, the green compacts obtained by uniaxial pressing were subjected to a sintering process in a conventional furnace under an argon atmosphere (see the illustration presented in Figure 1d).

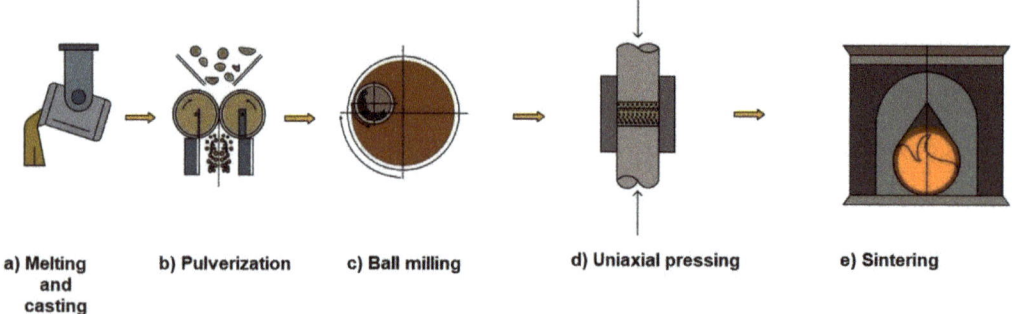

Figure 1. The processing route developed in the present work, involving melting and casting along with some powder metallurgy techniques of powder.

The processing parameters of each of the stages involved in the processing route developed in the present work are described as follows.

Materials. The elemental powders of metallic Al and intermetallic Fe40Al (99.9% purity) were blended in a mortar—three-particle volume fractions of 5, 10, and 15% of Fe40Al intermetallic particles, were added to the aluminum powder.

Composite powder. The powder mixture was introduced, under an argon atmosphere, to a hardened steel vial with balls of the same material. Analytical grade methyl alcohol, used as a process control agent, was added at a ratio of 0.003 mL per gram of mixture. Later, the mixture was mechanically milled by utilizing a planetary mill at a speed of 250 RPM during different milling times: 0.5, 1, 2, 4, 6, and 8 h. During the ball milling tests, a balls-to-powder weight ratio of 10:1 was employed.

Consolidation of composite powders. The powder mixture that was milled for 8 h was uniaxially pressed by using a pressure of 15 tons with a hydraulic press. The resulting compacts possessed a diameter of 3/8 in and $\frac{1}{4}$ in of height. Then, the green compacts were sintered at 500 °C for 2 h with subsequent cooling inside the furnace.

Characterizations. The evolution of the morphology, the particle size, and the distribution of the chemical elements of the powder mixture were analyzed in a scanning electron microscope (SEM, Stereoscan 440), which was connected to software for energy dispersive spectroscopy (EDS). X-ray diffraction analysis was performed to determine and identify the crystalline structure of the different phases that remained or were formed during milling, as well as to determine the crystallite size, crystalline state transitions, and the lattice parameters from the X-ray diffraction peaks. To perform this analysis, a Bruker D2 Phaser Diffractometer D500 was utilized, in which a voltage of 30 kV was applied with a current of 20 mA. The samples were scanned with a filter of CuKα radiation with a wavelength of λ = 1.5418 Å, using a step of $0.020°/0.6$ s in a range of 20° to 80°. Vickers microhardness measurements were performed in a Future-Tech Corp FM 700 microhardness tester to determine the microhardness of the sintered samples.

3. Results

3.1. Microstructural Analysis by Scanning Electron Microscopy

Figure 2 presents the morphology of the elemental starting powders before the ball milling process. These powders correspond to Al and Fe40Al, respectively. Figure 2a displays the irregular-shaped particles of Al (with a mean particle size of 2.7 microns ± 0.95). Figure 2b exhibits the irregular-shaped Fe40Al intermetallic particles (with a mean particle size of 27 microns ± 2.9).

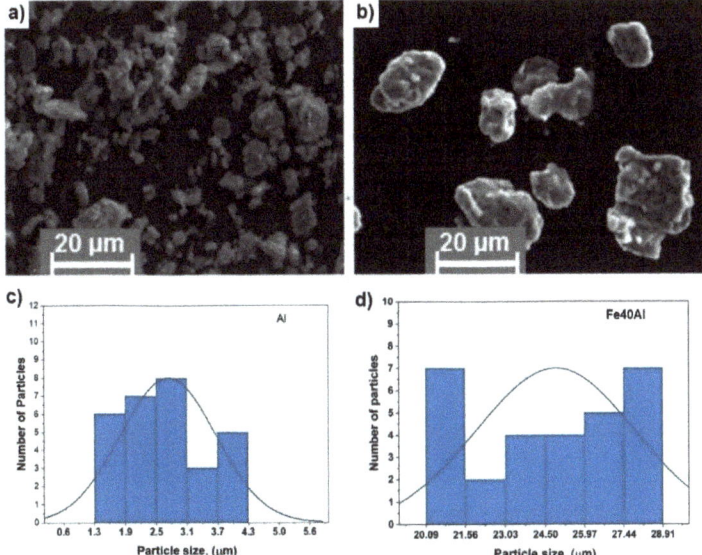

Figure 2. Scanning electron micrograph of (**a**) Al powder particles and (**b**) Fe40Al intermetallic powder particles. Particle size distributions of (**c**) Al and (**d**) Fe40Al.

Figure 3 presents the evolution of the particle morphology and size of the powder mixtures of Al-5Fe40Al, Al-10Fe40Al, and Al-15Fe40Al respectively. Aiming to improve the visibility and synthesize the results of the particle size, the micrographs corresponding to 0, 2, 4, and 8 h of milling taken from each composition (Al-5Fe40Al, Al-10Fe40Al, and Al-15Fe40Al), are presented in one figure. The scanning electron micrographs show that the particle size decreased as the milling time advanced from 0 to 8 h in the cases of the Al-10Fe40Al and Al-15Fe40Al powder mixtures. However, the particle size of Al-5Fe40Al underwent a progressive diminution from the beginning to the end of the processing. In addition, according to these micrographs, the particle morphology remained irregular.

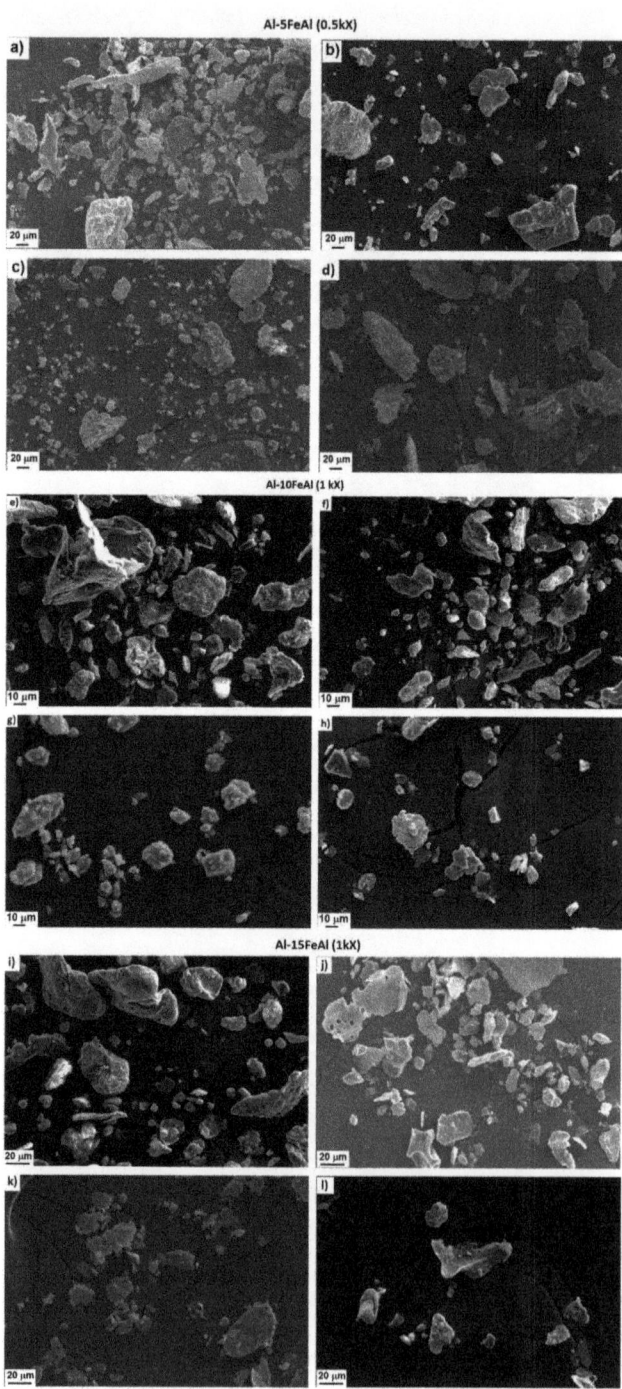

Figure 3. Microstructural evolution of the Al-5Fe40Al powder mixture composition as a function of milling time: (**a**) 0 h, (**b**) 2 h, (**c**) 4 h, (**d**) 8 h. Al-10Fe40Al: (**e**) 0 h, (**f**) 2 h, (**g**) 4 h, (**h**) 5 h. Al-15Fe40Al: (**i**) 0 h, (**j**) 2 h, (**k**) 4 h, (**l**) 6 h.

In Figure 4, the particle size variation of the Al-5Fe40Al composite powder is presented. It can be observed that the mean particle size decreased from around 4.7 microns to 1.6 microns as the powder mixture was milled from 1 h to 4 h. Between 4 h and the end of the milling process, the particle size increased from 1.6 μm to 2.4 μm. This behavior could be related to the predominance of welding among the particles. The size ranges of the powder composite mixtures exhibited a slight tendency to narrow with the advancement of the processing time.

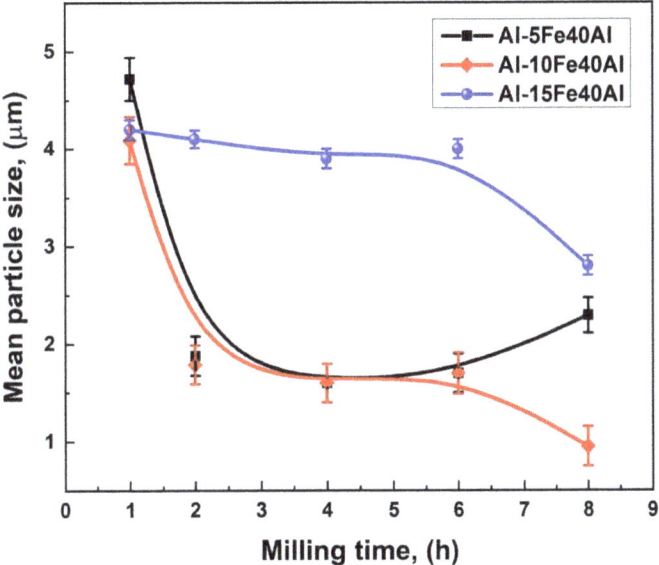

Figure 4. Variations of the mean particle size with the processing time of all powder mixtures (Al-5Fe40Al, Al-10Fe40Al, and Al-15Fe40Al) as a function of the ball milling period.

Additionally, in Figure 4, the particle size of the Al-10Fe40Al powder mixture and the standard deviation as a function of the milling period are presented. This plot shows that the particle size predominantly diminished from about 4.1 μm to 0.8 μm as the milling time progressed from 1 h to 8 h. The size ranges related to the standard deviation of the powder composite mixtures exhibited a very slight trend to decrease with the advancement of the ball milling time.

Figure 4 also shows the particle size of the Al-15Fe40Al powder mixture. In this plot, the particle size of the Al-15Fe40Al powder mixture decreased from 4.2 μm at 0 h to 2.8 μm at 8 h of ball milling.

Similarly, the particle size distributions were applied to determine the mean particle size and the size range of the powder particles in each specimen. The size ranges of the powder composite mixtures suggest an almost imperceptible propensity to decrease as the ball milling period progressed.

It is evident that the particle size variation as a function of milling time exhibited a similar trend for the processing of the three powder mixtures (Al-5Fe40Al, Al-10Fe40Al, and Al-15Fe40Al). This tendency indicates that extended milling times yield a more uniform size of powders and smaller mean particle sizes.

Since the constituents of a powder mixture are ductile–brittle components, in this case, at the beginning of milling, the irregular shapes of the ductile aluminum particles were transformed into a flat flake morphology by the ball–powder–ball collisions, while the fragile particles corresponding to the Fe40Al intermetallic compound were fragmented. These brittle Fe40Al particles tend to be occluded by ductile aluminum particles and settle in interlamellar spacings. As the mechanical milling time progresses, the ductile aluminum particles are hardened by the mechanical workings, and the lamellae become convoluted and refined. As the grinding process takes place, the lamellae become more refined, the interlamellar spacing decreases, and, in this case, the brittle intermetallic particles were uniformly distributed throughout the ductile aluminum matrix. However, if the brittle constituent is soluble, then the alloying process occurs between the brittle and ductile components, tending towards chemical homogeneity [19]. In this case, the Fe element from the Fe40Al intermetallic particles was solubilized in the Al matrix progressively during the milling process.

3.2. Microstructural Analysis by X-ray Diffraction

Figure 5a,b depicts the X-ray diffractograms of the pure Al and Fe40Al intermetallic phases, both in particulate powder aggregation. The X-ray diffractograms of Al exhibit the (111), (200), (220), and (311) diffraction peaks, indicating that the crystal orientations inside the powder particles were randomly distributed. In addition, Figure 5b illustrates the (110) and (200) diffraction peaks related to the intermetallic particulate phase. In this case, the presence of only two diffraction peaks could be due to a preferential crystallographic orientation since the intermetallic particulate powder was produced by a solid-state trituration and pulverization process from ingots with coarse grains.

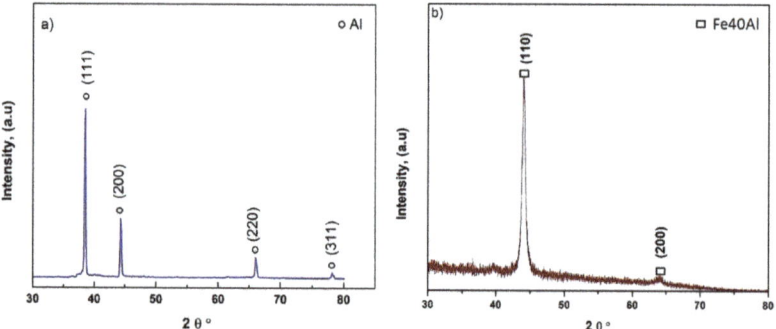

Figure 5. X-ray diffraction profiles of (**a**) pure Al and (**b**) Fe40Al intermetallic phase.

Figures 6–8 depict the X-ray diffraction patterns of the powder mixtures subjected to mechanical milling at various processing times. In accordance with the diffraction patterns, only the diffraction peaks of Al can be predominantly observed. The presence of the diffraction peaks of Fe-Al is negligible; this behavior could be related to the low content of the intermetallic phase and the possible overlap of the peaks of the intermetallic phase with those of aluminum. The highest peaks of Al pertain to the reflections produced by the (111) crystallographic planes. From 0 to 8 h of milling time, the little peak corresponding to the Fe40Al vanished progressively. This behavior could be related to a combined effect consisting of the disorder or amorphization of the Fe40Al intermetallic phase during the processing period.

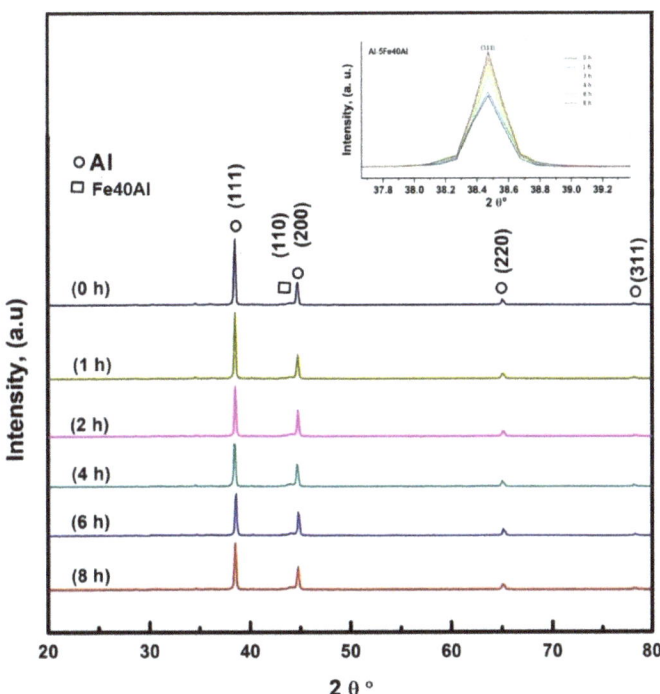

Figure 6. X-ray diffraction profiles of the powder mixture of Al-5Fe40Al at different processing times.

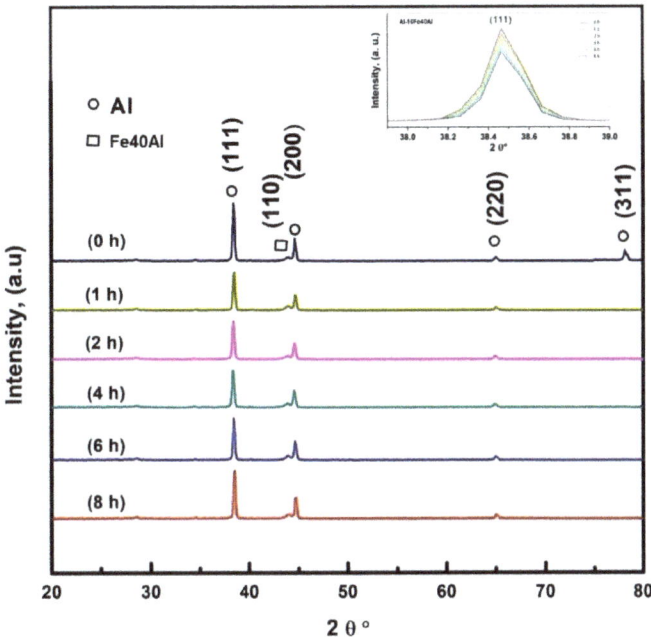

Figure 7. X-ray diffraction profiles of the powder mixture of Al-10Fe40Al at different processing times.

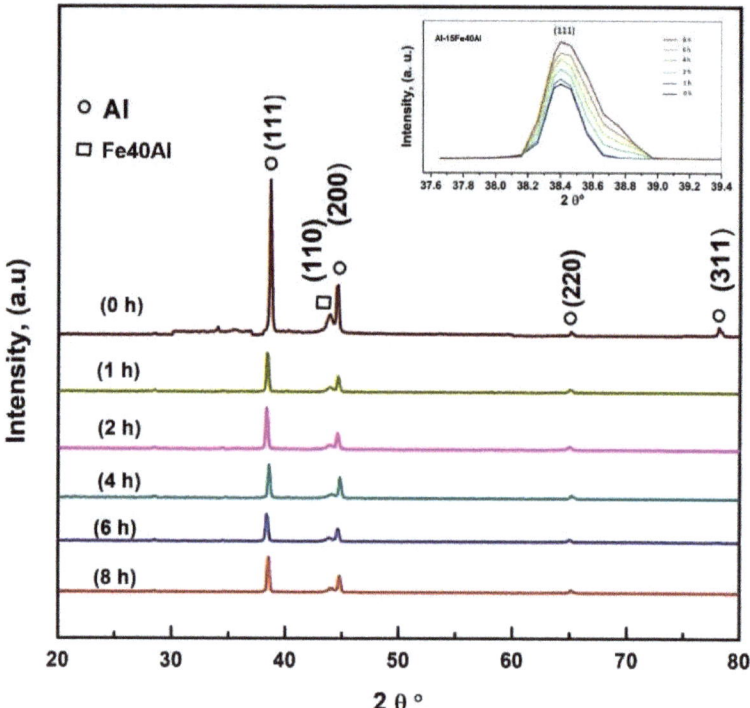

Figure 8. X-ray diffraction profiles of the powder mixture of Al-15Fe40Al at different processing times.

There is no evidence of the presence of secondary phases that could have been produced by mechanochemical reactions among Al and Fe40Al. It can be observed that the intensity of the Fe40Al diffraction peak increased as the content of the intermetallic phase increased in the Al-xFe40Al powder mixtures. All the diffractograms show that the width of peaks at half its height belonging to the Al exhibited a trend to increase as the milling time elapsed. It is well known that this tendency is related to a decrease in the crystallite size and the generation of lattice strain [20].

Figures 9–11 exhibit the dependence of the crystallite size as a function of the processing time of the powder mixtures composed of Al-5Fe40Al, Al-10Fe40Al, and Al-15 Fe40Al, respectively. In this case, the nanometric size of grains was determined from all the X-ray diffraction peaks, and a mean value of the crystallite size was calculated. The crystallite size (nm) was computed by the typical method that employs the Scherrer equation [21] by utilizing the width of the XRD diffraction peak at half its height. It is worth noticing that the trend of the decreasing crystallite size in the ball milling process has typically been reported in the literature where Al composite powders were subjected to this non-equilibrium processing technique [22–25].

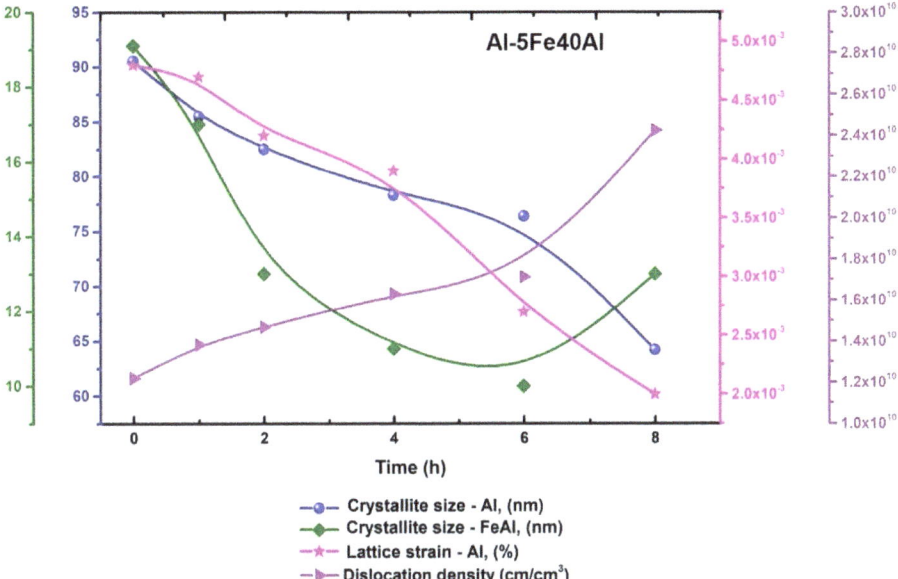

Figure 9. Variation of crystallite size of Al and Fe40Al, lattice strain (Al), and dislocation density (Al) as a function of processing time for Al-5Fe40Al powder mixture.

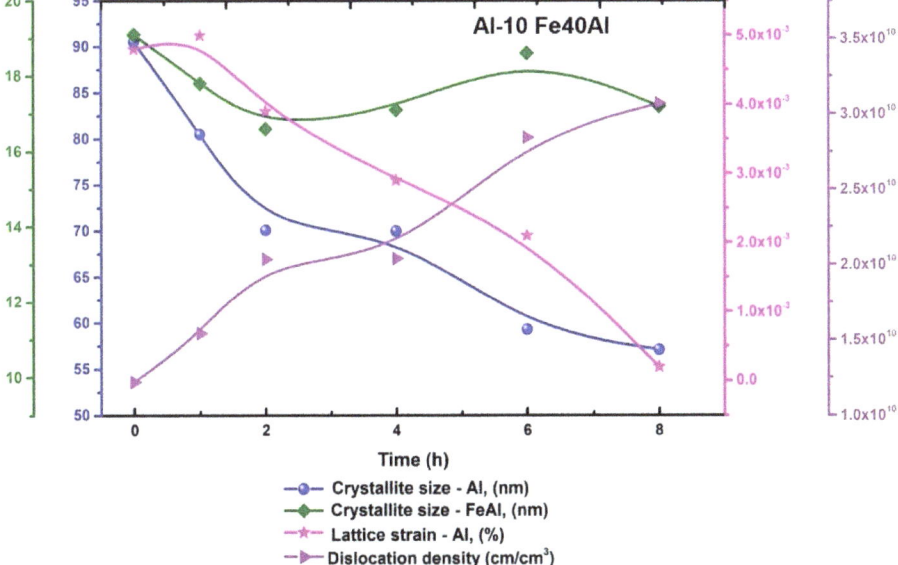

Figure 10. Variations of the crystallite sizes of Al and Fe40Al, lattice strain (Al), and dislocation density (Al) as a function of processing time for the Al-10Fe40Al powder mixture.

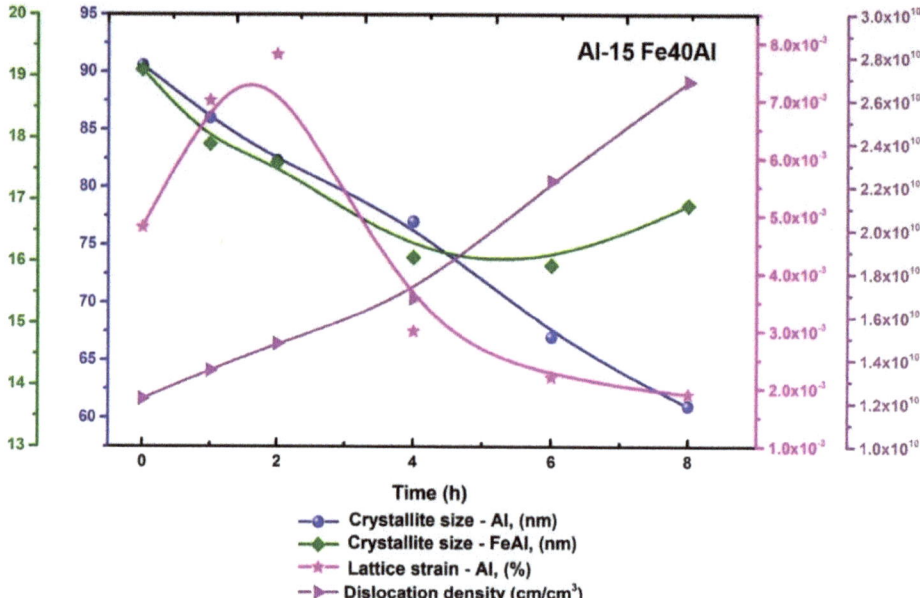

Figure 11. Variations of crystallite sizes of Al and Fe40Al, lattice strains (Al), and dislocation density (Al) as a function of processing time for Al-15Fe40Al powder mixture.

Crystallite sizes were utilized to compute dislocation density. In a preliminary approach, dislocation density was considered to be at least one dislocation per crystallite. Thus, in agreement with Equation (1), dislocation density (N) becomes the inverse square of the crystallite size (L_c) [26].

$$N = (L_c)^{-2} \qquad (1)$$

The variation of crystallite size for both the Al solid solution and Fe40Al intermetallic phase, along with the lattice strain, dislocation density, and the processing period of the Al-5Fe40Al powder mixture, is shown in Figure 9. This plot reveals that the nano-scaled grain size of the (alpha) Al phase displayed a tendency to decrease from about 90 nm at the onset of the process to around 65 nm at the end of mechanical milling (8 h). Concerning the Fe40Al intermetallic compound, its nano-metric grain size underwent a reduction from around 19 nm (at 0 h) to about 10 nm at 6 h; however, afterward, the powder mixture increase from 10 nm at 6 h to around 13 nm at 8 h. The lattice strain presented in Figure 9 exhibited a trend of decreasing from 1 h to 8 h of milling; however, after 8 h of processing, it displayed a trend of increasing until the end of milling. In this case, the fact that the dislocation density continuously increased from the start of the process to the end of 15 h of milling suggests that the lattice strain decreased during the first 8 h of processing because the Al phase predominantly experienced removal of vacancies, stacking failures, and punctual defects, among others.

The fact that the nano-scaled grain size of the intermetallic iron aluminide showed a decrease during the first 6 h and a subsequent increase up to 8 h of milling may be due to the dynamic recrystallization nature that it is typically observed during the hot deformation of ordered alloys. It was reported that during the deformation process within a temperature range of 600–700 °C, where <111> super-dislocations operate, dynamic recrystallization developed very slowly [27]. However, at temperatures within the interval of 750–900 °C, where <100> super-dislocations operate, dynamic recrystallization progressed rapidly [28] (Imayev et al., 1995).

The changes in the crystallite sizes for both Al and Fe40Al phases, along with the lattice strain, dislocation density, and the processing time of the Al-10Fe40Al powder mixture are shown in Figure 10. This graph indicates that the nano-scaled grain size of (alpha) Al phase displayed a decreasing trend from about (80 nm) at the beginning of the process to around (57 nm) at the end of the ball milling (8 h). Regarding the Fe40Al intermetallic phase, its crystallite size experienced a decrease from 19 nm at (0 h) to 16.5 nm at (2 h) of milling; however, after that period, the powder mixture increased from 16.5 nm at (2 h) to 18.5 nm at (6 h), and finally, decreased again after 6 h, until the nanometric grain-size reached a value of 17 nm at the end of processing. The lattice strain exhibited a decreasing trend from 1 h to 8 h of milling. In this case, the fact that the dislocation density continuously increased from the start of the process to the end of the 15 h of milling suggests that the lattice strain decreased during the first 8 h of processing because the Al phase predominantly experienced removal of vacancies, stacking failures, and punctual defects, among others.

The fact that the crystallite size of the intermetallic phase showed fluctuating behavior as a function of processing time may be due to the dynamic recrystallization character that is frequently observed in the hot deformation of ordered alloys. It could also be due to the fact that local high temperatures in the impact zone of the balls can reach up to 600 °C and the temperature also varies depending on the impact mode. Thus, it is expected that the dynamic recrystallization will be altered, considering as well that this process operates very slowly in the temperature range of 600 to 700 °C [29]; however, at high temperatures (750–900 °C), recrystallization develops rapidly [30]. This explains why the irregular variations in crystal size depend on the dynamic recrystallization processes for which the driving force is the temperature (which, in this case, varies from one point to another in the impact zones of the balls) and deformation.

The variations of nano-scaled grain sizes for both Al and Fe40Al phases, along with the lattice strain, dislocation density, and the ball milling time of the Al-15Fe40Al powder mixture are displayed in Figure 11. This plot depicts that the nano-metric grain size of the solid solution (alpha) Al showed a tendency to decrease from about (90 nm) at the beginning of the process to around (61 nm) at the end of the process (8 h). Regarding the Fe40Al intermetallic phase, its nanometric grain size decreased from around 19 nm at 0 h of milling to about 16.8 nm at the end of processing. The lattice strain exhibited a predominant decreasing trend for almost the entire processing period. Similarly, in this case, the curve of the dislocation density versus the milling time depicted a continuous rise from the beginning to the end of the non-equilibria processing; this behavior is reasonable since the increase in dislocations was generated by a plastic deformation process that took place during milling. In this way, during the deformation process, a hardening of the particles was induced by the increase in dislocation density, which, at the same time, contributed to inducing the fracture of particles. This tendency is in agreement with the continuous decrease in the size of particles as the milling time elapsed. A continuous increase in the dislocation density led to restricted dislocation motion, and afterward, led to their accumulation. When the dislocation density reaches a certain value, the crystal disintegrates into subgrains, which are separated at the beginning by low-angle grain boundaries. This process is repeated over and over until a final grain size is attained.

The crystallite size of the intermetallic compound decreased from the beginning of the process up to 6 h of milling, but after that period, the nanometric grain size decreased slightly from about 16 nm to 17 nm until the end of the mechanical milling. This behavior could be explained in terms of the complexity of the recovery and recrystallization process that develops in ordered alloys during milling. The intricate kinetics of these processes are related to the following observations: the drastic reduction in grain motion in ordered alloys severely delays recrystallization, a temperature interval is present where recrystallization does not develop, and the pre-recrystallization behavior relies on the annealing temperature and alloy composition.

Figure 12 shows Vegard's law plot calculated from the reticular constants obtained using the experimental results of the three Al-xFe40Al powder mixtures. This plot shows that the lattice parameter of Al or the solid solution Al-Fe varies mostly in a linear way from 0 to 8 h of ball milling. In this case, the physical origin of the deviations of the lattice constants from Vegard's law could be mainly due to the large size mismatch between Al and Fe [31]. Figure 12 exhibits that a solid solubility of about 0.02% was reached in the case of Al-5Fe40Al and Al-10Fe40Al powder mixtures. However, the Al-15Fe40Al powder mixture exhibited a solid solubility of about 0.012%.

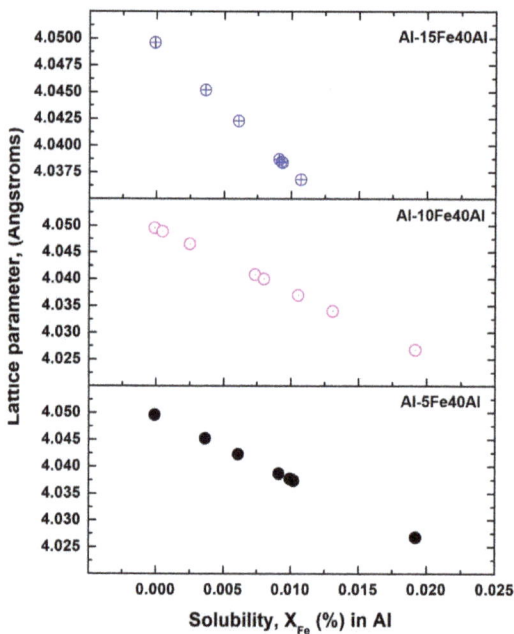

Figure 12. Variations of the sizes of lattice parameters of Al in Al-5Fe40Al, Al-10Fe40Al, and Al-15Fe40Al (vol. %) powder mixtures as a function of the milling period.

3.3. Microstructural Characterization by Microscopy of Consolidated Specimens

Figure 13 shows optical micrographs corresponding to the Al-5Fe40Al, Al-10Fe40Al, and Al-15Fe40Al consolidated samples. Figures 13a,b and 14c show bright phases (Fe40Al) uniformly distributed in a dark grey matrix (Al). The bright intermetallic phases presented an irregular morphology. The mean particle size corresponding to the intermetallic phase in Al-5Fe40Al composite was 9.2 microns with a standard deviation of ±3.6. Regarding the mean particle size of the Fe40Al particles in the Al-10Fe40Al composite, the computed value was 9.5 microns with a standard deviation of ±4.8. Concerning the mean particle size of the reinforcement Fe40Al particles in the consolidated composite Al-15Fe40Al, the calculation procedure gave a value of 10.8 microns with a standard deviation of ±4.1. This uniform size of intermetallic particles indicates that during milling, there was not a significant decrease in its size; however, the size of the Al particles exhibited a higher decrease percentage. Figure 13d–f exhibits the distribution of the precipitate size of the Fe40Al intermetallic particles. In the first composite composition, Al-5Fe40Al, most of the Fe40Al intermetallic particles are within the size interval of 4.4 to 8 µm. Concerning the composite composition of Al-10Fe40Al, most intermetallic particles are within the size interval of 8.8 to 20.8 µm.

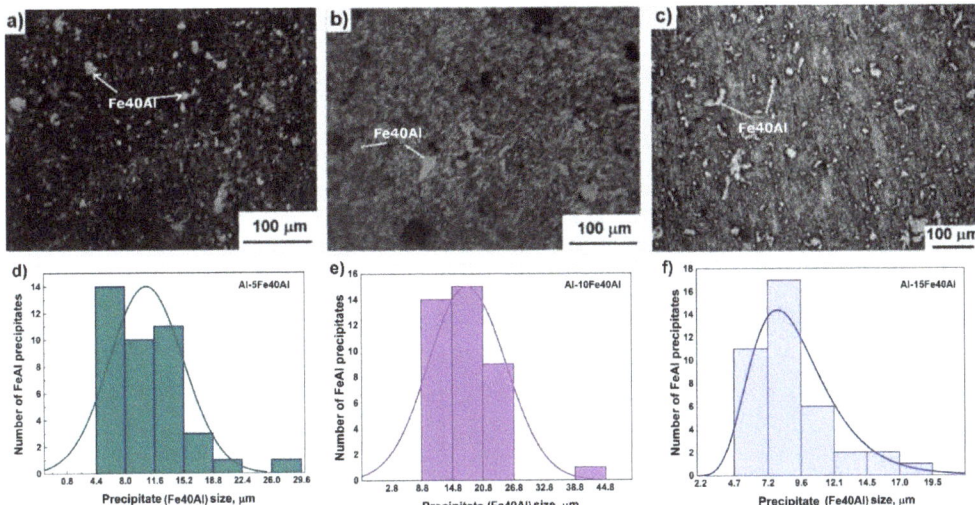

Figure 13. Optical micrographs of consolidated specimens, (**a**) Al-5Fe40Al, (**b**) Al-10Fe40Al, and (**c**) Al-15Fe40Al. Particle size distribution: (**d**) Al-5Fe40Al, (**e**) Al-10Fe40Al, and (**f**) Al-15Fe40Al.

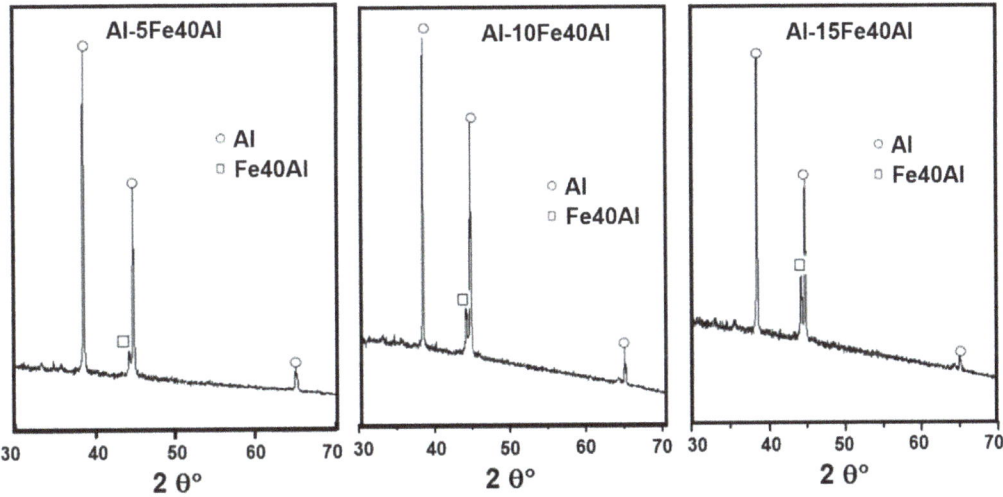

Figure 14. X-ray diffraction spectra of the sintered Al-5Fe40Al, Al-10Fe40Al, and Al-15Fe40Al specimens.

3.4. Microstructural Characterization by X-ray Diffraction of the Consolidated Specimens

Figure 14 exhibits the X-ray diffraction profiles of Al-xFe40Al specimens after the sintering treatment. The diffraction peak corresponding to the Fe40Al intermetallic phase grew after the sintering heat treatment. In this case, the intensity of the diffraction peak of the intermetallic phase was significantly greater than those observed in the X-ray diffraction peaks of the specimens that were mechanically milled. This behavior could be related to the crystallization of the Fe40Al intermetallic phase, which was induced by the sintering process. First, the Fe40Al intermetallic phase underwent amorphization during the ball milling process, and after, the compound crystallized because of sintering. Moreover, it is worth noticing that during the sintering, the crystallite size of the aluminum increased in accordance with the following percentages: in Al-5Fe40Al—29.6%, in Al-10Fe40Al—57.3%, and in Al-15Fe40Al—31.01%.

3.5. Microhardness of Sintered Samples

Figure 15 displays the three values of the hardness of the Al-5Fe40Al, Al-10Fe40Al, and Al-15Fe40Al composites determined in the present work, with values of 84.6, 101.1, and 107.2 VPN, respectively. In this case, the hardness also varied almost linearly with an increase in the volume fraction of the reinforcement phase [32]. In addition, the microhardness value of the pure Al utilized as the matrix is presented in the above-mentioned plot.

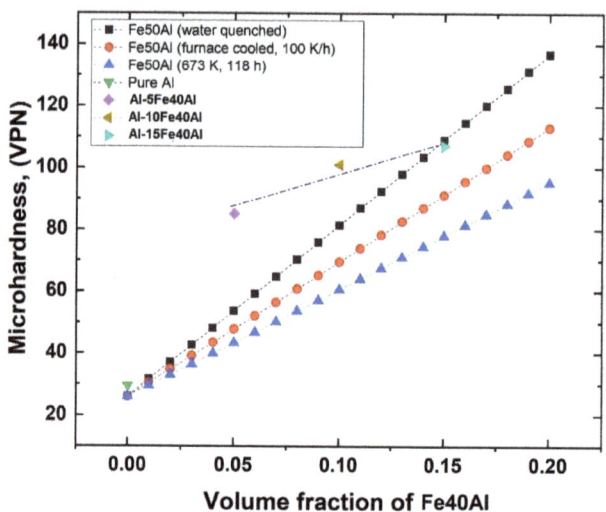

Figure 15. Vickers microhardness of composite materials Al-5Fe40Al, Al-10Fe40Al, and Al-15Fe40Al compared with the predicted values of hardness of various composites reinforced with Fe50Al intermetallics under different processing conditions.

Aiming to compare the hardness values measured in the present work with a theoretical basis, a model that describes the variation of hardness as a function of the reinforcement phase concentration was developed. In this theoretical model, the hardness values of pure Al and those of the Fe50Al intermetallic compounds (under different processing conditions) were taken from the literature [33]. The composition of the Fe50Al intermetallic phase was considered since its hardness is similar to that of the Fe40Al intermetallic phase. Then, the hardness values were modeled as a function of the volume fraction of the reinforcement phase (Fe50Al), in accordance with the rule of mixtures that is represented by the following equation:

$$p_c = \sum f_i p_i = f_1 p_1 + f_2 p_2 + \ldots + f_n p_n \qquad (2)$$

where p_c is the property of compound; $p_1, p_2, \ldots p_n$ are the properties of each of the constituents of composite material, and $f_1, f_2, \ldots f_n$ are the volume fractions of each of the components. However, the property modeled here was hardness.

Figure 15 displays the variation of hardness modeled as a function of the volume fraction of the reinforcement phase (Fe50Al), considering, in all cases, an aluminum (Al) matrix. The three curves presented in this plot were generated by considering the hardness of the Fe50Al intermetallic compounds reported in previous works that were processed under the following parameters: water quenched, furnace cooled at 100 k/h, and annealed at 673 K for 118 h [33]. In these three curves, the hardness of the Al matrix composite changed linearly with increases in the Fe50Al reinforcement phase [32].

It was observed that the hardness values of the Al-5Fe40Al, Al-10Fe40Al, and Al-15Fe40Al measured in the present research exhibited higher values than the hardness modeled as a function of the content of the Fe50Al reinforcement phase [33], as can be observed in Figure 15. The hardening effect observed in the composite produced in the present study is due to the hardening mechanism induced by the solid solution of solute Fe in solvent Al, which promotes the formation of strain fields around solute atoms, which, in turn, interact with the stress fields of dislocations and obstruct dislocation motion. Other reasons that promoted the hardness increase in the sintered specimens were the work hardening effect induced by the plastic deformation during milling, the continuous generation of dislocations, and the interruption of dislocation motion [34].

Figure 16 shows the dependence of hardness on dislocation density. In this case, the dislocation density was estimated from the Al-5Fe40Al, Al-10Fe40Al, and Al-15Fe40Al specimens that were subjected to the ball milling process for 8 h. In this plot, a linear relationship of both variables is depicted. For this analysis, the dislocation density was considered to remain constant during sintering since the predominant material transport mechanism occurs by surface, volume, or grain boundary diffusion, or even by vaporization and re-condensation [35]. In addition, Figure 16 displays a relatively linear increase in hardness with increasing dislocation density. This behavior is reasonable since, during the mechanical milling, the particles experienced a plastic deformation. This phenomenon is associated with the hardening of crystals due to an increase in the dislocation density and the mutual interaction of dislocations. When dislocations move, they interact and change their distribution and density, thereby causing hardening [36].

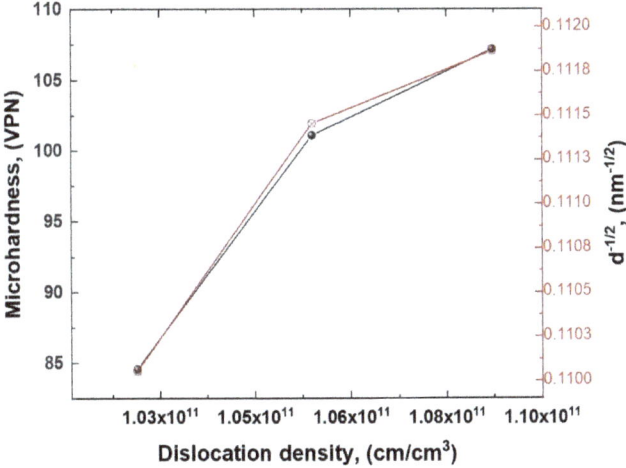

Figure 16. Variation of the Vickers microhardness with dislocation density and the inverse square root of the grain size, illustrating a typical Hall–Petch linear relationship.

The variations of microhardness as a function of the grain size for the three composites, Al-5Fe40Al, Al-10Fe40Al, and Al-15Fe40Al, are presented in Figure 16. In this plot, the microhardness values are related to the inverse square root of the grain size by a common Hall–Petch linear relationship. This correlation between the grain size and hardness was demonstrated by a variety of previous research studies on unreinforced alloys of Al [37–39].

4. Conclusions

During the stage of the mechanical milling process, the particle size of the initial powder mixtures of compositions Al-5Fe40Al, Al-10Fe40Al, and Al-15Fe40Al exhibited a predominant decreasing trend as the milling time progressed from the beginning to the end of the non-equilibria processing. For the case of the Al-5Fe40Al powder mixture, its particle size decreased from 4.7 μm at 0 h to 1.6 μm at 4 h, but after increased again up to 2.25 μm at 8 h. Regarding the Al-10Fe40Al powder mixture, its particle size decreased from 4.1 μm at 0 h to 0.9 μm at 8 h of processing. Finally, concerning the Al-15Fe40Al powder mixture, its particle size decreased from 4.2 μm at 0 h to 2.8 μm at 8 h of ball milling.

The X-ray diffraction analyses revealed that the crystallite size of the Al decreased during all of the processing periods and for all three compositions, and the crystallite size of the Fe40Al intermetallic phase tended to decrease during the first 4 h of milling. During the mechanical milling, there was no evidence of the formation of additional phases that could be formed by mechanochemical synthesis.

The crystallite size of the (α) Al phase in the Al-5Fe40Al powder mixture exhibited a decreasing trend from 90 nm at the onset of the process to 65 nm at the end of mechanical milling (8 h). Regarding the Al-10Fe40Al powder mixture, the nanometric grain size of the (α) Al phase displayed a decreasing trend from 90 nm at 0 h to 57.5 nm at 8 h. Finally, concerning the Al-15Fe40Al powder mixture, the nano-scaled grain size of the (α) Al phase showed a propensity to decrease from 90 nm at (0 h) to 61 nm at (8 h).

During ball milling, the Fe dissolved progressively in the Al matrix. In this case, the lattice parameter of the Al followed Vegard's Law since the reticular constant changed almost in a linear way with the percentage of the Fe dissolved in the Al matrix. Consolidation of the Al-xFe40Al powder mixtures produced solid and rigid pieces with low volume fractions of porosity, and optical micrographs revealed a uniform distribution of the Fe40Al intermetallic phase on the Al matrix. Moreover, XRD analyses of the consolidated specimens indicated that there was no evidence of the formation of secondary phases during the sintering process. During the mechanical milling, the intermetallic phase experienced amorphization; however, during sintering, the amorphous phase recrystallized again, as revealed by the definition of the diffraction peaks of the Fe40Al phase in the XRD patterns.

The addition of Fe40Al to the Al matrix effectively induced a hardening effect. This may have been due to the mechanisms of solid solution hardening, plastic deformation, and precipitation. Likewise, the observed hardening can be modeled in terms of the rule of mixtures as a fundamental mathematical expression of the properties of composite materials.

Microhardness (Vickers) of the Al-XFe40Al composites exhibited an increasing trend from 84 VPN with 5 vol. % of Fe40Al reinforcement up to 107 VPN when Fe40Al reinforcement was added in 15 vol. %.

The graph, representing the microhardness as a function of dislocations density and grain size, indicates that the hardness increased almost linearly with the increasing dislocations density and decreasing grain size.

Author Contributions: Conceptualization, R.A.R.D., A.M.O. and S.R.G.S.; methodology, R.A.R.D., S.R.G.S. and A.M.O.; validation, R.A.R.D.; formal analysis, J.P.C. and J.L.R.B.; investigation, R.A.R.D., J.P.C., S.R.G.S., H.C.M. and J.I.B.F.; writing—original draft preparation, R.A.R.D. and J.L.R.B.; writing—review and editing, R.A.R.D., J.L.R.B. and J.P.C.; supervision, C.A.G.R., H.C.M., J.I.B.F. and V.R.V.; project administration, R.A.R.D., C.A.G.R., V.R.V. and A.M.O. All authors have read and agreed to the published version of the manuscript.

Funding: This research received no external funding.

Institutional Review Board Statement: Not applicable.

Informed Consent Statement: Not applicable.

Acknowledgments: The authors thank our affiliated institutions for the facilities granted for carrying out this research work. The authors also acknowledge Eng. Jaime Agustin Ramírez Ibarra for his technical support.

Conflicts of Interest: The authors declare no conflict of interest.

References

1. Kainer, K.U. Basics of Metal Matrix Composites. In *Metal Matrix Composites*; John Wiley & Sons, Ltd.: Hoboken, NJ, USA, 2006; pp. 1–54, ISBN 978-3-527-60811-9.
2. Köhler, E.; Niehues, J. Aluminum-matrix Composite Materials in Combustion Engines. In *Metal Matrix Composites*; John Wiley & Sons, Ltd.: Hoboken, NJ, USA, 2006; pp. 95–109, ISBN 978-3-527-60811-9.
3. Weinert, K.; Buschka, M.; Lange, M. Machining Technology Aspects of Al-MMC. In *Metal Matrix Composites*; John Wiley & Sons, Ltd.: Hoboken, NJ, USA, 2006; pp. 147–172, ISBN 978-3-527-60811-9.
4. Tan, M.J.; Zhang, X. Powder Metal Matrix Composites: Selection and Processing. *Mater. Sci. Eng.* **1998**, *244*, 80–85. [CrossRef]
5. Fogagnolo, J.B.; Velasco, F.; Robert, M.H.; Torralba, J.M. Effect of Mechanical Alloying on the Morphology, Microstructure and Properties of Aluminium Matrix Composite Powders. *Mater. Sci. Eng.* **2003**, *342*, 131–143. [CrossRef]
6. Ozdemir, I.; Ahrens, S.; Mücklich, S.; Wielage, B. Nanocrystalline Al–Al2O3p and SiCp Composites Produced by High-Energy Ball Milling. *J. Mater. Process. Tech.* **2008**, *205*, 111–118. [CrossRef]
7. Lu, L.; Lai, M.O.; Zhang, S. Preparation of Al-Based Composite Using Mechanical Alloying. *Key Eng. Mat.* **1995**, *104–107*, 111–124. [CrossRef]
8. Scudino, S.; Liu, G.; Sakaliyska, M.; Surreddi, K.B.; Eckert, J. Powder Metallurgy of Al-Based Metal Matrix Composites Reinforced with β-Al3Mg2 Intermetallic Particles: Analysis and Modeling of Mechanical Properties. *Acta. Mater.* **2009**, *57*, 4529–4538. [CrossRef]
9. Hashim, J. The Production of Metal Matrix Composites Using the Stir Casting Technique. Ph.D. Thesis, Dublin City University, Dublin, Ireland, 1999.
10. George, E.P.; Yamaguchi, M.; Kumar, K.S.; Liu, C.T. Ordered Intermetallics. *Annu. Rev. Mater. Sci.* **1994**, *24*, 409–451. [CrossRef]
11. Mitra, R.; Wanhill, R.J.H. Structural Intermetallics. In *Aerospace Materials and Material Technologies; Volume 1: Aerospace Materials*; Prasad, N.E., Wanhill, R.J.H., Eds.; Springer: Singapore, 2017; pp. 229–245, ISBN 978-981-10-2134-3.
12. Palm, M.; Stein, F.; Dehm, G. Iron Aluminides. *Annu. Rev. Mater. Res.* **2019**, *49*, 297–326. [CrossRef]
13. Chaubey, A.K.; Scudino, S.; Mukhopadhyay, N.K.; Khoshkhoo, M.S.; Mishra, B.K.; Eckert, J. Effect of Particle Dispersion on the Mechanical Behavior of Al-Based Metal Matrix Composites Reinforced with Nanocrystalline Al–Ca Intermetallics. *J. Alloys Compd.* **2012**, *536*, S134–S137. [CrossRef]
14. Varin, R.A. Design of a Low-Melting Point Metal Matrix Composite Reinforced with Intermetallic Ribbons/Design Eines Niedrig-Schmelzenden Metall-Matrix-Verbundwerkstoffes Mit Verstärkenden Intermetallischen Bändern. *Int. J. Mater. Res.* **1990**, *81*, 373–379. [CrossRef]
15. Pour, H.A.; Lieblich, M.; López, A.J.; Rams, J.; Salehi, M.T.; Shabestari, S.G. Assessment of Tensile Behaviour of an Al–Mg Alloy Composite Reinforced with NiAl and Oxidized NiAl Powder Particles Helped by Nanoindentation. *Compos. Part A Appl. Sci. Manuf.* **2007**, *38*, 2536–2540. [CrossRef]
16. Muñoz-Morris, M.A.; Rexach, J.I.; Lieblich, M. Comparative Study of Al-TiAl Composites with Different Intermetallic Volume Fractions and Particle Sizes. *Intermetallics* **2005**, *13*, 141–149. [CrossRef]
17. Da Costa, C.E.; Velasco, F.; Torralba, J.M. Mechanical, Intergranular Corrosion, and Wear Behavior of Aluminum-Matrix Composite Materials Reinforced with Nickel Aluminides. *Metall. Mater. Trans. A* **2002**, *33*, 3541–3553. [CrossRef]
18. Wolf, B.; Bambauer, K.O.; Paufler, P. On the Temperature Dependence of the Hardness of Quasicrystals. *Mater. Sci. Eng.* **2001**, *298*, 284–295. [CrossRef]
19. Suryanarayana, C. Mater Sci Forum: A Novel Technique to Synthesize Advanced Materials. *Research* **2019**, *2019*, 1–17. [CrossRef]
20. Ahmad, Z.; Abu, M.; Rahman, M.; Ain, M.F.; Ahmadipour, M. Assessment of Crystallite Size and Strain of CaCu3Ti4O12 Prepared via Conventional Solid-State Reaction. *Micro. Nano Lett.* **2016**, *11*, 147–150. [CrossRef]
21. Cherepanov, P. High Intensity Ultrasound Processing of AlNi (50 Wt.% Ni) Particles for Electrocatalytic Water Splitting. Ph.D. Thesis, Bayreuth University, Bayreuth, Germany, 2015.
22. Erturun, V.; Sahin, O. Investigation of Microstructural Evolution in Ball-Milling of SiC Reinforced Aluminum Matrix Composites. *Powder Metall. Met. Ceram.* **2019**, *57*, 687–696. [CrossRef]
23. Travessa, D.N.; Silva, M.J.; Cardoso, K.R. Niobium Carbide-Reinforced al Matrix Composites Produced by High-Energy Ball Milling. *Metall. Mater. Trans. A* **2017**, *48*, 1754. [CrossRef]
24. Tamizi Junqani, M.; Madaah Hosseini, H.R.; Azarniya, A. Comprehensive Structural and Mechanical Characterization of In-Situ Al–Al3Ti Nanocomposite Modified by Heat Treatment. *Mater. Sci. Eng.* **2020**, *785*, 139351. [CrossRef]
25. Tomiczek, B.; Pawlyta, M.; Adamiak, M.; Dobrzański, L.A. Effect of milling time on microstructure of AA6061 composites fabricated via mechanical alloying. *Arch. Metall. Mater.* **2015**, *60*, 789–793. [CrossRef]
26. Raghu, T.; Sundaresan, R.; Ramakrishnan, P.; Mohan, T.R. Synthesis of Nanocrystalline Copper–Tungsten Alloys by Mechanical Alloying. *Mater. Sci. Eng.* **2001**, *304*, 438–441. [CrossRef]
27. Baker, I.; Munroe, P.R. Mechanical Properties of Fe40Al. *Int. Mater. Rev.* **1997**, *42*, 181–205. [CrossRef]
28. Imayev, R.; Evangelista, E.; Tassa, O.; Stobrawa, J. Relationship between Mechanism of Deformation and Development of Dynamic Recrystallization in Fe40Al Intermetallic. *Mater. Sci. Eng.* **1995**, *202*, 128–133. [CrossRef]
29. Zhan, Z.; He, Y.; Wang, D.; Gao, W. Low-Temperature Processing of Fe–Al Intermetallic Coatings Assisted by Ball Milling. *Intermetallics* **2006**, *14*, 75–81. [CrossRef]

30. Siegel, R.W. Nanophase materials: Synthesis, structure, and properties. In *Physics of New Materials*; Springer: Heidelberg/Berlin, Germany, 1994; pp. 65–105. [CrossRef]
31. Magomedov, M.N. On the Deviation from the Vegard's Law for the Solid Solutions. *Solid State Commun.* **2020**, *322*, 114060. [CrossRef]
32. Kaczmar, J.W.; Pietrzak, K.; Włosiński, W. The Production and Application of Metal Matrix Composite Materials. *J. Mater. Process. Tech.* **2000**, *106*, 58–67. [CrossRef]
33. Nagpal, P.; Baker, I. Effect of cooling rate on hardness of FeAl and NiAl. *Met. Trans. A* **1990**, *21*, 2281–2282. [CrossRef]
34. Suryanarayana, C. Mechanical Alloying and Milling. *Prog. Mater. Sci.* **2001**, *46*, 1–184. [CrossRef]
35. Thümmler, F.; Oberacker, R. An Introduction to Powder Metallurgy. Book/The Institute of Materials; Institute of Materials: London, UK, 1993; ISBN 978-0-901716-26-2.
36. Hull, D.; Bacon, J.D. Introduction to Dislocations, 5th ed.Butterworth-Heinemann: Oxford, UK, 2011; pp. 1–253.
37. Rajinikanth, V.; Venkateswarlu, K.; Sen, M.K.; Das, M.; Alhajeri, S.N.; Langdon, T.G. Influence of Scandium on an Al–2% Si Alloy Processed by High-Pressure Torsion. *Mater. Sci. Eng. A* **2011**, *528*, 1702–1706. [CrossRef]
38. Xu, C.; Horita, Z.; Langdon, T.G. The Evolution of Homogeneity in an Aluminum Alloy Processed Using High-Pressure Torsion. *Acta Mater.* **2008**, *56*, 5168–5176. [CrossRef]
39. Loucif, A.; Figueiredo, R.B.; Baudin, T.; Brisset, F.; Langdon, T.G. Microstructural Evolution in an Al-6061 Alloy Pro-cessed by High-Pressure Torsion. *Mater. Sci. Eng. A* **2010**, *527*, 4864–4869. [CrossRef]

Article

Using the Instrumented Indentation Technique to Determine Damage in Sintered Metal Matrix Composites after High-Temperature Deformation

Alexander Smirnov *, Evgeniya Smirnova, Anatoly Konovalov and Vladislav Kanakin

Institute of Engineering Science, UB RAS, 34 Komsomolskaya St., 620049 Ekaterinburg, Russia; evgeniya@imach.uran.ru (E.S.); avkonovalov@mail.ru (A.K.); kanakin.v.s@google.com (V.K.)
* Correspondence: smirnov@imach.uran.ru

Citation: Smirnov, A.; Smirnova, E.; Konovalov, A.; Kanakin, V. Using the Instrumented Indentation Technique to Determine Damage in Sintered Metal Matrix Composites after High-Temperature Deformation. *Appl. Sci.* **2021**, *11*, 10590. https://doi.org/10.3390/app112210590

Academic Editor: Myoung-Gyu Lee

Received: 18 October 2021
Accepted: 8 November 2021
Published: 10 November 2021

Publisher's Note: MDPI stays neutral with regard to jurisdictional claims in published maps and institutional affiliations.

Copyright: © 2021 by the authors. Licensee MDPI, Basel, Switzerland. This article is an open access article distributed under the terms and conditions of the Creative Commons Attribution (CC BY) license (https://creativecommons.org/licenses/by/4.0/).

Abstract: The paper shows the applicability of data on the evolution of the elastic modulus measured by the instrumented microindentation technique to the determination of accumulated damage in metal matrix composites (MMCs) under high temperature deformation. A composite with a V95 aluminum alloy matrix (the Russian equivalent of the 7075 alloy) and SiC reinforcing particles is used as the research material. The metal matrix composite was produced by powder technology. The obtained results show that, under macroscopic compression at temperatures ranging between 300 and 500 °C, the V95\10% SiC MMC has the best plasticity at 300 °C. At a deformation temperature of 500 °C, the plastic properties are significantly lower than those at 300 and 400 °C.

Keywords: damage; elastic modulus; microindentation; metal matrix composite; high temperatures; aluminum; 7075 alloy

1. Introduction

Practical problems of materials processing by plastic deformation often require damage criteria and damage models, with the help of which it would be possible to assess the deformation ability of materials in various technological processes [1–5]. After thermomechanical deformation, damage models are most often experimentally identified and verified on the basis of density measurements of deformable material specimens [6–9], metallographic analysis [6,8,10,11], computed tomography [12,13], as well as on the basis of measurements of physical characteristics implicitly dependent on metal imperfection, such as the elastic modulus [8,10,13], hardness [10,14], electrical resistance [8,10,15], acoustic emission characteristics [8,16], and heat dissipation [17–19].

Under severe plastic deformation of specimens, an inhomogeneous stress-strain state is formed due to the loss of specimen stability. A neck is formed during tensile tests, and a barrel is formed during compression tests. This is especially evident at high temperatures, when lubricant effectiveness is greatly reduced, thus making the specimen barrel-shaped. In this case, the stress state and accumulated strain in the specimen strongly depend on the area under consideration. The smallest strain is concentrated at the specimen–die contact, the largest strain being found in the specimen center. Under certain conditions, the sign of the stress-state factors on the specimen side surface can reverse. Therefore, the level of accumulated damage in a specimen is significantly heterogeneous in volume, and the use of macro-level tests for determining the relation of accumulated material damage to external thermomechanical effects may lead to a significant error. As a result, micro-level testing is required to find the relationships between the stress-strain state history and material damage.

Methods for determining damage at the micro-level are based on data of metallographic analysis, computed tomography, hardness and elasticity measurements. The former two methods are direct methods for measuring damage, and they are theoretically the most accurate. However, their practical application is inconvenient. In the metallographic method, it is rather difficult to prepare a thin section where there would be no particles of the composite matrix or the alloy phase, which result from mechanical grinding and polishing. Besides, microcracks and micropores are polished out during thin section preparation. Another methodological difficulty of the metallographic technique is the lack of reliable methods for detecting all submicropores and submicrocracks, as well as all micropores and microcracks. At the moment, the only drawback of the technique based on computed tomography data is the resolution of computer tomographs, which allow us to detect reliably defects with a linear size of above 15 μm in real measurements of metal-based specimens [12,20].

Methods based on measuring hardness and elasticity by instrumented indentation at the micro-level avoid the disadvantages of the metallographic technique; however, they have a number of other significant problems [13,21]. As shown in [13], the technique based on hardness measurements cannot be used to calculate damage in metallic materials due to the presence of competitive processes associated with hardening and softening as a result of increasing material damage by defects. Damage determination based on elastic modulus measurements is devoid of the disadvantages of the metallographic technique caused by polishing out submicroscopic and microscopic material defects, as well as by their detection methods. Besides, the elastic modulus is weakly affected by relaxation processes and by grain shape and size [22,23]. Nevertheless, as shown in [13,21], the determination of the elastic modulus in metallic materials at the micro-level by the Oliver-Pharr method has a number of methodological difficulties, such as the effect of texture evolution on the elastic modulus, phase transformations during deformation and the appearance of residual stresses in materials, as well as the necessity to produce specimens with a surface plane perpendicular to the indenter. The latter problem is fairly easy to solve, but the issues related to the effect of residual stress formation, as well as texture and phase evolution, on the elastic modulus cannot be easily resolved since these characteristics represent the state of the material after deformation. The authors of [24] solved these problems for the DP600 steel using heat treatment of the specimen, which included recrystallization and recovery. Due to the thermal effect on the steel, they were able to restore the initial grain microstructure and texture and to remove residual stresses. At the same time, based on the metallographic method, the authors found that the density of specimen defects remained unchanged. This allowed them to determine specimen damage using the instrumented indentation technique. It was reported in [24] that this technique is laborious. Besides, thermal effects reduce material damage [9,25], thus generally making it impossible to use heat treatment to determine specimen damage proceeding from data on elastic modulus evolution.

At high deformation temperatures, relaxation processes associated with dynamic recovery and recrystallization occur in metal materials. This forms new grains and decreases texturization and residual stresses [26–29]. Thus, the deformation of metal materials under conditions of dynamic structure formation can form material microstructure suitable for the determination of damage in metallic materials from microindentation data.

Numerous studies have reported on the effect of thermomechanical conditions on damage accumulation [9,25,30–33] and damage healing [9,25,34,35] in metals and alloys in terms of both the mechanics of solids [9,25,31–33] and physical material science [25,30,34]. Unlike the case with alloys, interface boundaries between the phase constituents of metal matrix composites act more clearly as sources for the formation of discontinuities and barriers for micro- and macrocrack growth. As a result, the models relating the history of changes in temperature and the stress-strain state to damage accumulated in a metal matrix composite due to the presence of reinforcing particles may in general differ from previously constructed models for alloys. A limited number of studies deal with this problem [36–39], and this makes it relevant to study damage evolution in metal matrix composites under various thermomechanical conditions. The goal of this article is to evaluate the applicability of data on elastic modulus evolution at the micro-level to the study of damage in metal matrix composites under high-temperature deformation conditions. The best type of metal matrix composites used to estimate the damage measuring method is a composite produced by powder technology since microcracks can be much more easily detected in it than in a composite made by liquid-phase technology. This is what determined the choice of the method of composite manufacturing.

2. Material and Research Methods

The V95\10% SiC aluminum matrix composite and the V95\0% SiC material were manufactured using powder metallurgy technologies with and without the addition of reinforcing particles, respectively. The V95\10% SiC and V95\0% SiC materials were not extruded after sintering. The size of the used SiC particles with an average diameter of 3.0 ± 0.5 µm corresponded to F1200 (FEPA-Standard 42-2:2006). The V95 alloy powder particles had a shape close to a solid sphere. The particle diameter ranged between 1 and 60 µm. The V95 alloy (similar to the 7075 aluminum alloy in the chemical composition) had the following chemical composition, wt%: Al 89.73, Zn 5.72, Mg 1.95, Cu 1.71, Mn 0.34, Fe 0.26, Cr 0.14, Ni 0.06, Si 0.05, Ti 0.05, and Zn 0.02. Raw material constituents of the V95\10% SiC MMC and the V95\0% SiC material were mixed by a vibromixer. Sintering was carried out at 470 °C for 60 min. The sintering pressure was 30 MPa. There were no additives for modifying the surface of the SiC particles.

Particle examination and fracture surface analysis after specimen deformation were performed using a Vega II Tescan scanning electron microscope. The microstructure of the specimens before deformation was studied by the electron backscatter diffraction (EBSD) technique using a Vega II Tescan scanning electron microscope with an EBSD Oxford HKL NordlysF+ analysis accessory. Figure 1 shows the SEM image and the EBSD image of the matrix of the V95\10% SiC metal matrix composite and the V95\0% SiC material after sintering. When constructing EBSD images, we assumed that high-angle boundaries had a misorientation of more than 15° and that the low-angle boundaries were in the range from 2 to 15°. The grain boundaries in the EBSD images are black, and the low-angle boundaries are shown in gray. Dark spots on the EBSD images correspond to places where it was impossible to determine the orientation of the matrix lattice, and they generally correspond to areas with a large accumulation of SiC particles.

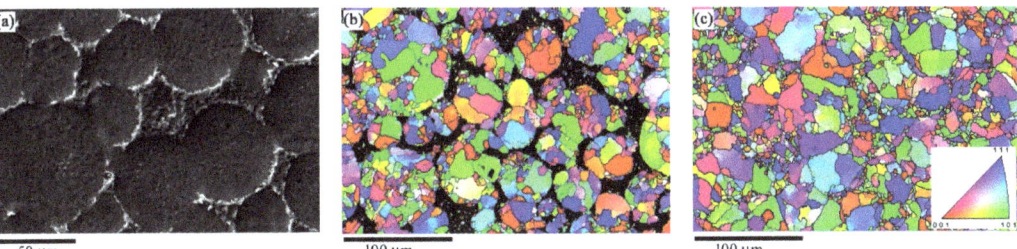

Figure 1. An SEM image of the microstructure of the V95\10% SiC metal matrix composite (**a**); An EBSD image of V95\10% SiC (**b**) and V95\0% SiC sintered powder (**c**).

The scanning pitch in the EBSD analysis was 0.3 µm. The size of the scanned area was 300 × 200 µm. For EBSD analysis and search for pores and microcracks, the specimens were first mechanically polished. Then, they were ion-polished by a Linda SEMPrep2 ion milling device for 30 min at an accelerating voltage of 10 kV with the angle of specimen inclination to the ion beam equal to 7°.

Cylindrical specimens were tested for compression and tension in the Instron 8801 servohydraulic experimental system. The compression specimen had a diameter of $d_0 = 6 \pm 0.05$ mm and a height of $h_0 = 9 \pm 0.05$ mm. The cylindrical tensile specimen had the following dimensions: $d_0 = 5 \pm 0.05$ mm and the length of the working part $l_0 = 25 \pm 0.1$ mm. In compression tests, a graphite-based lubricant was used to reduce friction between the punch and the specimen. The lubricant provided the coefficient of Coulomb friction between the punch and the aluminum alloys equal to 0.09 at 300 °C; at temperatures of 400 and 500 °C, the coefficient of friction was 0.1 and 0.13, respectively. After the test, despite lubrication, the specimen became barrel-shaped (Figure 2) in the entire deformation temperature range. In order to determine the stress-strain state evolution in the specimen during the test, its computational model was constructed and solved by the finite element method in the DEFORM program. An isothermal viscoplastic model with isotropic strain hardening was adopted for the specimen material. The flow stress was set in tabular form based on cylindrical specimen compression testing (see Figure 3a). The calculations were performed under the assumption of the axisymmetric stress-strain state in the deformation center and deformation symmetry about the horizontal geometric symmetry axis of the specimen (see Figure 3b). Thus, only a quarter of the specimen cross-section was simulated. The simulation was performed in accordance with experimental conditions of specimen loading, when the punch movement speed $V_y(t)$ was constant and equal to 1.19 mm/s. It was assumed that the friction between the specimen and the punch followed Coulomb's friction law and depended on temperature. In the simulation of specimen deformation at a temperature of 300 °C, the Coulomb friction coefficient was set to 0.09, and at temperatures of 400 and 500 °C it was equal to 0.1 and 0.13, respectively.

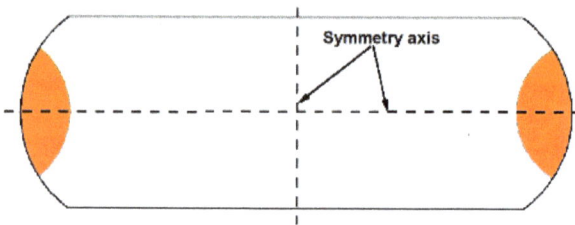

Figure 2. The longitudinal section of the cylindrical specimen after compression: orange—the indentation zone.

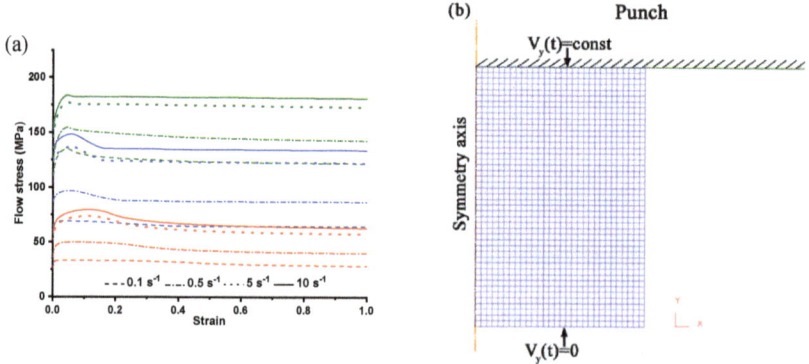

Figure 3. The dependences of flow stresses on strain (**a**) for 300 °C (green curve), 400 °C (blue curve), 500 °C (red curve) and strain rates ranging between 0.1 and 10 s^{-1}, as well as the finite element grid used for the simulations (**b**).

Indentation experiments were made in a Hysitron TI 950 nanomechanical testing device using a Berkovich indenter. A restriction was set on the movement of the indenter into the specimen. The indenter penetration and lifting rate was 0.5 µm/s, the holding time under load was 0 s. At this loading form, the device was tested on standard calibration samples made of fused quartz and monocrystalline aluminum.

After compressions, the specimens were indented in the zone close to the side surface on the thin sections after mechanical and ion polishing. The plane of the thin section was parallel to the direction of the compression axis, and it passed through the specimen symmetry axis (see Figure 2).

Material density was determined by the hydrostatic method according to ASTM B311-13 by weighing the specimens in air and in distilled water. The weighing was made on an Ohaus Pioneer PA 214 analytical balance. Table 1 shows the densities of the materials, as well as their porosity for the V95\10% SiC metal matrix composite and the V95\0% SiC material. The theoretical density ρ_{MMC}^{th} of the V95\10% SiC MMC was calculated according to the mixture rule. The porosity P_{MMC} of the V95\10% SiC MMC was calculated by the formula

$$P_{MMC} = \left(1 - \frac{\rho_{MMC}^{exp}}{\rho_{MMC}^{th}}\right) \cdot 100\%,$$

where ρ_{MMC}^{exp} is the experimentally determined density of the V95\10% SiC metal matrix composite; ρ_{MMC}^{th} is the theoretically calculated density of V95\10% SiC.

The porosity P_{CM} of the V95\0% SiC material was calculated by the formula

$$P_{CM} = \left(1 - \frac{\rho_{CM}^{exp}}{\rho_{M}^{exp}}\right) \cdot 100\%,$$

where ρ_{CM}^{exp} is the experimentally determined density of the V95\0% SiC material; ρ_{M}^{exp} is the experimentally determined density of the V95 alloy.

Table 1. The density and porosity of the materials.

Material	Density, g/cm³		Porosity, %
	Experimental	Calculated	
V95\10% SiC	2.85	2.86	0.30
V95\0% SiC	2.81	-	0.12
V95 alloy	2.82	-	-

3. Methodology for Measuring Damage in a Metal Matrix Composite and Discussion of Results

3.1. The Mechanism of Fracture in the Sintered V95\10% SiC Metal Matrix Composite

In order to determine the effect of reinforcing particles on the fracture mechanism in the V95\10% SiC metal matric composite, cylindrical specimens made from the V95\0% SiC material and the V95\10% SiC metal matrix composite were tensile tested at room temperature. The results of the mechanical tests are shown in Figure 4a. The figure shows that the strain to fracture of the composite is higher than that of the sintered material under external tensile stresses at room temperature. The fracture of both materials occurs along the interface boundaries rather than in the material matrix (see Figure 4b,c). A similar fracture of the composite occurs under compressive stresses, this being clearly demonstrated by the composite fracture surface in compression tests of the cylindrical specimens at room temperature (see Figure 5). Figures 4c and 5b show SEM images of the fracture surface with a superimposed maps of the distribution of SiC particles and the composite matrix, which were constructed based on EDS analysis. In these figures, the SiC particles are highlighted in green and the matrix is orange.

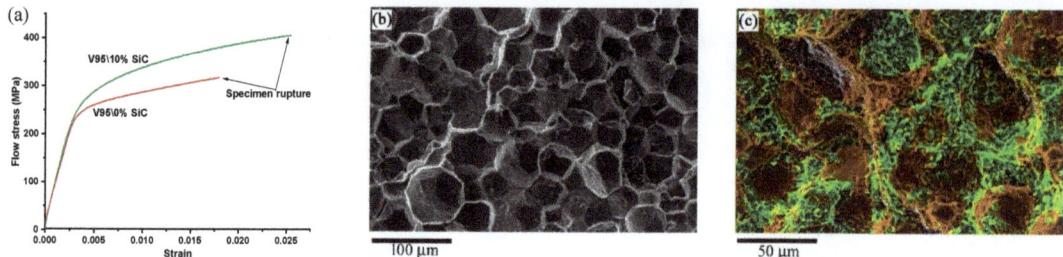

Figure 4. The strain dependence of flow stress under tension at room temperature (**a**); the fracture surface after the tension of the V95\0% SiC sintered material (**b**) and the V95\10% SiC metal matrix composite (**c**): orange—the matrix; green—SiC particles.

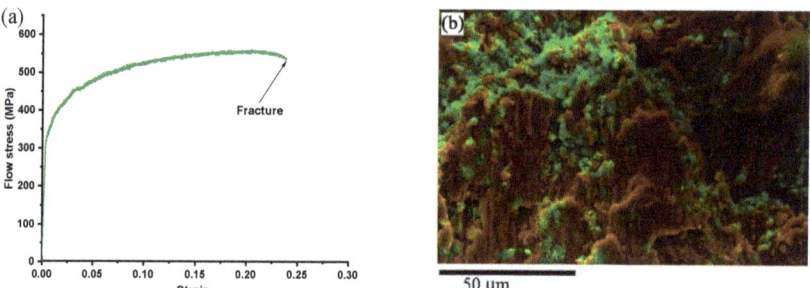

Figure 5. The strain dependence of flow stress under compression (**a**) and the fracture surface of the compressed V95\10% SiC metal matrix composite at room temperature (**b**): orange—the matrix; green—SiC particles.

The revealed deformation behavior of the metal matrix composite can be explained by the fact that the interface boundaries between the matrix and a reinforcing particle, as well as the interface boundaries among the matrix particles, were the main centers of fracture. The fracture of the composite did not occur in one stage; it resulted from gradual damage accumulation in the material leading to the appearance of macrocracks and subsequent fracture. According to the general ideas of the mechanisms of metallic material fracture [9,33,40], the fracture of the composite under study can be divided into three stages. At the first stage of the fracture of the V95\10% SiC MMC, submicropores and submicrocracks occur both inside the composite matrix and at the particle interface (see Figure 6, stage I). The composite matrix is plastically deformable and fails by ductile fracture (see Figure 7); therefore, in addition to the formation of new submicropores and submicrocracks in the matrix, the submicropores coalesce and previously formed submicrocracks become blunted, thus turning into submicropores. Then, with a significant deformation of the matrix, the submicropores coalesce, with the subsequent appearance of micropores and ductile microcracks. In parallel with increasing damage in the matrix, the stage of isolated development of microcracks occurs at the interfaces (see Figure 6, stage II). At this stage, sharp wedge-shaped microcracks quickly reach a critical value, above which their development requires no increase in external stresses (self-similarity). The growing microcrack runs into a barrier represented by a reinforcing particle or a matrix particle. A microcrack can overcome this barrier only with an increase in external stresses (see Figure 6, stage III); the microcracks coalesce thereafter to form a crack, and this leads to the final fracture of the composite along the interface boundary.

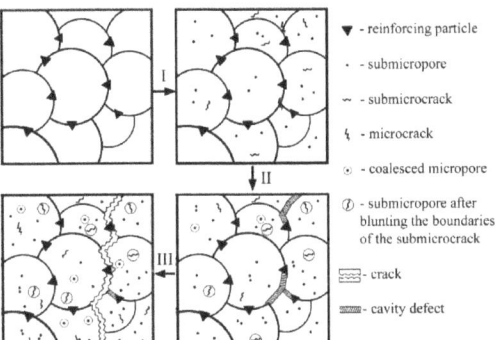

Figure 6. Schematic damage accumulation in the sintered metal matrix composite during plastic deformation.

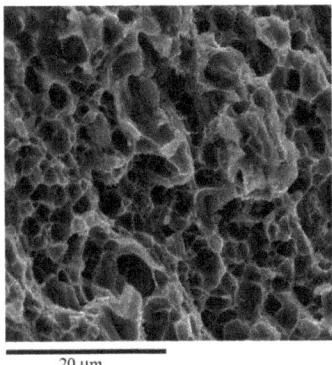

Figure 7. The fracture surface of the V95 alloy after specimen tension at room temperature.

3.2. Damage Measurement Technique

The main non-damage-induced influences on the elastic modulus in metallic materials are texture change, residual stress formation, and phase composition evolution. In order to determine whether an oriented texture was formed in the composite during deformation, EBSD analysis was performed on the V95\10% SiC MMC specimens after deformation at temperatures ranging from 300 to 500 °C. Figure 8 shows images of {100} pole shapes, as well as EBSD images of the specimens before and after deformation in the presence of macrocracks on the side surface. This figure shows that the shape of the grains inside the matrix particles remains close to equiaxed although the matrix particles as wholes stretch along the material flow direction. This behavior of the material matrix is attributable to the active process of dynamic recrystallization, which causes the formation of new grains. In its turn, the dynamic recrystallization process in the matrix is responsible for the absence of a predominant lattice orientation. Thus, we can say that the formed microstructure of the matrix will have no effect on the measured elastic modulus.

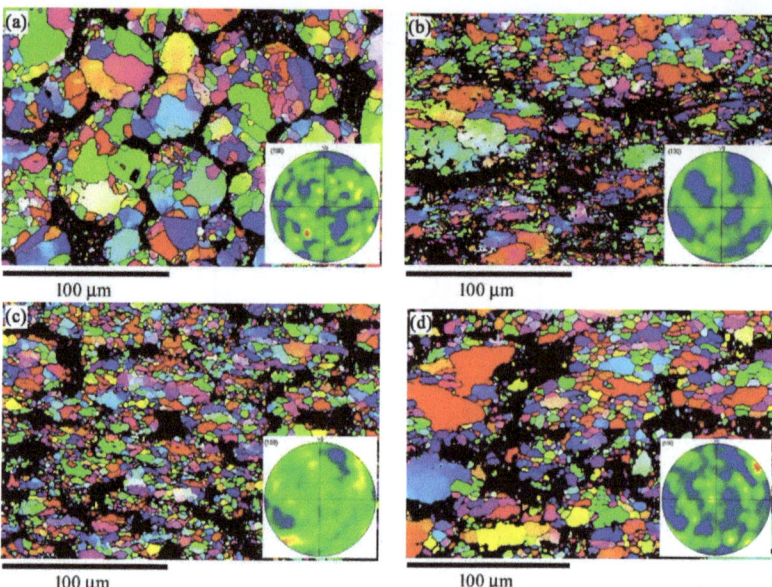

Figure 8. The initial state (**a**), the state after deformation at 300 (**b**), 400 (**c**), and 500 °C (**d**) for the V95\10% SiC metal matrix composite.

Figure 9 shows microcracks formed at the interparticle boundary in the V95\10% SiC MMC before the appearance of macrocracks on the side surface of the specimen due to compressive plastic deformation. The elastic modulus of the material must decrease due to microcracking, and this can be detected via instrumented indentation by pressing a diamond indenter into the material.

When the indenter is pressed into the metal matrix composite, the indentation diagram is affected by both the composite matrix and the reinforcing particles. Under low loads, the indenter penetrates either only into the matrix or only into the reinforcing material, and this causes a large scatter in the indentation diagrams [41,42]. As the load gradually increases, this scatter decreases and reaches a certain threshold value, which is determined by the variation of the macroscopic properties of the specimen and by the measurement error of the device. The correctness of using the technique for determining damage by the indentation technique requires that the volume of the deformed material under the indenter exceed the minimum material volume (representative volume) containing the number of the *carriers* of the structural phase state sufficient for a statistical description of the material state. Reinforcing particles, matrix phases and interphase boundaries, matrix grains and subgrains, etc. act as such *carriers* in metal matrix composites. The representative volume during indentation, as well as the determined value of the elastic modulus, depends on the indenter depth. The variation coefficient δ was chosen as a measure of the relative variation of the reduced elasticity modulus, and the penetration depth dependence of the variation coefficient δ was plotted thereafter. Such dependences for the V95\10% SiC MMC are shown in Figure 10a. This figure demonstrates that the detected reduced elastic modulus for the composite remains unchanged at indenter penetration depth exceeding 12 μm. Thus, for the correct determination of the local value of the reduced elastic modulus for the whole composite, not its structural components, the indenter penetration depth must exceed 12 μm. In order to meet this requirement, all further micromechanical tests aimed at the determination of the reduced elastic modulus were performed at an indenter penetration depth of 15 μm. Typical images of indents on the specimens at this penetration depth are shown in Figure 10b.

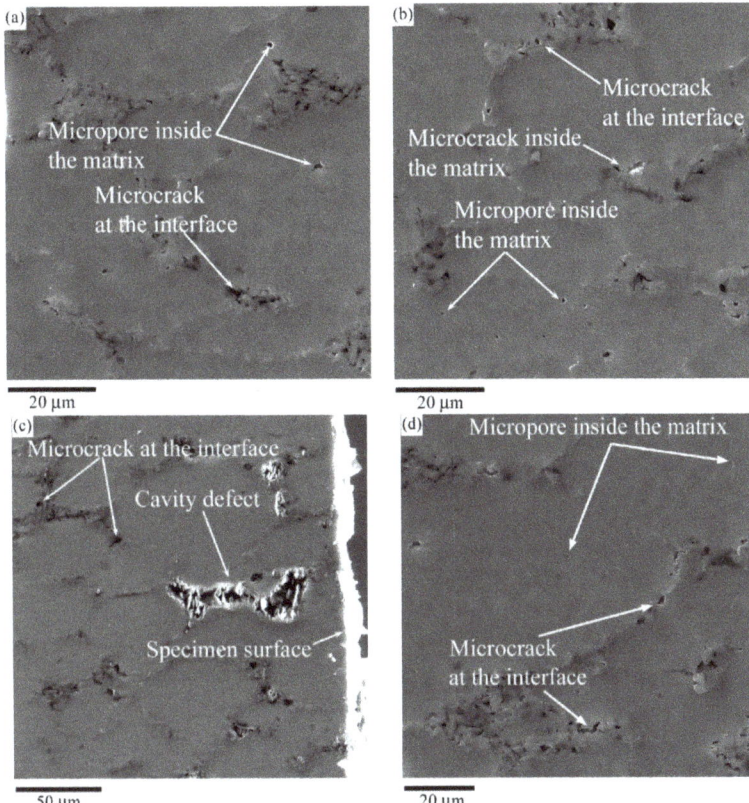

Figure 9. The microstructure near the side surface of the sintered V95/10% SiC MMC specimen after compression at temperatures of 300 °C and $D_E = 0.28$ (**a**), 400 °C and $D_E = 0.27$ (**b**), 500 °C and $D_E = 0.28$ (**c**,**d**). The images were obtained from a longitudinal thin section with a plane passing through the symmetry axis of the specimen (see Figure 2).

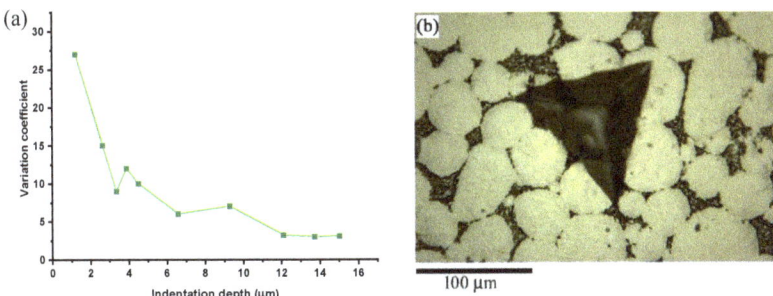

Figure 10. The penetration depth dependence of the variation coefficient δ for the V95\10% SiC MMC (**a**) and a typical image of an indent on the specimens (**b**) at an indenter penetration depth of 15 μm.

Figure 11 shows the behavior of the reduced elastic modulus $E*$ on the side surface of the metal matrix composite specimen depending on specimen strain ε at test temperatures of 300, 400 and 500 °C. The strain at each time t is calculated by the formula $\varepsilon = \sqrt{\frac{2}{3} e_{ij} e_{ij}}$, where e_{ij} denotes strain tensor deviator components determined from the results of the finite element simulation of cylindrical specimen compression (see Section 2). The data in Figure 11 show that the value of $E*$ decreases with increasing strain. For temperatures of 300 and 400 °C, as soon as the strain reaches the value $\varepsilon = 0.65$, there is a sharp decrease in the reduced elastic modulus $E*$. At a temperature of 500 °C, this occurs as early as at $\varepsilon = 0.4$. Figure 9 shows the microstructures of the composite specimens in the region close to the side surface. These microstructures were obtained after compression to the strain $\varepsilon = 0.76$ at temperatures of 300 and 400 °C and $\varepsilon = 0.59$ at a temperature of 500 °C. As seen from this figure, the microstructures of the specimens before fracture are similar to each other. For all the test temperatures, the specimens have plenty of microcracks at the interparticle boundaries and micropores inside the matrix. Further deformation of the specimens leads to the appearance of macrocracks on the side surface at $\varepsilon = 0.78$ for temperatures of 300 and 400 °C. For a temperature of 500 °C, macrocracks appear at $\varepsilon = 0.63$. Figure 9c shows a cavity defect on the specimen side surface, which may come out to the surface and develop into a macrocrack with further specimen compression.

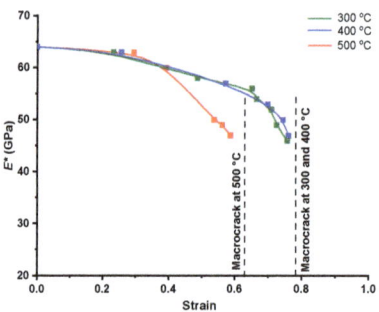

Figure 11. The strain (ε) dependence of the reduced elastic modulus $E*$ for the V95\10% SiC MMC specimen at 300, 400 and 500 °C, the dependence being valid for the zone near the side surface of the specimen (see Figure 2).

Material damage D_E is here considered to mean a measure that characterizes the degree of strain-induced material defectiveness. At the same time, the value of material damage is equal to 0 prior to deformation. At the time of macrocrack formation, damage takes the value equal to 1. In order to determine the moment of macrocracking, the specimens were deformed to different strains, and the moment of macrocrack appearance was determined iteratively. The presence of a macrocrack was detected by means of an optical microscope with ×20 magnification.

The parameter D_E is calculated by the formula found in [7],

$$D_E = 1 - \frac{E_{cur}}{E_{in}} \quad (1)$$

where E_{in} is the elastic modulus of the specimen in the initial (undeformed) state; E_{cur} is the elastic modulus of the specimen before macrocracking, i.e., at a current value of strain ε. The data obtained by the instrumented indentation technique do not allow us to determine the elastic modulus directly. The reduced elastic modulus $E*$ is calculated from the results of instrumented indentation, which is related to the elastic modulus E by the following formula (ISO 14577-1-2002):

$$E = \frac{1-(\nu_s)^2}{\frac{1}{E*} - \frac{1-(\nu_i)^2}{E_i}} \quad (2)$$

where ν_s is Poisson's ratio for the test specimen; ν_i is Poisson's ratio for the indenter, which is equal to 0.07 for diamond (ISO 14577-1-2002); $E*$ is the reduced elastic modulus of the test specimen; E_i is the elastic modulus of the indenter, which is equal to $1.14 \cdot 10^6$ N/mm^2

for diamond (ISO 14577-1-2002). Using the relation between the elastic modulus and the reduced elastic modulus represented by Equation (2) and then substituting it into Equation (1), we obtain the following formula for calculating material damage:

$$D_E = \frac{E_i \cdot (E_{in}^* - E_{cur}^*)}{E_{in}^* \cdot (E_i + (E_{cur}^* \cdot \nu_i^2 - E_{cur}^*))} \quad (3)$$

According to the technique described above, the $D_E - \varepsilon$ dependencies for the V95\10% SiC metal matrix composite deformed at temperatures of 300, 400, and 500 °C were obtained for the zone near the side surface of the specimen (see Figure 12). The last point on each $D_E - \varepsilon$ curve is marked with a cross; it corresponds to the state of the specimen with a macrocrack detected on the side surface, and this corresponds to damage equal to 1.

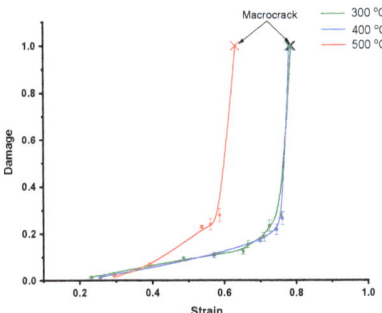

Figure 12. The strain (ε) dependence of accumulated damage D_E for the V95\10% SiC MMC in the zone near the side surface at 300, 400, and 500 °C (see Figure 2).

Figure 12 shows that the beginning of damage accumulation in the composite belongs to the strain range from 0.2 to 0.3 and increases with deformation temperature. The results of damage accumulation indicate that the slowest damage accumulation occurs at 300 °C, that it is more rapid at 400 °C and the most rapid at 500 °C. The history of changes in the stress state is one of the factors affecting damage accumulation. In this paper, the stress state coefficient k and the Lode-Nadai coefficient μ_σ are used as the stress state indices. They are calculated by the following formulas:

$$k = \frac{\sigma}{T} \text{ and } \mu_\sigma = 2\frac{\sigma_{22} - \sigma_{33}}{\sigma_{11} - \sigma_{33}} - 1,$$

where $\sigma = \frac{1}{3}(\sigma_{11} + \sigma_{22} + \sigma_{33})$ is the average normal stress; $T = \sqrt{0.5\sigma'_{ij}\sigma'_{ij}}$ is the intensity of tangential stresses; σ'_{ij} are the stress deviator components; σ_{11}, σ_{22}, σ_{33} are principal stresses. The stress state indices are calculated from finite element method computer simulation of specimen compression under conditions of the axisymmetric stress-strain state (see Section 2).

Figure 13a,b show changes in the stress state indices k and μ_σ on the side surface of the specimens during testing at three test temperatures. The difference from the values $k = -\frac{1}{\sqrt{3}}$ and $\mu_\sigma = +1$, which are typical for uniaxial compression conditions, is caused by the curvature of the side surface of the specimens under compression due to the action of friction forces on the contact surfaces. It is obvious from these figures that, even though the most favorable conditions of the stress state at the time of macrocracking on the lateral surface of the specimen exist at 500 °C, the composite demonstrates significantly lower plasticity at this temperature. At this temperature, the strain to fracture on the side surface of the specimen is the lowest and equal to 0.63, and it is 0.78 for test temperatures of 300 and 400 °C (see Figure 13c). Despite the same accumulated strain to fracture at 300 and 400 °C, the composite was deformed at a temperature of 300 °C under worse stress conditions (see Figure 13). This allows us to assume that the V95\10% SiC metal matrix composite has better plastic properties at 300 °C than at the other two temperatures.

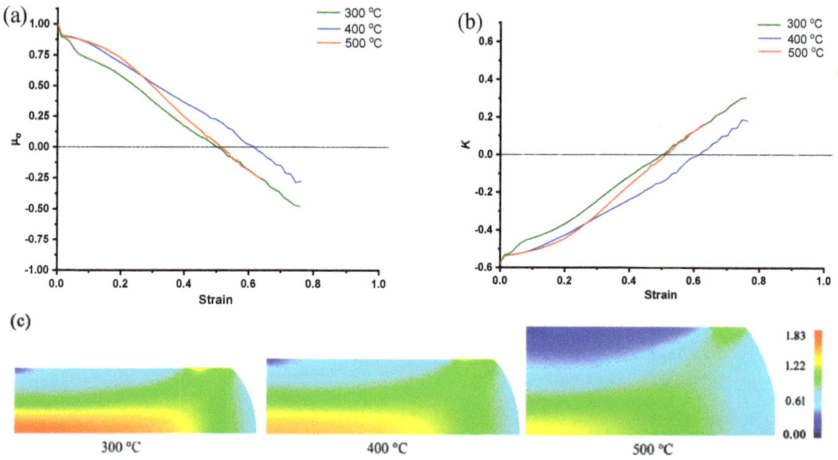

Figure 13. The strain (ε) dependence of the stress state indices μ_σ (**a**) and k (**b**) in the zone near the side surface of the specimen (see Figure 2) for a temperature range of 300 to 500 °C; the accumulated strain field. ε (**c**) at the time of macrocracking on the side surface of the V95\10% SiC MMC specimens at temperatures ranging from 300 to 500 °C.

As shown above, the fracture in the composite under study occurs along the interphase boundary. Reinforcing particles and matrix zones with high interparticle contact strength are barriers for the growth of microcracks (see Section 3.1). As the temperature increases, the strength of the interparticle contact in the matrix decreases due to the temperature softening of the V95 alloy. This fact may explain the identical strains of the composite at 300 and 400 °C despite the more rigid stress state at 300 °C. The same reason can explain a significant decrease in plastic properties at 500 °C.

4. Conclusions

1. A mechanism of damage accumulation in a sintered V95\10% SiC metal matrix composite has been proposed. The mechanism consists of a three-stage accumulation of defects on the interphase boundaries and inside the composite matrix.
2. The applicability of data on elastic modulus evolution obtained from instrumented microindentation in order to determine damage in sintered metal matrix composites after high-temperature deformation has been exemplified by the V95\10% SiC metal matrix composite. The correctness of using the technique for determining damage by indentation requires that the possible effect of residual stresses and texture be taken into account and that the volume of the deformed material under the indenter exceed the minimum material volume (the representative volume) containing a sufficient number of *carriers* of the structural phase state for a statistical description of the material state.
3. The strain dependences of damage for temperatures ranging from 300 to 500 °C have been obtained for the V95\10% SiC MMC. It has been shown that the best plasticity of the composite under compression conditions at the macro-level is observed at a temperature of 300 °C. At a deformation temperature of 500 °C, the plastic properties significantly decrease from those at 300 and 400 °C.

Author Contributions: Conceptualization, A.S.; Validation, E.S.; Project Administration, A.K.; investigation, A.S., E.S. and V.K. All authors have read and agreed to the published version of the manuscript.

Funding: The study was partially financially supported by the RFBR (project 19-08-00765) in the development of a damage measurement technique for metal matrix composites; the work on studying the rheological properties of the V95\10% SiC MMC was performed within the research conducted by the Institute of Engineering Science, Ural Branch of the Russian Academy of Sciences, project No. AAAA-A18-118020790140-5.

Acknowledgments: The facilities of the Plastometriya shared access center at the IES UB RAS and a shared access center at the B. N. Yeltsin Ural Federal University were used.

Conflicts of Interest: The authors declare no conflict of interest.

References

1. Tang, B.; Wu, F.; Guo, N.; Liu, J.; Ge, H.; Bruschi, S.; Li, X. Numerical modeling of ductile fracture of hot stamped 22MnB5 boron steel parts in three-point bending. *Int. J. Mech. Sci.* **2020**, *188*, 105951. [CrossRef]
2. Modanloo, V.; Alimirzaloo, V.; Elyasi, M. Optimal Design of Stamping Process for Fabrication of Titanium Bipolar Plates Using the Integration of Finite Element and Response Surface Methods. *Arab. J. Sci. Eng.* **2020**, *45*, 1097–1107. [CrossRef]
3. Imran, M.; Afzal, M.J.; Buhl, J.; Bambach, M.; Dunlap, A.; Schwedt, A.; Aretz, A.; Wang, S.; Lohmar, J.; Hirt, G. Evaluation of process-induced damage based on dynamic recrystallization during hot caliber rolling. *Prod. Eng.* **2020**, *14*, 5–16. [CrossRef]
4. Zhang, K.; Badreddine, H.; Hfaiedh, N.; Saanouni, K.; Liu, J. Enhanced CDM model accounting of stress triaxiality and Lode angle for ductile damage prediction in metal forming. *Int. J. Damage Mech.* **2021**, *30*, 260–282. [CrossRef]
5. Smirnov, S.; Vichuzhanin, D.; Nesterenko, A.; Smirnov, A.; Pugacheva, N.; Konovalov, A. A fracture locus for a 50 volume-percent Al/SiC metal matrix composite at high temperature. *Int. J. Mater. Form.* **2017**, *10*, 831–843. [CrossRef]
6. Meya, R.; Kusche, C.F.; Löbbe, C.; Al-Samman, T.; Kerzel, S.K.; Tekkaya, A.E. Global and high-resolution damage quantification in dual-phase steel bending samples with varying stress states. *Metals* **2019**, *9*, 319. [CrossRef]
7. Schmitt, J.H.; Jalinier, J.M.; Baudelet, B. Analysis of damage and its influence on the plastic properties of copper. *J. Mater. Sci.* **1981**, *16*, 95–101. [CrossRef]
8. Lemaitre, J.; Dufailly, J. Damage measurements. *Eng. Fract. Mech.* **1987**, *28*, 643–661. [CrossRef]
9. Smirnov, S.V. Accumulation and healing of damage during plastic metal forming: Simulation and experiment. *Key Eng. Mater.* **2013**, *528*, 61–69. [CrossRef]
10. Alves, M. Measurement of ductile material damage. *Mech. Struct. Mach.* **2001**, *29*, 451–476. [CrossRef]
11. Tasan, C.C.; Hoefnagels, J.P.M.; Geers, M.G.D. A brittle-fracture methodology for three-dimensional visualization of ductile deformation micromechanisms. *Scr. Mater.* **2009**, *61*, 20–23. [CrossRef]
12. Hild, F.; Bouterf, A.; Roux, S. Damage measurements via DIC. *Int. J. Fract.* **2015**, *191*, 77–105. [CrossRef]
13. Tasan, C.C.; Hoefnagels, J.P.M.; Geers, M.G.D. A critical assessment of indentation-based ductile damage quantification. *Acta Mater.* **2009**, *57*, 4957–4966. [CrossRef]
14. Mkaddem, A.; Gassara, F.; Hambli, R. A new procedure using the microhardness technique for sheet material damage characterisation. *J. Mater. Process. Technol.* **2006**, *178*, 111–118. [CrossRef]
15. Saijun, Z.; Chi, Z.; Qinxiang, X.; Songmao, C. Measurement of full-field ductile damage based on resistance Method. *Procedia Eng.* **2014**, *81*, 1055–1060. [CrossRef]
16. Harizi, W.; Anjoul, J.; Acosta Santamaría, V.A.; Aboura, Z.; Briand, V. Mechanical behavior of carbon-reinforced thermoplastic sandwich composites with several core types during three-point bending tests. *Compos. Struct.* **2021**, *262*, 113590. [CrossRef]
17. Kumar, J.; Sundara Raman, S.G.; Kumar, V. Analysis and Modeling of Thermal Signatures for Fatigue Damage Characterization in Ti-6Al-4V Titanium Alloy. *J. Nondestruct. Eval.* **2016**, *35*, 1–10. [CrossRef]
18. Sharkeev, Y.P.; Vavilov, V.P.; Skripnyak, V.A.; Klimenov, V.A.; Belyavskaya, O.A.; Nesteruk, D.A.; Kozulin, A.A.; Tolmachev, A.I. Evolution of the temperature field during deformation and fracture of specimens of coarse-grained and ultrafine-grained titanium. *Russ. J. Nondestruct. Test.* **2011**, *47*, 701–706. [CrossRef]
19. Fedorova, A.Y.; Bannikov, M.V.; Plekhov, O.A. A study of the stored energy in titanium under deformation and failure using infrared data. *Frat. ed Integrita Strutt.* **2013**, *24*, 81–88. [CrossRef]
20. Tasan, C.C.; Hoefnagels, J.P.M.; Geers, M.G.D. Identification of the continuum damage parameter: An experimental challenge in modeling damage evolution. *Acta Mater.* **2012**, *60*, 3581–3589. [CrossRef]
21. Xu, Z.H.; Li, X. Effect of sample tilt on nanoindentation behaviour of materials. *Philos. Mag.* **2007**, *87*, 2299–2312. [CrossRef]
22. Kim, H.S.; Bush, M.B. Effects of grain size and porosity on the elastic modulus of nanocrystalline materials. *Nanostruct. Mater.* **1999**, *11*, 361–367. [CrossRef]
23. Shen, T.D.; Koch, C.C.; Tsui, T.Y.; Pharr, G.M. On the elastic moduli of nanocrystalline Fe, Cu, Ni, and Cu-Ni alloys prepared by mechanical milling/alloying. *J. Mater. Res.* **1995**, *10*, 2892–2896. [CrossRef]

24. Tasan, C.C.; Hoefnagels, J.P.M.; Geers, M.G.D. Indentation-based damage quantification revisited. *Scr. Mater.* **2010**, *63*, 316–319. [CrossRef]
25. Zapara, M.A.; Tutyshkin, N.D.; Müller, W.H.; Wille, R. Experimental study and modeling of damage of Al alloys using tensor theory. *Contin. Mech. Thermodyn.* **2010**, *22*, 99–120. [CrossRef]
26. Degtyarev, M.V.; Chashchukhina, T.I.; Voronova, L.M. Grain growth in dynamically recrystallized copper during annealing above and below the temperature of thermally activated nucleation. *Diagn. Resour. Mech. Mater. Struct.* **2016**, *5*, 15–29. [CrossRef]
27. Chaudhuri, A.; Sarkar, A.; Kapoor, R.; Chakravartty, J.K.; Ray, R.K.; Suwas, S. Understanding the Mechanism of Dynamic Recrystallization During High-Temperature Deformation in Nb-1Zr-0.1C Alloy. *J. Mater. Eng. Perform.* **2019**, *28*, 448–462. [CrossRef]
28. Gourdet, S.; Montheillet, F. An experimental study of the recrystallization mechanism during hot deformation of aluminium. *Mater. Sci. Eng. A* **2000**, *283*, 274–288. [CrossRef]
29. Rollett, A.D.; Rohrer, G.S.; Humphreys, F.J. *Recrystallization and Related Annealing Phenomena*; Elsevier: Oxford, UK, 2004; ISBN 9780080982359.
30. Zamani, M.; Seifeddine, S.; Jarfors, A.E.W. High temperature tensile deformation behavior and failure mechanisms of an Al-Si-Cu-Mg cast alloy—The microstructural scale effect. *Mater. Des.* **2015**, *86*, 361–370. [CrossRef]
31. Bannikov, M.; Bilalov, D.; Oborin, V.; Naimark, O. Damage evolution in the AlMg6 alloy during high and very high cycle fatigue. *Frat. Integrita Strut.* **2019**, *13*, 383–395. [CrossRef]
32. Zhang, K.; Badreddine, H.; Yue, Z.; Hfaiedh, N.; Saanouni, K.; Liu, J. Failure prediction of magnesium alloys based on improved CDM model. *Int. J. Solids Struct.* **2021**, *217–218*, 155–177. [CrossRef]
33. Saanouni, K.; Devalan, P. *Damage Mechanics in Metal Forming: Advanced Modeling and Numerical Simulation*; John Wiley & Sons Inc.: London, UK, 2012; ISBN 9781118562192.
34. Karamyshev, A.P.; Nekrasov, I.I.; Nesterenko, A.V.; Parshin, V.S.; Smirnov, S.V.; Fedulov, A.A.; Shveikin, V.P. Studying the damage of ingots under plastic deformation on a lever-type radial forging machine. *AIP Conf. Proc.* **2016**, *1785*, 030010.
35. Versteylen, C.D.; Sluiter, M.H.F.; van Dijk, N.H. Modelling the formation and self-healing of creep damage in iron-based alloys. *J. Mater. Sci.* **2018**, *53*, 14758–14773. [CrossRef]
36. Junaedi, H.; Ibrahim, M.F.; Ammar, H.R.; Samuel, A.M.; Soliman, M.S.; Almajid, A.A.; Samuel, F.H. Effect of testing temperature on the strength and fracture behavior of Al-B4C composites. *J. Compos. Mater.* **2016**, *50*, 2871–2880. [CrossRef]
37. Kurşun, A.; Bayraktar, E.; Enginsoy, H.M. Experimental and numerical study of alumina reinforced aluminum matrix composites: Processing, microstructural aspects and properties. *Compos. Part B Eng.* **2016**, *90*, 302–314. [CrossRef]
38. Wang, H.; Zhang, H.; Cui, Z.; Chen, Z.; Chen, D.; Wang, H. Investigation on the high-temperature ductility and fracture mechanisms of an in-situ particle reinforced Al matrix composite 7075Al/TiB2. *Mater. Sci. Eng. A* **2019**, *764*, 138263. [CrossRef]
39. Weck, A.; Wilkinson, D.S.; Maire, E. Observation of void nucleation, growth and coalescence in a model metal matrix composite using X-ray tomography. *Mater. Sci. Eng. A* **2008**, *488*, 435–445. [CrossRef]
40. Tekkaya, A.E.; Bouchard, P.O.; Bruschi, S.; Tasan, C.C. Damage in metal forming. *CIRP Ann.* **2020**, *69*, 600–623. [CrossRef]
41. Smirnov, A.S.; Shveikin, V.P.; Smirnova, E.O.; Belozerov, G.A.; Konovalov, A.V.; Vichuzhanin, D.I.; Muizemnek, O.Y. Effect of silicon carbide particles on the mechanical and plastic properties of the AlMg6/10% SiC metal matrix composite. *J. Compos. Mater.* **2018**, *52*, 3351–3363. [CrossRef]
42. Konovalov, D.A.; Veretennikova, I.A.; Bykova, T.M.; Michurov, N.S. Development of an Approach to Determining the Representative Volume Element of the Al/SiC Metal Matrix Composite Material Fabricated by Squeeze Casting. *Russ. Metall.* **2020**, *2020*, 738–745. [CrossRef]

Article

Electrochemical Synthesis of Nano-Sized Silicon from KCl–K$_2$SiF$_6$ Melts for Powerful Lithium-Ion Batteries

Timofey Gevel [1,2], Sergey Zhuk [1,2], Natalia Leonova [1], Anastasia Leonova [1], Alexey Trofimov [1,2], Andrey Suzdaltsev [1,2,*] and Yuriy Zaikov [1,2]

[1] Scientific Laboratory of the Electrochemical Devices and Materials, Ural Federal University, Mira St. 28, 620002 Yekaterinburg, Russia; Timofey.gevel@urfu.ru (T.G.); zhuksi83@mail.ru (S.Z.); n.m.leonova@urfu.ru (N.L.); a.m.leonova@urfu.ru (A.L.); a.a.trofimov@urfu.ru (A.T.); i.p.zaikov@urfu.ru (Y.Z.)

[2] Institute of High-Temperature Electrochemistry, Ural Branch, The Russian Academy of Sciences, Academicheskaya St. 20, 620137 Yekaterinburg, Russia

* Correspondence: suzdaltsev_av@ihte.uran.ru or a.v.suzdaltsev@urfu.ru

Abstract: Currently, silicon and silicon-based composite materials are widely used in microelectronics and solar energy devices. At the same time, silicon in the form of nanoscale fibers and various particles morphology is required for lithium-ion batteries with increased capacity. In this work, we studied the electrolytic production of nanosized silicon from low-fluoride KCl–K$_2$SiF$_6$ and KCl–K$_2$SiF$_6$–SiO$_2$ melts. The effect of SiO$_2$ addition on the morphology and composition of electrolytic silicon deposits was studied under the conditions of potentiostatic electrolysis (cathode overvoltage of 0.1, 0.15, and 0.25 V vs. the potential of a quasi-reference electrode). The obtained silicon deposits were separated from the electrolyte residues, analyzed by scanning electron microscopy and spectral analysis, and then used to fabricate a composite Si/C anode for a lithium-ion battery. The energy characteristics of the manufactured anode half-cells were measured by the galvanostatic cycling method. Cycling revealed better capacity retention and higher coulombic efficiency of the Si/C composite based on silicon synthesized from KCl–K$_2$SiF$_6$–SiO$_2$ melt. After 15 cycles at 200 mA·g^{-1}, material obtained at 0.15 V overvoltage demonstrates capacity of 850 mAh·g^{-1}.

Keywords: lithium-ion battery; silicon; halide melt; electrodeposition; silicon fibers; nanotubes; nanoneedles; composite anode

1. Introduction

Currently, silicon and silicon-based composite materials are widely used in microelectronics, solar energy devices, as well as in the manufacture of portable energy storage devices [1–3]. In particular, a new lithium-ion battery (LIB) with a silicon-based anode is being actively developed. The theoretical capacity of such an anode (4200 mAh·g^{-1}) is an order of magnitude higher than the widely used graphite anode [3,4]. However, during cycling, silicon is subject to huge volumetric expansion (up to 400% [5]), which leads to mechanical destruction of the electrode and significant loss of the original capacity. To avoid this, it is recommended to use anodes based on nanosized silicon particles. Nanosized silicon cannot be obtained by the existing methods of the carbothermal reduction of quartz to silicon with subsequent purification of the silicon by hydrogenation–dehydrogenation [6]. The use of nanosized silicon particles, mainly fibers and tubes within composite materials, can significantly reduce the problem of volumetric expansion. Suitable polymer binders also reduce mechanical degradation of the electrode, but they do not affect the silicon expansion [7,8]. Another method that can improve silicon's performance is the coating (mainly carbon based) of silicon particles and creation of a "core-shell" structure [9,10]. Doped silicon anodes have also shown improved electrochemical performance and cyclic resistance in comparison with pure silicon [11–13].

Methods for preparing silicon nanofibers can be divided into two groups. The first group comprises methods based on removing excess material-laser ablation [14], ion etching [15], chemical etching [16,17]. These methods allow one to obtain high-purity components with the required sizes. However, they require expensive equipment, catalysts (gold is most often used) and are poorly suited for industrial scale production. The second group of methods includes silicon deposition on the substrates. Among these, the most commonly used method is the vapor–liquid–crystal deposition [18,19]. Silane is usually used as a source of silicon. The disadvantages of these methods also include the use of gold as a catalyst and the difficulty in scaling and hardware design of the process. To obtain silicon suitable for microelectronics, several stages of purification are needed, one of which is represented by a group of methods based on zone crystallization from molten silicon [20].

One of the cheapest and promising methods for producing nanosized silicon is electrolytic reduction from molten salts. To date, many studies have been carried out aimed at the electrolytic production of silicon from various molten electrolytes with K_2SiF_6, Na_2SiF_6, SiO_2, and $SiCl_4$ additives in a temperature range mainly from 550 °C to 750 °C [21–39]. All these studies show the fundamental possibility of using electroreduction for the production of silicon deposits of various morphologies, including continuous coatings up to 1 mm thick, submicron films (0.5–1 µm), micro-sized dendrites, nano- and micro-sized disordered fibers.

One of the most promising electrolytes for silicon production is the KCl–KF melt with a KF molar fraction of up to 66% [27–31], which is a good solvent for K_2SiF_6 and SiO_2 and is water soluble. However, this system also has a number of disadvantages, including relatively high aggressiveness of KF to reactor materials, the need to remove impurities such as H_2O and HF from KF when preparing a molten KCl–KF mixture, and a good solubility of oxides. Despite the positive results of the work performed, all of the above factors can impede the deposition of high-purity silicon, especially without oxygen inclusions.

To eliminate these drawbacks, alternative media with the reduced or zero fluoride content for the production of silicon are being investigated. In particular, this concerns the KCl–KI–KF–K_2SiF_6 [32,33] and $CaCl_2$–CaO–SiO_2 melts [34–37], although their use also implies the absence of moisture in the reactor and increased requirements for the preparation of molten electrolytes and instrumentation. In previous works, we have shown the fundamental possibility of obtaining nanosized silicon fibers by electrolytic refining of silicon in a low-fluoride KCl–K_2SiF_6 system [38,39]. The advantage of this system is that its anionic composition virtually does not change. That provides stabilization of the energy characteristics of silicon-containing electroactive ions and possibility of controlling the morphology of silicon deposits. Preliminary electrolysis tests [38,39] show that it is difficult to obtain nanosized silicon deposits from molten salts. In particular, this is due to the varied concentration of silicon-containing ions in the melt, the semiconducting nature of silicon, and increased sensitivity of the deposit morphology to oxygen impurities.

Despite the active interest in Si-based anodes for LIBs [3,4,6–19,40], the works devoted to the utilization of electrolytic silicon deposits as the LIB anode material are limited [41–46]. The highest characteristics for silicon obtained by electrodeposition are on carbon cloth. The capacity was more than 1100 mAh·g^{-1} by the 200th cycle at a discharge current of 500 mA·g^{-1}, while the silicon nanopowder itself, without the use of carbon fabric, demonstrates very low capacity—less than 500 mAh·g^{-1} by the 5th cycle and less than 200 mAh·g^{-1} by 200th cycle [40]. In recent years, most researchers are focused on silicon electrodeposition from $CaCl_2$-based melts. In work [42] carried out by researchers from the USA and China, electrodeposited nanofibers in LIB demonstrate a capacity of 714 mAh·g^{-1} after 500 cycles at C/2 current. According to [45], China launched a large-scale test production of electrodeposited silicon with capacity more than 2000 mAh·g^{-1} after 70 cycles at a current of 0.1 C; capacity at a current of 5 C is about 1600 mAh·g^{-1}. It is noted that for practical use of the produced silicon they made Si/graphite mixture with a total capacity of 650 mAh·g^{-1} [47].

In this regard, in this work, we studied the effect of SiO$_2$ on the morphology of nanosized silicon deposits during the electrolysis of KCl–K$_2$SiF$_6$ melts. For this, silicon deposits were obtained from the KCl–K$_2$SiF$_6$ and KCl–K$_2$SiF$_6$–SiO$_2$ melts at different cathode overvoltage. Obtained deposits were analyzed and tested on their electrochemical characteristics as a part of an Si-based anode of a LIB.

2. Materials and Methods

Melt preparation. The investigated KCl electrolytes with additions of K$_2$SiF$_6$ and SiO$_2$ were prepared by mixing individual salts of a chemically pure grade (Reakhim, Russia) and then melting them in a glassy carbon crucible immediately before the experiments. In a previous study [48], it was shown that the purity of KCl used is close to that of the recrystallized salt; therefore, additional purification of the prepared melt from electropositive impurities was not carried out. The K$_2$SiF$_6$ salt was preliminarily purified from oxygen impurities by HF fluorination. For this, K$_2$SiF$_6$ was mixed with NH$_4$F and heated in stages to a temperature of 450 °C according to a previously described procedure [49]. SiO$_2$ was added to the melt in an amount of 0.34 wt.%. The KCl–K$_2$SiF$_6$ and KCl–K$_2$SiF$_6$–SiO$_2$ melts were held for an hour at the operating temperature to establish equilibrium between silicon ions [50,51]. Then, the melt samples were taken for analysis, and electrolysis was started.

Experimental setup. Electrolysis tests were carried out in a sealed quartz retort with a high-purity argon atmosphere at a temperature of 790 °C (Figure 1). A glassy carbon crucible with the melt was placed at the bottom of the retort, which was hermetically closed with a fluoroplastic cap. The inner walls of the retort were additionally shielded with nickel plates to protect against fluorine-containing sublimates. Holes with fittings were made in the cap, in which a working electrode (glassy carbon), a glassy carbon counter electrode, and a silicon QRE were placed. In addition, holes were provided for an additional working electrode, thermocouple, and a tube for gas supply and loading of silicon-containing additives. During the experiments, the temperature was controlled within ±2 °C by a Pt/PtRh thermocouple and a thermocouple module USB-TC01 (National Instruments, Austin, TX, USA).

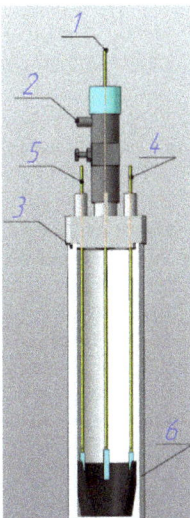

Figure 1. Experimental setup. 1—working electrode; 2—gas inlet; 3—quartz cell; 4—counter electrode; 5—quasi-reference electrodes; 6—glassy carbon crucible.

Electrochemical measurements and electrodeposition. To establish the effect of SiO_2 on the kinetics of silicon electrodeposition, voltammograms were obtained by the cyclic chronovoltammetry. Before measurement, the electrodes were kept in the melt for 30 min to establish a stable (within ± 5 mV) potential difference between the working electrode and the quasi-reference electrode (QRE). In order to determine and compensate the ohmic voltage drop in the measuring circuit, impedance spectroscopy and current interruption (I-Interrupt) technique were used.

The electrodeposition of silicon was carried out on glassy carbon plates with an area of 2 cm^2 at different cathode overvoltage for one hour. The electrolysis time was selected based on the results of preliminary electrolysis tests [24,25]. For electrochemical measurements and electrolysis, PGSTAT AutoLAB 302N with Nova 1.11 software (MetrOhm, Herisau, Switzerland) was used. Glassy carbon plates were used as anodes.

At the end of the electrolysis, the electrode was raised above the melt for 20–30 min in order to remove excess salts included in the cathode deposit. Then, the electrode was raised to the cold zone of the retort and, after cooling in an inert atmosphere, was removed.

Separation of the cathode deposit. The obtained silicon deposits were washed in acidic (HCl) water solution; the pH of the solution for washing varies in the range of 2 to 4. An ultrasonic homogenizer SONOPULS UW mini 20 was used to disperse the deposits. Dispersion was carried out in a periodic mode at a given power of 0.995 kJ with a pulse duration of 90 s. After washing, the deposits were dried in a vacuum oven at 80 °C for 18 h.

Analysis of the morphology and composition. The elemental composition of the melt and the obtained silicon deposits was analyzed by atomic emission spectroscopy with inductively coupled plasma (AES-ICP) using an iCAP 6300 Duo Spectrometer (Thermo Scientific, Waltham, MA, USA). The morphology and elemental composition of the obtained deposits were studied using a Tescan Vega 4 (Tescan, Kohoutovice, Czech Republic) scanning electron microscope with an Xplore 30 EDS detector (Oxford, UK).

Electrochemical performance. Silicon performance was investigated in a 3-electrode half-cell. The anode composition was 10 wt.% polyvinylidene difluoride dissolved in N-methyl-2-pyrollidone, 10 wt.% carbon black, and 80 wt.% silicon. LIB fabrication was performed in an argon-filled glove box (O_2, H_2O < 0.1 ppm). Stainless steel mesh with applied composite anode was used as the working electrode and two separate lithium strips as the counter and reference electrodes. All electrodes were divided by two layers of separator and tightly placed in the cell. The cell was flooded with 1 mL of electrolyte—1 M $LiPF_6$ in a mixture of ethylene carbonate/dimethyl carbonate/diethyl carbonate (1:1:1 by volume). Cycling experiments were performed using a Zive-SP2 potentiostat (WonATech, Seoul, Korea).

3. Results

3.1. Electrolysis Test Results

According to ICP, the silicon contents in the $KCl-K_2SiF_6$ and $KCl-K_2SiF_6-SiO_2$ melts before electrolysis tests were 0.05 and 0.15 wt.%, respectively. The deposits were obtained at cathode overvoltage of 0.1, 0.15, and 0.25 V relative to the QRE potential. For the $KCl-K_2SiF_6$ melt (0.05 wt.% of silicon), the obtained deposits are shown in Figure 2. It can be seen that the growth of deposits occurs with the formation of silicon-salt mixtures. In this case, with a decrease in the cathode overvoltage, the deposits change color from gray to light brown. This can be associated both with a change in the crystalline structure of silicon to amorphous, and with a change in the size of silicon particles [31,38].

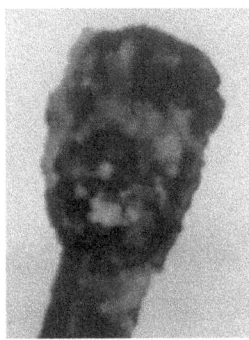 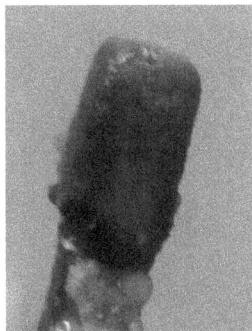 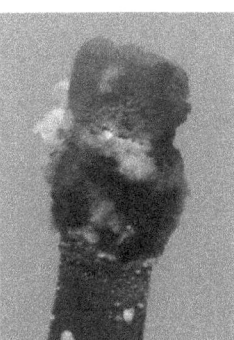

Figure 2. Photographs of the cathode deposits obtained after the electrolysis of the KCl–K$_2$SiF$_6$ melt at a temperature of 790 °C and cathode overvoltage of 0.25, 0.15, and 0.1 V vs. QRE potential.

Electrolysis tests in the KCl–K$_2$SiF$_6$ melt with the addition of SiO$_2$ (0.15 wt.% of silicon) at the same cathode overvoltages of 0.1, 0.15, and 0.25 V vs. QRE potential were performed. As a result, cathodic deposits were obtained, photographs of which are shown in Figure 3. It can be noted that during electrolysis of the melt with the addition of SiO$_2$, more volumetrically uniform deposits are obtained, while the color of the precipitates remains the same in all tests.

  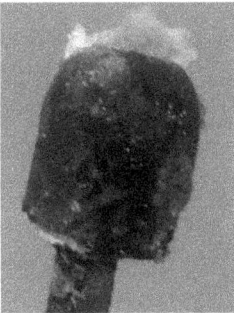

Figure 3. Photographs of the cathode deposits obtained after the electrolysis of the KCl–K$_2$SiF$_6$–SiO$_2$ melt at a temperature of 790 °C and cathode overvoltage of 0.25, 0.15, and 0.1 V vs. QRE potential.

3.2. Effect of SiO$_2$ on the Morphology of Silicon Deposits

For more detailed analysis of the effect of SiO$_2$ on the morphology of the silicon, SEM images were made. Figure 4 shows the corresponding micrographs of the deposits. It can be seen that in the KCl–K$_2$SiF$_6$ system, silicon was deposited in the form of disordered fibers with an average diameter of 300–400 nm, while the effect of cathode overvoltage on the morphology of the deposit is not clearly observed. Such a result is probably due to the limiting stage of the preceding chemical reaction of silicon-containing ion dissociation.

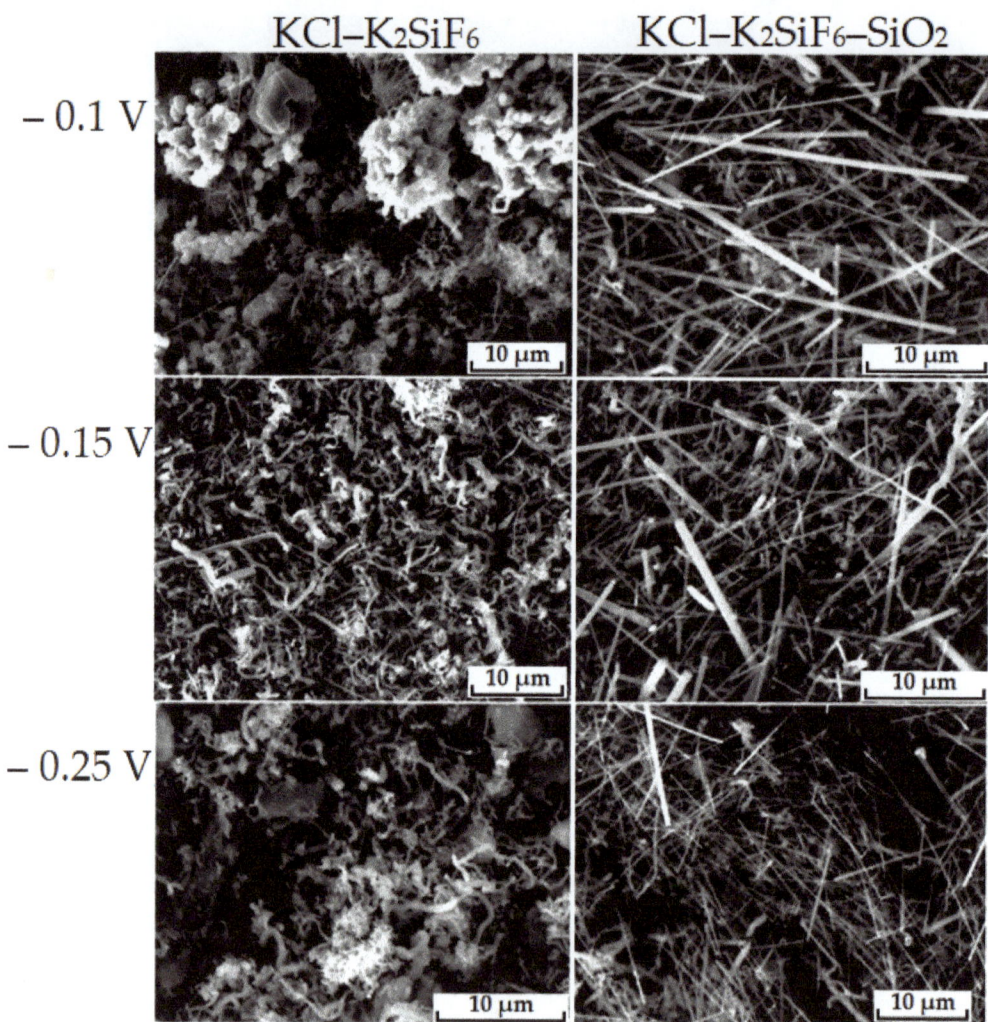

Figure 4. Micrographs of the cathode deposits obtained during the electrolysis of KCl–K$_2$SiF$_6$ and KCl–K$_2$SiF$_6$–SiO$_2$ melts at a temperature of 790 °C and cathode overvoltage of 0.25, 0.15, and 0.1 V vs. QRE.

When SiO$_2$ is added in the melt, silicon is predominantly deposited in the form of nano-sized tubes and needles, the length of which varies in the range of 30–60 μm, and the diameter varies in the range of 200 nm to 400 nm. This effect can be interpreted as the mutual influence of several factors:

(1) a change in the composition of silicon-containing electroactive particles, namely the inclusion of oxygen atoms in their composition, leading to a decrease in binding energy of silicon atoms [51] and facilitation of the charge transfer stage;

(2) a decrease in electrical conductivity (increase in resistance) of the near-cathode layer melt, which will redistribute the current lines among the cathode surface. Namely, silicon electrodeposition will occur at the tips of the deposit, rather than on the lateral surfaces;

(3) a change in the composition of silicon-containing electroactive ions, as a result the ions activity and the rate constant of the possible preceding chemical reaction in the melt change.

Taking into account the weak effect of the cathode overvoltage on the morphology of silicon deposits, the effect of SiO_2 addition can be associated with the last two factors. It can be noted that the average size of silicon particles decreases with the addition of SiO_2 to the melt. A similar effect was observed in other works [31,41].

According to EDX analysis, the average oxygen content in the fibers obtained during the electrolysis of the $KCl-K_2SiF_6$ melt was 2.1–4.3 wt.%, and the average oxygen content in nano-sized tubes and needles obtained during the electrolysis of this melt with the addition of SiO_2 was 2.7–5.1 wt.%. The appearance of oxygen in the deposits can be caused both by the use of a quartz retort as a reactor container and after-electrolysis silicon treatment. It is important to note that the addition of SiO_2 does not significantly increase oxygen content in silicon deposits, which is important from the point of the electrochemical characteristics as the anode of LIB.

3.3. Effect of SiO_2 on the Kinetics of Silicon Electrodeposition

To check the effect of SiO_2 addition on the silicon electroreduction, voltammograms were recorded in the $KCl-K_2SiF_6$ and $KCl-K_2SiF_6-SiO_2$ melts at a temperature of 790 °C. Figure 5 shows that the electroreduction of silicon-containing electroactive ions from the $KCl-K_2SiF_6$ melt occurs at a potential more negative than 0.1 V, with the formation of a cathodic peak **Si** [27,32] at a potential of −0.1 V, relative to the QRE potential (Figure 5, black line).

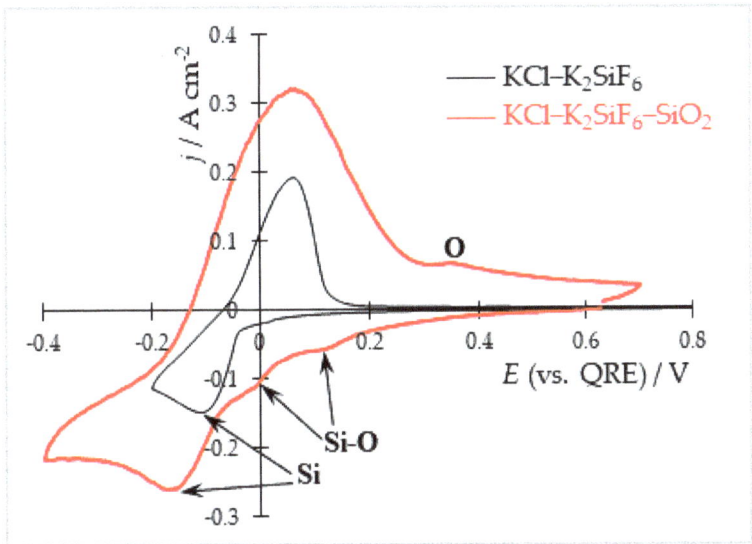

Figure 5. Cyclic voltammograms for the $KCl-K_2SiF_6$ and $KCl-K_2SiF_6-SiO_2$ melts with a total silicon content of 0.05 and 0.15 wt.%, respectively. Potential sweep rate is 0.4 V s^{-1}.

For the SiO_2-containing melt, additional **Si-O** cathode waves appeared on the voltammogram at the potentials of about 0.1 and 0 V relative to the QRE potential (Figure 5, red line). Those waves can be associated with the discharge of different oxygen-containing electroactive Si-O-Cl-F anions, which appear in the melt during SiO_2 dissolution [50,51]. A similar effect was observed in the works, aimed at the electroreduction of silicon and other elements from halide-oxide melts [52–54]. In the $KCl-K_2SiF_6-SiO_2$ melt, an increase of the silicon electroreduction current was observed, since the silicon content has been increased. In the anode region, peak **Si'** and a wave **(Si-O)'** at a potential of 0.1 and 0.33 V relative to the QRE potential were detected. We assume that the peak **Si'** is associated with silicon dissolution [27,32], and the additional wave **(Si-O)'** can be caused by oxidation of a reduced form of the silicon electroactive ion. To obtain a more detailed picture, further study of the cathodic process in KCl–K$_2$SiF$_6$ and KCl–K$_2$SiF$_6$–SiO$_2$ melts is required.

3.4. Electrochemical Characteristics of the Obtained Silicon Deposits

The obtained silicon deposits were studied as a composite Si/C anode for a lithium-ion battery. In the first charge/discharge cycle (Figure 6), the capacity of Si/C composite prepared on the base of silicon electrodeposited from KCl–K$_2$SiF$_6$ was between 1450 and 3000 mAh·g^{-1} during lithiation, and between 950 and 1670 mAh·g^{-1} during delithiation. Despite of the similar morphology and size of silicon fibers synthesized at different over-potentials, their energy characteristics are different. As can be seen from Figure 6, samples 2 and 3 synthesized at 0.15 and 0.25 V show similar charging times, although the initial period associated with electrolyte reduction and solid electrolyte interphase formation is quite different. Silicon synthesized at 0.1 V demonstrates 60% lower capacity, and the highest initial coulombic efficiency of 65.5%.

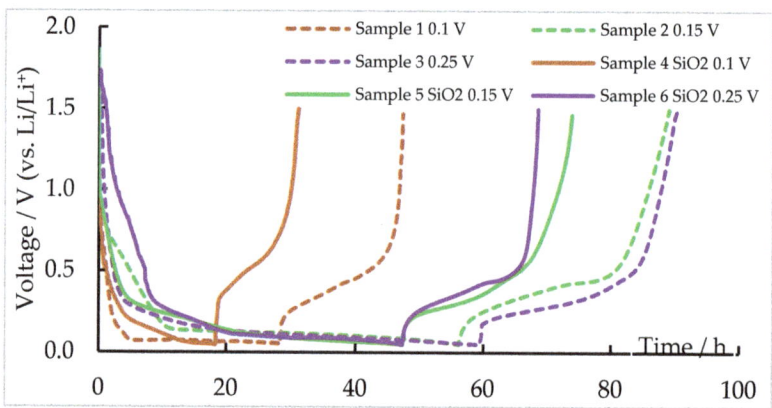

Figure 6. First charge/discharge cycle curves of Si/C composites based on electrodeposited silicon at 50 mA g^{-1}. Solid lines represent silicon deposited from molten KCl–K$_2$SiF$_6$; dashed lines represent silicon deposited from molten KCl–K$_2$SiF$_6$–SiO$_2$.

After 15 cycles at 200 mA·g^{-1} (Figure 7a), the capacitance for the best sample of silicon fibers was 750 mAh·g^{-1} for sample 3 synthesized at 0.25 V overvoltage. The coulombic efficiency of the anode half-cell with silicon fibers in this case increased from 59% up to 92%. Silicon synthesized at 0.15 V (sample 2) demonstrates a similar capacity fade rate as sample 3, but a lower overall capacity. However, these samples have different coulombic efficiency during the first 10 cycles. Sample 1, while having a much lower capacity, demonstrates better capacity retention, with −50.7% of the initial capacity after 15 cycles.

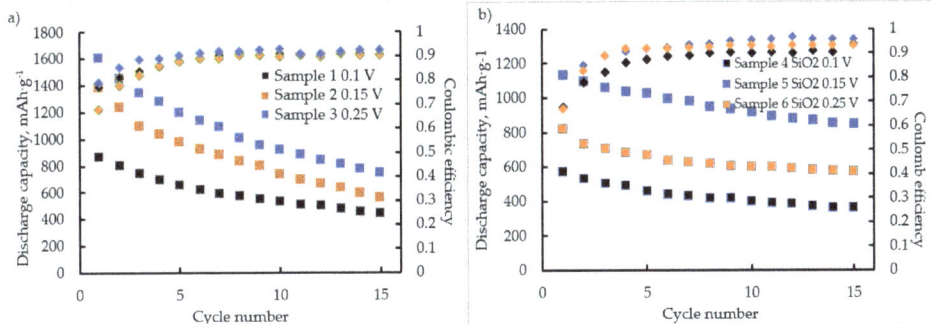

Figure 7. Cycling performance and the coulombic efficiency of the Si/C made from silicon electrodeposited from (**a**) KCl–K$_2$SiF$_6$ (samples 1–3) and (**b**) KCl–K$_2$SiF$_6$–SiO$_2$ (samples 4–6) at the cathode overvoltages of –0.1, –0.15, and –0.25 V. Diamonds-coulombic efficiency; squares-capacity.

For silicon samples obtained from the KCl–K$_2$SiF$_6$–SiO$_2$ melt, the anode capacity was between 920 and 2430 mAh·g^{-1} for charging and between 600 and 1300 mAh·g^{-1} during discharge. As can be seen from Figure 6, the behavior of these samples is similar to the ones synthesized without the addition of SiO$_2$ to the melt, but lower capacities are achieved for all the samples. The highest capacity value was observed for the sample 5 synthesized at 0.15 V overvoltage. The initial coulombic efficiency of the silicon needles/tubes was 44–65%, lower than for the samples 1–3. Sample 4 demonstrates much higher delithiation potential than all other samples, probably due to the higher C-rate in comparison to other samples. After 15 cycles (Figure 7b), samples synthesized from the melt with SiO$_2$ addition demonstrate higher coulombic efficiency—between 93.5% and 96.1%—and better capacity retention rate. Remaining capacity of the sample 5 is 850 mAh·g^{-1}, which is 74.5% of the initial capacity.

It is probable that such a difference in capacity fade rate is caused by different silicon structures or surface layer composition; however, the reasons for drastic capacity difference among the samples with similar morphology synthesized at different deposition overpotential is not clear. The results obtained indicate that further research is promising, aimed both at optimizing the conditions for obtaining silicon deposits from the investigated molten electrolytes, and at improving the performance of a lithium-ion battery with a composite anode based on the obtained silicon.

4. Conclusions

In this work, we studied the effect of cathode overvoltage and the addition of SiO$_2$ on the morphology of silicon electrolytic deposits obtained from KCl–K$_2$SiF$_6$ and KCl–K$_2$SiF$_6$–SiO$_2$ melts at a temperature of 790 °C. Using scanning electron microscopy, it was shown that during electrolysis of the KCl–K$_2$SiF$_6$ melt, nano-sized disordered fibers (300–400 nm in diameter) are formed on the cathode, while the addition of SiO$_2$ to the melt leads to an ordered growth of silicon deposit in the form of nano-sized needles and tubes (diameter 200–400 nm, length 30–60 µm). At the same time, it was noted that an increase in cathode overvoltage during the electrolysis of the studied molten electrolytes has virtually no effect on the morphology and size of the deposit, which may be associated with the course of the process under the conditions of a slowed-down preceding chemical reaction of the silicon-containing ion dissociation.

According to EDX analysis, the average oxygen content in silicon deposits obtained during the electrolysis of the KCl–K$_2$SiF$_6$ melt was 2.1–4.3 wt.%, and in the deposits obtained during the electrolysis of this melt with the addition of SiO$_2$, the average oxygen content was 2.7–5.1 wt.%.

The resulting deposits were used for the manufacture of composite Si/C anode half-cells of lithium-ion batteries. All the synthesized samples have similar initial charge/discharge curves, although the capacity is different. Samples synthesized at 0.1 V show much lower capacity than the ones synthesized at 0.15 and 0.25 V overvoltage. Although the first cycle capacity of silicon synthesized with SiO_2 addition is lower than that of silicon obtained without additive, further cycling results in higher capacity and much higher capacity retention for all samples. For silicon deposited from $KCl–K_2SiF_6$, the highest capacity value is 750 mAh·g^{-1} with capacity retention of 46.6 % and coulombic efficiency of 92.7%. The best results are achieved for silicon synthesized from $KCl–K_2SiF_6–SiO_2$ at 0.15 V overvoltage; its capacity was 850 mAh·g^{-1} after 15 cycles at a current of 200 mA·g^{-1} and capacity retention of 74.5% with coulombic efficiency of 96.1%. Such differences in the capacity retention for the samples synthesized with and without SiO_2 addition is not clear and requires further investigation.

Author Contributions: Conceptualization, T.G. and A.S.; methodology; T.G., S.Z., N.L., A.L. and A.T.; validation, T.G. and A.S.; formal analysis, A.S. and A.T.; investigation, T.G., S.Z., N.L., A.L. and A.T.; writing—original draft preparation, T.G. and A.S.; writing—review and editing, T.G. and A.S.; supervision, A.S. and Y.Z.; project administration, A.T. and A.S. All authors have read and agreed to the published version of the manuscript.

Funding: This research received no external funding.

Institutional Review Board Statement: Not applicable.

Informed Consent Statement: Not applicable.

Acknowledgments: This work is performed in the frame of the State Assignment number 075-03-2020-582/1, dated 18 February 2020 (the theme number 0836-2020-0037).

Conflicts of Interest: The authors declare no conflict of interest.

References

1. Cohen, U. Some prospective applications of silicon electrodeposition from molten fluorides to solar cell fabrication. *J. Electron. Mater.* **1977**, *6*, 607–643. [CrossRef]
2. Liu, Z.; Sofia, S.E.; Laine, H.S.; Woodhouse, M.; Wieghold, S.; Peters, I.M.; Buonassisi, T. Revisiting thin silicon for photovoltaics: A technoeconomic perspective. *Energy Environ. Sci.* **2020**, *13*, 12–23. [CrossRef]
3. Feng, K.; Li, M.; Liu, W.; Kashkooli, A.G.; Xiao, X.; Cai, M.; Chen, Z. Silicon-based anodes for lithium-ion batteries: From fundamentals to practical applications. *Small* **2018**, *14*, 1702737. [CrossRef] [PubMed]
4. Galashev, A.Y.; Suzdaltsev, A.V.; Ivanichkina, K.A. Design of the high performance microbattery with silicene anode. *Mater. Sci. Eng.* **2020**, *261*, 114718. [CrossRef]
5. Schmidt, H.; Jerliu, B.; Hüger, E.; Stahnc, J. Volume expansion of amorphous silicon electrodes during potentiostatic lithiation of Li-ion batteries. *Electrochem. Commun.* **2020**, *115*, 106738. [CrossRef]
6. Wu, J.J.; Chen, Z.; Ma, W.; Dai, Y. Thermodynamic estimation of silicon tetrachloride to trichlorosilane by a low temperature hydrogenation technique. *Silicon* **2017**, *9*, 69–75. [CrossRef]
7. Zhang, C.; Wang, F.; Han, J.; Bai, S.; Tan, J.; Liu, J.; Li, F. Challenges and recent progress on silicon-based anode materials for next-generation lithium-ion batteries. *Small Struct.* **2021**, *2*, 2100009. [CrossRef]
8. Zhao, Y.; Yue, F.; Li, S.; Zhang, Y.; Tian, Z.; Xu, Q.; Xin, S.; Guo, Y. Advances of polymer binders for silicon-based anodes in high energy density lithium-ion batteries. *InfoMat* **2021**, *3*, 460–501. [CrossRef]
9. Liu, X.; Zhu, X.; Pan, D. Solutions for the problems of silicon–carbon anode materials for lithium-ion batteries. *R. Soc. Open Sci.* **2018**, *5*, 172370. [CrossRef]
10. Azam, M.A.; Safie, N.E.; Ahmad, A.S.; Yuza, N.A.; Adilah Zulkifli, N.S. Recent advances of silicon, carbon composites and tin oxide as new anode materials for lithium-ion battery: A comprehensive review. *J. Energy Storage* **2021**, *33*, 102096. [CrossRef]
11. Salah, M.; Hall, C.; Murphy, P.; Francis, C.; Kerr, R.; Stoehr, B.; Rudd, S.; Fabretto, M. Doped and reactive silicon thin film anodes for lithium ion batteries: A review. *J. Power Sources* **2021**, *506*, 230194. [CrossRef]
12. Cho, S.; Jung, W.; Jung, G.Y.; Eom, K.S. High-performance boron-doped silicon micron-rod anode fabricated using a mass-producible lithography method for a lithium ion battery. *J. Power Sources* **2020**, *454*, 227931. [CrossRef]
13. Chen, M.; Li, B.; Liu, X.; Zhou, L.; Yao, L.; Zai, J.; Qian, X.; Yu, X. Boron-doped porous Si anode materials with high initial coulombic efficiency and long cycling stability. *J. Mater. Chem.* **2018**, *6*, 3022. [CrossRef]
14. Fukata, N.; Oshima, T.; Tsuruid, T.; Itod, S.; Murakami, K. Synthesis of silicon nanowires using laser ablation method and their manipulation by electron beam. *Sci. Techn. Adv. Mater.* **2005**, *6*, 628–632. [CrossRef]

15. Hung, Y.-J.; Lee, S.-L.; Thibeault, B.J.; Coldren, L.A. Fabrication of highly-ordered silicon nanowire arrays with controllable sidewall profiles for achieving low surface reflection. *IEEE J. Sel. Top. Quantum Electron.* **2011**, *17*, 869–877. [CrossRef]
16. Epur, R.; Hanumantha, P.J.; Datta, M.K.; Hong, D.; Gattu, B.; Kumta, P.N. A simple and scalable approach to hollow silicon nanotube (h-SiNT) anode architectures of superior electrochemical stability and reversible capacity. *J. Mater. Chem.* **2015**, *3*, 11117–11129. [CrossRef]
17. Wendisch, F.J.; Rey, M.; Vogel, N.; Bourret, G.R. Large-scale synthesis of highly uniform silicon nanowire arrays using metal-assisted chemical etching. *Chem. Mater.* **2020**, *32*, 9425–9434. [CrossRef]
18. Demami, F.; Ni, L.; Rogel, R.; Salaun, A.C.; Pichon, L. Silicon nanowires synthesis for chemical sensor applications. *Procedia Eng.* **2010**, *5*, 351–354. [CrossRef]
19. Prosini, P.P.; Rufoloni, A.; Rondino, F.; Santoni, A. Silicon nanowires used as the anode of a lithium-ion battery. *AIP Conf. Proc.* **2015**, *1667*, 020008. [CrossRef]
20. Parr, N.L.; Mech, E. *Zone Refining and Allied Techniques*; George Newnes Ltd.: London, UK, 1960.
21. Zaykov, Y.P.; Zhuk, S.I.; Isakov, A.V.; Grishenkova, O.V.; Isaev, V.A. Electrochemical nucleation and growth of silicon in the KF-KCl-K$_2$SiF$_6$ melt. *J. Solid State Electrochem.* **2015**, *19*, 1341–1345. [CrossRef]
22. Rao, G.M.; Elwell, D.; Feigelson, R.S. Electrodeposition of silicon onto graphite. *J. Electrochem. Soc.* **1981**, *128*, 1708–1711. [CrossRef]
23. Bieber, A.L.; Massot, L.; Gibilaro, M.; Cassayre, L.; Taxil, P.; Chamelot, P. Silicon electrodeposition in molten fluorides. *Electrochim. Acta* **2012**, *62*, 282–289. [CrossRef]
24. Kuznetsova, S.V.; Dolmatov, V.S.; Kuznetsov, S.A. Voltammetric study of electroreduction of silicon complexes in a chloride–fluoride melt. *Russ. J. Electrochem.* **2009**, *45*, 742–748. [CrossRef]
25. Cai, Z.; Li, Y.; Tian, W. Electrochemical behavior of silicon compound in LiF–NaF–KF–Na$_2$SiF$_6$ molten salt. *Ionics* **2011**, *17*, 821–826. [CrossRef]
26. Juzeliunas, E.; Fray, D.J. Silicon electrochemistry in molten salts. *Chem. Rev.* **2020**, *120*, 1690–1709. [CrossRef]
27. Maeda, K.; Yasuda, K.; Nohira, K.T.; Hagiwara, R.; Homma, T. Silicon electrodeposition in water-soluble KF–KCl molten salt: Investigations on the reduction of Si(IV) ions. *J. Electrochem. Soc.* **2015**, *162*, D444–D448. [CrossRef]
28. Zhuk, S.I.; Isakov, A.V.; Apisarov, A.P.; Grishenkova, O.V.; Isaev, V.A.; Vovkotrub, E.G.; Zaykov, Y.P. Electrodeposition of continuous silicon coatings from the KF–KCl–K$_2$SiF$_6$ melts. *J. Electrochem. Soc.* **2017**, *164*, H5135–H5138. [CrossRef]
29. Zhuk, S.I.; Isaev, V.A.; Grishenkova, O.V.; Isakov, A.V.; Apisarov, A.P.; Zaykov, Y.P. Silicon electrodeposition from chloride-fluoride melts containing K$_2$SiF$_6$ and SiO$_2$. *J. Serb. Chem. Soc.* **2017**, *82*, 51–62. [CrossRef]
30. Yasuda, K.; Maeda, K.; Hagiwara, R.; Homma, T.; Nohira, T. Silicon electrodeposition in a water-soluble KF–KCl molten salt: Utilization of SiCl$_4$ as Si source. *J. Electrochem. Soc.* **2017**, *164*, D67–D71. [CrossRef]
31. Zaykov, Y.P.; Isakov, A.V.; Apisarov, A.P.; Chemezov, O.V. Production of silicon by electrolysis of halide and oxide-halide melts. *Tsvetnye Met.* **2013**, *2013*, 58–62.
32. Laptev, M.V.; Isakov, A.V.; Grishenkova, O.V.; Vorob'ev, A.S.; Khudorozhkova, A.O.; Akashev, L.A.; Zaikov, Y.P. Electrodeposition of thin silicon films from the KF–KCl–KI–K$_2$SiF$_6$ melt. *J. Electrochem. Soc.* **2020**, *167*, 042506. [CrossRef]
33. Khudorozhkova, A.O.; Isakov, A.V.; Kataev, A.A.; Redkin, A.A.; Zaykov, Y.P. Density of KF–KCl–KI melts. *Rus. Met. (Metally)* **2020**, *2020*, 918–924. [CrossRef]
34. Cho, S.K.; Lim, T. Catalyst-mediate d doping in electrochemical growth of solar silicon. *Electrochim. Acta* **2021**, *367*, 137472. [CrossRef]
35. Xie, H.; Zhao, H.; Liao, J.; Yin, H.; Bard, A.J. Electrochemically controllable coating of a functional silicon film on carbon materials. *Electrochim. Acta* **2018**, *269*, 610–616. [CrossRef]
36. Zou, X.; Ji, L.; Yang, X.; Lim, T.; Yu, E.T.; Bard, A.J. Electrochemical formation of a p−n junction on thin film silicon deposited in molten salt. *J. Amer. Chem. Soc.* **2017**, *139*, 16060–16063. [CrossRef]
37. Yasuda, K.; Nohira, T.; Hagiwara, R.; Ogata, Y.H. Direct electrolytic reduction of solid SiO$_2$ in molten CaCl$_2$ for the production of solar grade silicon. *Electrochim. Acta* **2007**, *53*, 106–110. [CrossRef]
38. Gevel, T.A.; Zhuk, S.I.; Ustinova, Y.A.; Suzdaltsev, A.V.; Zaikov, Y.P. Silicon electroreduction from the KCl–K$_2$SiF$_6$ melt. *Rasplavy* **2021**, *2021*, 187–198. [CrossRef]
39. Gevel, T.; Zhuk, S.; Suzdaltsev, A.; Zaikov, Y. Silicon electrodeposition from the KCl–K$_2$SiF$_6$ melt. *SSRN Electron. J.* **2021**. [CrossRef]
40. Liang, B.; Liu, Y.; Xu, Y. Silicon-based materials as high capacity anodes for next generation lithium ion batteries. *J. Power Sources* **2014**, *267*, 469–490. [CrossRef]
41. Chemezov, O.V.; Isakov, A.V.; Apisarov, A.P.; Brezhestovsky, M.S.; Bushkova, O.V.; Batalov, N.N.; Zaikov, Y.P.; Shashkin, A.P. Electrolytic production of silicon nanofibers from the KCl–KF–K$_2$SiF$_6$–SiO$_2$ melt for composite anodes of lithium-ion batteries. *Electrochem. Energetics* **2013**, *13*, 201–204.
42. Dong, Y.; Slade, T.; Stolt, M.J.; Li, L.; Girard, S.N.; Mai, L.; Jin, S. Low-temperature molten-salt production of silicon nanowires by the electrochemical reduction of CaSiO$_3$. *Angew. Chem.* **2017**, *129*, 14645–14649. [CrossRef]
43. Yuan, Y.; Xiao, W.; Wang, Z.; Fray, D.J.; Jin, X. Efficient nanostructuring of silicon by electrochemical alloying / dealloying in molten salts for improved lithium storage. *Angew. Chem.* **2018**, *130*, 15969–15974. [CrossRef]

44. Hamedani, A.A.; Ow-Yang, C.W.; Soytas, S.H. Mechanisms of Si nanoparticle formation by molten salt magnesiothermic reduction of silica for lithium-ion battery anodes. *ChemElectroChem* **2021**, *8*, 3181–3191. [CrossRef]
45. Yu, Z.; Wang, N.; Fang, S.; Qi, X.; Gao, Z.; Yang, J.; Lu, S. Pilot-plant production of high-performance silicon nanowires by molten salt electrolysis of silica. *Ind. Eng. Chem. Res.* **2020**, *59*, 1–8. [CrossRef]
46. Yu, Z.; Fang, S.; Wang, N.; Shi, B.; Hu, Y.; Shi, Z.; Shi, D.; Yang, J. In-situ growth of silicon nanowires on graphite by molten salt electrolysis for high performance lithium-ion batteries. *Mater. Lett.* **2020**, *273*, 127846. [CrossRef]
47. Hou, Y.; Yang, Y.; Meng, W.; Lei, B.; Ren, M.; Yang, X.; Wang, Y.; Zhao, D. Core-shell structured Si@Cu nanoparticles encapsulated in carbon cages as high-performance lithium-ion battery anodes. *J. Alloys Compd.* **2021**, *874*, 159988. [CrossRef]
48. Nikolaev, A.Y.; Suzdaltsev, A.V.; Zaikov, Y.P. High temperature corrosion of ZrN powder in molten LiCl with additions of $PbCl_2$, KCl, Li_2O, H_2O, and LiOH. *J. Electrochem. Soc.* **2019**, *166*, C147–C152. [CrossRef]
49. Nikolaev, A.; Suzdaltsev, A.; Zaikov, Y. Cathode process in the $KF–AlF_3–Al_2O_3$ melts. *J. Electrochem. Soc.* **2019**, *166*, D784–D791. [CrossRef]
50. Zaykov, Y.P.; Isakov, A.V.; Zakiryanova, I.D.; Reznitskikh, O.G.; Chemezov, O.V.; Redkin, A.A. Interaction between SiO_2 and a $KF–KCl–K_2SiF_6$ melt. *J. Phys. Chem.* **2014**, *118*, 1584–1588. [CrossRef]
51. Vorob'ev, A.S.; Isakov, A.V.; Kazakovtseva, N.A.; Khudorozhkova, A.O.; Galashev, A.E.; Zaikov, Y.P. Calculations of silicon complexes in $KF–KCl–KI–K_2SiF_6$ and $KF–KCl–KI–K_2SiF_6–SiO_2$ molten electrolytes. *AIP Conf. Proc.* **2019**, *2174*, 020072. [CrossRef]
52. Li, H.; Liang, J.; Xie, S.; Reddy, R.G.; Wang, L. Electrochemical and phase analysis of Si(IV) on Fe electrode in molten $NaCl–NaF–KCl–SiO_2$ system. *High Temp. Mater. Proc.* **2018**, *37*, 921–928. [CrossRef]
53. Pershin, P.; Suzdaltsev, A.; Zaikov, Y. Synthesis of silumins in $KF–AlF_3–SiO_2$ melt. *J. Electrochem. Soc.* **2016**, *163*, D167–D170. [CrossRef]
54. Dolmatov, V.S.; Kuznetsov, S.A. Electroreduction of tantalum oxofluoride complexes in equimolar mixture of sodium and potassium chlorides. *Rasplavy* **2016**, *2016*, 322–332. [CrossRef]

Article

A Porous Tungsten Substrate for Catalytic Reduction of Hydrogen by Dealloying of a Tungsten–Rhenium Alloy in an Aqueous Solution of Hydrochloric Acid

Aleksander A. Chernyshev [1,2,*] and Evgenia V. Nikitina [1,2]

[1] Department of Technology of Electrochemical Production, Institute of Chemical Technology, Ural Federal University, 620002 Yekaterinburg, Russia; neekeetina@mail.ru
[2] The Institute of High Temperature Electrochemistry of the Ural Branch of the Russian Academy of Sciences, 620137 Yekaterinburg, Russia
* Correspondence: aac-vp@ihte.uran.ru

Citation: Chernyshev, A.A.; Nikitina, E.V. A Porous Tungsten Substrate for Catalytic Reduction of Hydrogen by Dealloying of a Tungsten–Rhenium Alloy in an Aqueous Solution of Hydrochloric Acid. *Appl. Sci.* **2022**, *12*, 1029. https://doi.org/10.3390/app12031029

Academic Editors: Suzdaltsev Andrey and Oksana Rakhmanova

Received: 29 October 2021
Accepted: 17 January 2022
Published: 19 January 2022

Publisher's Note: MDPI stays neutral with regard to jurisdictional claims in published maps and institutional affiliations.

Copyright: © 2022 by the authors. Licensee MDPI, Basel, Switzerland. This article is an open access article distributed under the terms and conditions of the Creative Commons Attribution (CC BY) license (https://creativecommons.org/licenses/by/4.0/).

Abstract: Selective dissolution of a tungsten (85 wt.%)–rhenium (15 wt.%) alloy with rhenium in hydrochloric acid at the temperature of 298 K and anodic polarization modes was carried out to develop a porous catalytic substrate and to recycle rare metals. The parameters of the effective selective anodic dissolution of the tungsten–rhenium alloy, including the differences in applied potentials and electrolyte composition, were found. It was established that samples of the tungsten–rhenium alloy possess the smallest average pore size after being exposed for 6000 s. The obtained porous tungsten samples were characterized by X-ray diffraction and scanning electron spectroscopy. A thermodynamic description of the processes occurring during the anodic selective dissolution of a binary alloy was proposed. In the course of the work, the selectivity coefficient was determined using an X-ray fluorescence wave-dispersion spectrometer XRF-1800. The existence of a bimodal structure on the tungsten surface after dealloying was proved.

Keywords: tungsten; rhenium; dealloying; anodic polarization; developed surface; catalyst for HER

1. Introduction

The search for renewable, environmentally friendly energy sources, which is triggered by the depletion of fossil energy and environmental pollution [1,2], has determined the important role of hydrogen energy [3]. Despite the high specific energy release (160 MJ/g), its significant disadvantage and limitation include the significant consumption of electricity required for hydrogen electrolysis [4]. The main task in the development of hydrogen energy is to reduce the overvoltage of the hydrogen evolution reaction (HER), which will reduce the total energy consumption. Therefore, new materials should be based on metals with high electrocatalytic activity of cathodic hydrogen release that will lead to a decrease in the overvoltage of HER and, consequently, to a reduction in the hydrogen production cost.

Recently, base metals such as Ni, Co, W, or Mo [2,5–12]; their alloys [2,5–13]; and base metal-containing materials such as sulfides, nitrides, phosphides, and carbides of transition metals have been recognized as effective HER catalysts [14–19]. This is due to their chemical stability; their low cost; and, most importantly, to the low overvoltage of hydrogen release [20].

Tungsten, having unique properties such as a high melting point and chemical resistance, is a promising material for the HER from alkaline electrolytes [21,22]. The potential use of a porous layer of tungsten obtained in the process of dealloying accompanied with the removal of the second metal has determined the interest in the study of tungsten–rhenium alloys. Due to its thermodynamic properties, a tungsten–rhenium alloy is promising for the formation of a highly porous surface by anodic dissolution. The mechanism of anodic formation of such structures has received considerable scientific and technological interest.

A large number of works are devoted to the synthesis and study of nanostructured mesoporous materials obtained by various methods at ordinary temperatures. The current state of the problem is analyzed elsewhere [23–25]. There are several common methods for synthesizing mesoporous materials. For example, one of them is based on the use of nanotemplates such as silicon dioxide or organic polymers as structure-forming agents for metallic nanolayer formation. After chemical removal of the template, nanopores are formed. We would also like to mention here the method of thermal decomposition of unstable nitrogenous metal compounds, which are used to form nanofoams with a record low density. The removal of a more electronegative alloy component (so-called dealloying) is less widespread and is mainly used to obtain precious metals in the mesoporous state—gold, silver, and platinoids [26,27]. Thus, Hakamada and Mabuchi [28] presented a method for the preparation of mesoporous palladium in an aqueous sulfuric acid solution.

The existing works on the electrochemical behavior of anodes made of binary alloys in low-temperature electrolytes are scattered and, as a rule, do not pay much attention to the formed near-surface structures and the development of the electrode surface. The situation is somewhat better in low-temperature electrolytes [29].

Currently, the research in the field of mesoporous materials is shifting its focus toward the materials' synthesis. Apart from impressive specific surface areas, these materials demonstrate a sufficiently high electronic conductivity and high reaction capacity. The possibility of extracting expensive rhenium and simultaneous formation of a porous tungsten structure, which can be used as a substrate for catalyst deposition, from a tungsten–rhenium alloy is of considerable interest. The main task of this work is to evaluate the change in the structure of the tungsten–rhenium alloy depending on the electrolysis conditions and the composition of the solution at constant temperature.

Modern electrochemical materials science, using advancements in the field of nanomaterials, opens up new opportunities in the development of electric and hybrid vehicles, power supplies, and portable electronics. Various mesoporous materials are being intensively studied as electrodes for supercapacitors, biosensors, and catalysts.

Within the framework of this research, the main attention is focused on the structure of a metallic material synthesized by the selective dissolution of a tungsten and rhenium alloy. In further studies, it is planned to establish the patterns that govern the formation, structure, and properties of the tungsten structures.

The purpose of this work is to study the regularities of the formation of a porous structure on the surface of tungsten during selective anodic dissolution of a tungsten (85 wt. %)–rhenium (15 wt. %) alloy in a solution of hydrochloric acid at 298 K; to develop the technology for obtaining nanostructured electrode materials with a developed surface; and to study their physicochemical properties. There is a lack of scientific literature on the tungsten-containing systems of this type. There are only works on electrochemical dealloying of other materials in aqueous solutions. Tungsten is a promising catalyst and/or substrate. Production of nanoporous tungsten is also of great practical interest. Processing of tungsten–rhenium alloys is important to prove the possibility of extracting more expensive rhenium [30–32]. The method of selective dissolution, in addition to the synthesis of a bimodal structure at the surface of tungsten, provides the extraction of rhenium in the form of chlorrenic acid or potassium chlorrenate.

2. Materials and Methods

Electrochemical measurements were carried out using an Autolab PGStat 302N electrochemical workstation (Metrohm, Switzerland).

Experimental measurements were carried out in a standard three-electrode two-chamber glass cell using the following working electrodes: high-purity rhenium with a surface area of 1.06 cm^2, high-purity tungsten (0.72 cm^2), and a tungsten–rhenium alloy BP-20 (Russia) containing 15 wt. % rhenium with a surface of 2.5 cm^2. A rod made of spectrally pure graphite was used as a counter electrode, and a silver chloride electrode (SCE) was used as the reference electrode. All potentials in this work are given with respect to SCE. A Luggin capillary was used to eliminate the ohmic drop through the electrolyte solution.

The surfaces of the working electrodes were prepared by cleaning SiC paper with different grain sizes, followed by polishing with diamond powder. Linear voltammetry data were obtained for all electrodes from hydrochloric acid solutions; their compositions are shown in Table 1.

Table 1. Compositions of solutions for electrochemical studies.

Component	Solution 1	Solution 2	Solution 3	Solution 4
HCl, g L^{-1}	200	200	350	350
KCl, g L^{-1}	0	340	0	340

All solutions were prepared based on distilled water with the use of potassium chloride and hydrochloric acid of high purity. Electrochemical studies were carried out in the ambient air at an environment temperature of 298 K and a rate of potential change of 1 mV/s.

Based on linear voltammetry, the potential of the rhenium separation from a tungsten–rhenium alloy was determined. To confirm the selective separation of rhenium from the alloy, chronoamperometry was performed at a constant potential (Econst = 600 mV (SCE)) within 30,000 s.

The surface of porous tungsten samples obtained as a result of anodic potentiodynamic polarization (chronoamperometry) was analyzed by the micro-X-ray spectral method using the TESCAN MIRA 3 LMU auto-emission electron microscope (TESCAN, Czechia) and the X-ray phase method using the Rigaku D/MAX-2200VL/PC automatic X-ray diffractometer (Rigaku, Japan).

3. Results

Linear voltammetry was used to study the anodic process occurring on individual rhenium, tungsten, and tungsten–rhenium electrodes. Figure 1 shows that rhenium was dissolved in all solutions according to the mechanism of the delayed ionization. The active dissolution stage began at the potential of 600 mV relative to the silver chloride electrode. At current densities below 20 A/m^2, the addition of potassium chloride practically did not affect the course of the polarization curve. However, at high current densities above 2500 A/m^2, the addition of potassium chloride caused electrode salt passivation and the formation of potassium chlorrenate.

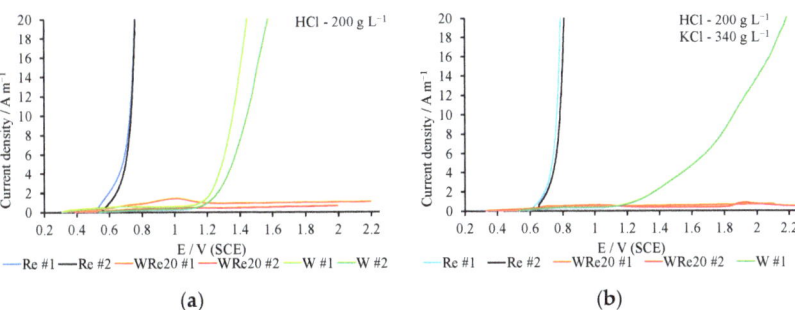

Figure 1. Cont.

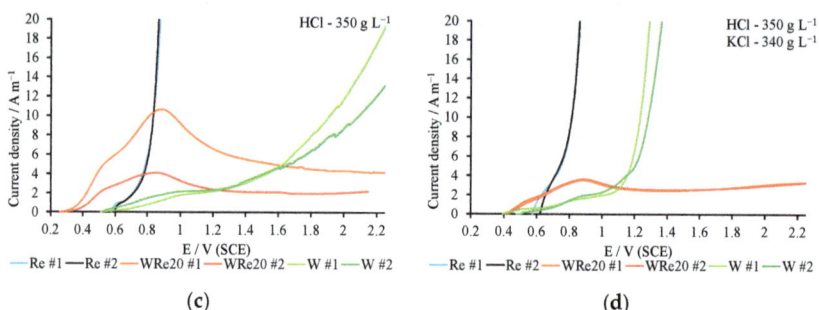

Figure 1. The data of linear voltammetry in different solutions: (**a**) Solution 1; (**b**) Solution 2; (**c**) Solution 3; (**d**) Solution 4.

The process taking place on the tungsten electrode was accompanied with significant polarization. The active stage began at the potential of 1100 mV (SCE) and was accompanied by abundant gas release. At the same time, there was no decrease in the mass of the electrode itself. Therefore, tungsten dissolution and the active release of gaseous oxygen were not observed.

No pronounced processes were registered in solutions 1 and 2 for the tungsten–rhenium alloy. Pronounced peaks associated with the selective dissolution of rhenium were observed in solutions 3 and 4.

The potential of rhenium release in hydrochloric acid solutions was determined on the basis of the obtained voltage dependences. The active dissolution stage of rhenium began at a potential of 600 mV (SCE) for both individual rhenium and the tungsten–rhenium alloy.

The extraction of rhenium from a tungsten–rhenium alloy depending on the processing time was carried out using chronoamperometry at a constant potential (Figure 2).

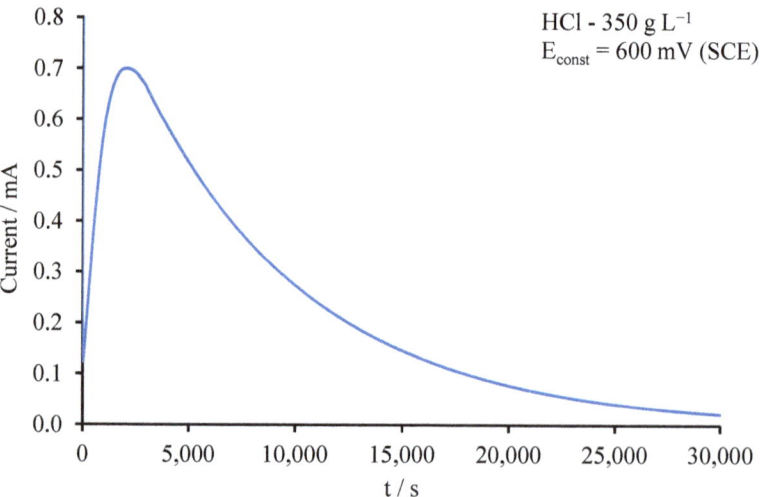

Figure 2. Chronoamperogram of tungsten–rhenium alloy dissolution.

The samples of the tungsten–rhenium alloy after potentiostatic dealloying were subjected to X-ray phase analysis using a Rigaku D/MAX-2200VL/PC X-ray diffractometer (Rigaku, Japan) (Figure 3). It was found that after dealloying, new phases (for example, an oxide film) were not formed on the surface of the samples. This indicates the absence of the process passivation and favorably affects the depth of extraction of rhenium from the alloy.

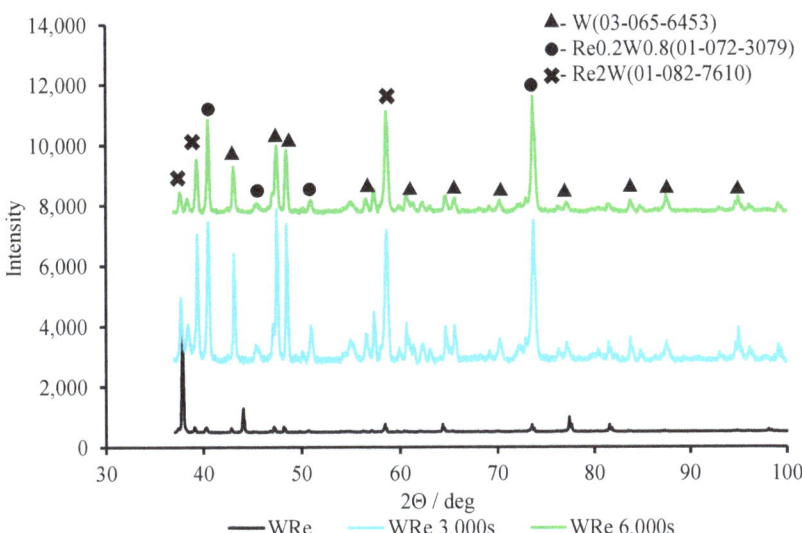

Figure 3. Diffractograms of initial alloy, alloy after 3000 s exposure in the solution, and alloy after 6000 s exposure in the solution.

To determine the change in the alloy composition, SEM studies of the samples' surfaces were carried out using energy-dispersive X-ray spectroscopy (EDS) with the TESCAN MIRA 3 LMU (Figure 4).

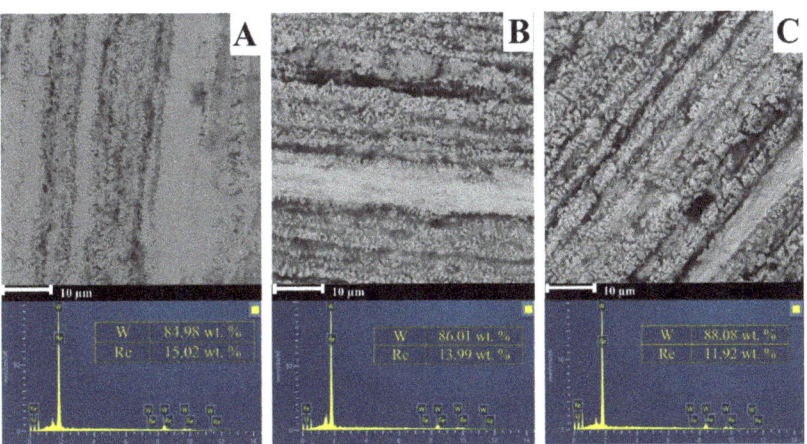

Figure 4. Microstructure of initial alloy (**A**), alloy after 3000 s exposure in solution (**B**), and alloy after 6000 s exposure in the solution (**C**).

Figure 4 shows that with the increase in the duration of the electrolysis, the sample surface dissolved selectively, forming a rough surface. The EDS spectra confirmed by linear voltammetry demonstrate that at a given potential, selective dissolution of rhenium occurs. In the first 3000 s, the process of active dissolution of rhenium proceeds and a strong change in the surface occurs. With the continuation of the electrolysis, the diffusion of rhenium ions from the electrode into the electrolyte bulk is hampered by the emerging bimodal structures, which leads to a slowdown but not to an interruption in the dealloying process. The EDS spectra of the sample after 6000 s exposure confirm this. To confirm the EDS data, the wavelengths of the samples were also analyzed using an XRF-1800 dispersive X-ray fluorescence spectrometer (Shimadzu, Japan). According to the results of the analysis, the concentration of rhenium in the alloy changed from 15 wt.% in the initial electrode up to 14 wt.% after 3000 s, to 12 wt.% after 6000 s, and to 6.8 wt.% after 30,000 s.

The dealloying penetration depth was estimated by SEM images of the sample cross-section using EDS with the TESCAN MIRA 3 LMU (Figure 5 and Table 2).

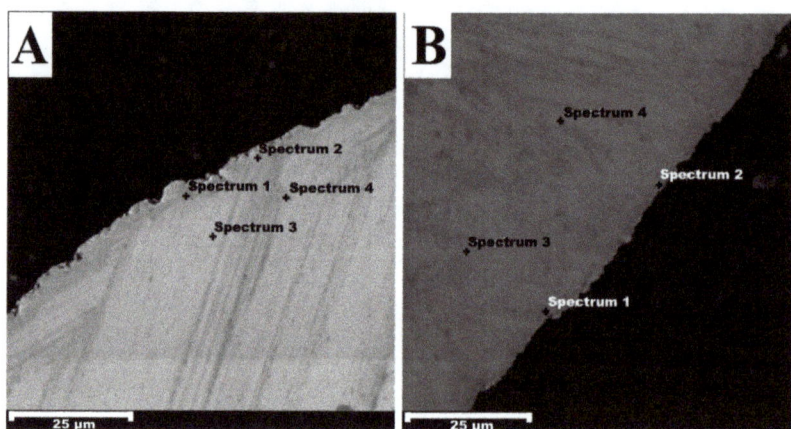

Figure 5. SEM images of cross-section of the sample after dealloying: (**A**) alloy after 3000 s; (**B**) alloy after 6000 s.

Table 2. EDS analysis of cross-section of the sample after dealloying.

Figure 5A	Component	Spectrum 1	Spectrum 2	Spectrum 3	Spectrum 4
	W, wt.%	88.25	88.75	86.22	86.31
	Re, wt.%	11.75	11.25	13.78	13.69
Figure 5B	Component	Spectrum 1	Spectrum 2	Spectrum 3	Spectrum 4
	W, wt.%	90.66	90.73	88.63	88.36
	Re, wt.%	9.34	9.27	11.37	11.64

The change in the concentration of rhenium from the volume to the front of the corrosion attack (dealloying) indicates that the process occurs not only on the surface of the sample but also in the pores formed at the initial moment of time.

A porous bimodal structure formed at the surface of tungsten during selective anodic dissolution of the tungsten (85 wt.%)–rhenium (15 wt.%) alloy in a solution of hydrochloric acid at 298 K. The specific morphology of the resulting electrode materials based on mesoporous tungsten was studied.

4. Discussion

The criterion of thermodynamic probability of anodic processes in binary alloys is applicable for the process of dissolution of tungsten alloys with a more electronegative component (rhenium). The anode reaction involving a component with a more negative potential has the highest rate. For a tungsten–rhenium alloy, the process scheme is as follows:

$$W(\text{alloy } W - Re) - ne^- = W^{n+} \tag{1}$$

$$Re(\text{alloy } W - Re) - ne^- = Re^{n+} \tag{2}$$

This process is accompanied by selective dissolution of rhenium from the surface layer of the alloy and the formation of a developed surface structure. The linear decrease in the thickness of the electrode and the difference in the partial diffusion coefficients of the alloy components were taken into account.

Selective anodic dissolution of one of the alloy components is quantitatively characterized by the coefficient of selective dissolution.

$$z_A = \frac{(C_W/C_{Re})_{solution}}{(C_W/C_{Re})_{alloy}} \quad (3)$$

A thermodynamic vacancy mechanism of this process has been developed for low-temperature electrolytes [33]. Based on the Darken dependence of the logarithm of the mutual diffusion coefficient from the inverse temperature, the value of the self-diffusion coefficient at 298K is estimated.

$$D = D_0 \cdot \exp\left(-\frac{E}{RT}\right) \quad (4)$$

The external flow is caused by two effects: a change in the concentration gradient C_{Re} and a decrease in the thickness of the electrode, causing the dissolution of the electronegative component.

The rate of dissolution of the alloy q_{Re}^0 (mol/cm^{2}s) is given by the following equation:

$$-q_{Re}^0 = -D\frac{\partial C_{Re}}{\partial x}(0,t) - C_A(0,t)U_L(t) \quad (5)$$

where D is the diffusion coefficient, x is the initial flat surface of the electrode, t is the time, and U_L is the rate of linear loss of the alloy (cm/s).

$$U_L = q_{Re}V_{alloy} \quad (6)$$

where V_{alloy} is the molar volume of the alloy.

The anodic dissolution of rhenium at the alloy–solution boundary, which is a powerful source of vacancies, where vacancies are absorbed by dislocations, and its diffusion from the depth of the alloy to the boundary cause the formation of vacancies in the alloy volume. There are two sources of vacancies with a total flow:

$$J_V = J_V' \uparrow\downarrow + J_V'' \uparrow \quad (7)$$

where $J_V' \uparrow\downarrow$ is the flow of excess or lack of vacancies in comparison with the equilibrium concentration in the alloy volume, and $J_V'' \uparrow$ is the flow of vacancies formed directly due to the process of selective dissolution of rhenium.

The change in the total volume of the formed voids (negative crystals) as a result of selective dissolution can be calculated:

$$\Delta V = C^{-1} \int_0^t J_V(t)dt \quad (8)$$

The excess of vacancies at the boundary causes the linear loss of the alloy and the decrease in the thickness of the electrode. Vacancies can both move to the border and merge into the pores.

It is the supersaturation of vacancies in the alloy volume that causes them to merge into pores and form the so-called diffusion porosity or Frenkel effect. As the temperature increases, the alloy dissolves (decreasing in thickness) faster, the defectiveness decreases, and the boundary diffusion coefficient increases faster than the volume diffusion coefficient. With an increase in temperature, the relative contribution of diffusion along the boundaries decreases; at a lower temperature, the process of formation of sufficiently small pores prevails.

The result of selective dissolution of the alloy is the formation of a surface layer with a modified composition. In the process of anodic dissolution, both the surface relief of the solid electrode and the proportion of the alloy components change and electrochemical destruction of the starting material occurs.

Studies of the regularity of selective dissolution are important for corrosiology and for the kinetics of electrochemical processes in general, as well as for electrochemical materials science and the production of structured materials for catalysis since it is necessary to take into account the real heterogeneity of solid metal surfaces.

To quantitatively characterize the type of destruction of intermetallic phases, the coefficient of selective dissolution Z_A was used, which shows how many times the ratio of the amounts of tungsten to rhenium in the electrolyte solution was greater than the corresponding ratio in the alloy. With uniform dissolution, this coefficient is equal to unity, and with selective dissolution, it tends to infinity (tungsten transfers into the solution) or to zero (rhenium transfers into the solution) [33]. According to the obtained experimental data, the selectivity coefficient and the rate of linear loss of the alloy were calculated (Table 3).

Table 3. Compositions of solutions for electrochemical studies.

Time, s	3000	6000	30,000
$Z_A(W)$	1.08	1.29	2.35
$Z_A(Re)$	0.92	0.77	0.41
U_L, sm/s	1.4×10^{-9}	2.9×10^{-9}	7.2×10^{-8}

Simultaneous occurrence of reactions (1) and (2) is thermodynamically possible at a value of applied anode potential exceeding the equilibrium potential of both alloy components. At an applied potential of 600 mV (SCE), only the dissolution of rhenium is thermodynamically possible, which is consistent with the data of linear chronovoltammetry (Figure 1).

During the anodic dissolution of a tungsten alloy with a more electronegative component, such as rhenium, the alloy dissolves in the active state at a relatively low overvoltage of the anodic metal dissolution reaction. Under the conditions of anodic dissolution of a two-phase alloy, a structural-selective dissolution is most often observed.

5. Conclusions

The regularities of anodic dissolution of a two-component tungsten–rhenium alloy in an aqueous solution of hydrochloric acid were studied.

The parameters of the effective selective anodic dissolution of the tungsten–rhenium alloy, including the differences in applied potentials and electrolyte composition, were found. The existence of a bimodal structure on the tungsten surface after dealloying was proved.

The technological parameters such as the composition of the electrolyte, the process temperature, and the modes of anodic polarization, which provide the most efficient separation of tungsten and rhenium, were determined.

Author Contributions: Data curation, A.A.C.; Investigation, A.A.C.; Writing—review & editing, E.V.N. All authors have read and agreed to the published version of the manuscript.

Funding: This research received no external funding.

Institutional Review Board Statement: Not applicable.

Informed Consent Statement: Not applicable.

Conflicts of Interest: The authors declare that they have no known competing financial interests or personal relationships that could have appeared to influence the work reported in this paper.

References

1. Pu, Z.; Zhao, J.; Amiinu, I.S.; Li, W.; Wang, M.; He, D.; Mu, S. A universal synthesis strategy for P-rich noble metal diphosphide-based electrocatalysts for the hydrogen evolution reaction. *Energy Environ. Sci.* **2019**, *12*, 952–957. [CrossRef]
2. Ashraf, M.A.; Liu, Z.; Zhang, D. Novel 3-D urchin-like Ni–Co–W porous nanostructure as efficient bifunctional superhydrophilic electrocatalyst for both hydrogen and oxygen evolution reactions. *Int. J. Hydrogen Energy* **2020**, *45*, 17504–17516. [CrossRef]
3. Kotkondawar, A.V.; Mangrulkar, P.; Wanjari, S.; Maddigapu, P.R.; Rayalu, S. Photothermal hydrogen production from oxidative hydrolysis of electrochemically synthesized nano-sized zinc. *Int. J. Hydrogen Energy* **2019**, *44*, 6514–6524. [CrossRef]
4. Schmidt, O.; Gambhir, A.; Staffell, I.; Hawkes, A.; Nelson, J.; Few, S. Future cost and performance of water electrolysis: An expert elicitation study. *Int. J. Hydrogen Energy* **2017**, *42*, 30470–30492. [CrossRef]
5. Yang, Y.; Zhu, X.; Zhang, B.; Yang, H.; Liang, C. Electrocatalytic properties of porous Ni-Co-WC composite electrode toward hydrogen evolution reaction in acid medium. *Int. J. Hydrogen Energy* **2019**, *44*, 19771–19781. [CrossRef]
6. Xu, Y.; Xiao, X.; Ye, Z.; Zhao, S.; Shen, R.; He, C.; Zhang, J.; Li, Y.; Chen, X. Cage-Confinement Pyrolysis Route to Ultrasmall Tungsten Carbide Nanoparticles for Efficient Electrocatalytic Hydrogen Evolution. *J. Am. Chem. Soc.* **2017**, *139*, 5285–5288. [CrossRef] [PubMed]
7. Nady, H.; Negem, M. Ni–Cu nano-crystalline alloys for efficient electrochemical hydrogen production in acid water. *RSC Adv.* **2016**, *6*, 51111. [CrossRef]
8. Ngamlerdpokin, K.; Tantavichet, N. Electrodeposition of nickel–copper alloys to use as a cathode for hydrogen evolution in an alkaline media. *Int. J. Hydrogen Energy* **2014**, *39*, 2505–2515. [CrossRef]
9. Chang, Z.; Zhu, L.; Zhao, J.; Chen, P.; Chen, D.; Gao, H. NiMo/Cu-nanosheets/Ni-foam composite as a high performance electrocatalyst for hydrogen evolution over a wide pH range. *Int. J. Hydrogen Energy* **2020**, *46*, 3493–3503. [CrossRef]
10. Lotfi, N.; Shahrabi, T.; Yaghoubinezhad, Y.; Barati Darband, G.H. Electrodeposition of cedar leaf-like graphene Oxide@Ni–Cu@Ni foam electrode as a highly efficient and ultra-stable catalyst for hydrogen evolution reaction. *Electrochim. Acta* **2019**, *326*, 134949. [CrossRef]
11. Sun, Q.; Zhou, M.; Shen, Y.; Wang, L.; Ma, Y.; Li, Y.; Bo, X.; Wang, Z.; Zhao, C. Hierarchical nanoporous Ni(Cu) alloy anchored on amorphous NiFeP as efficient bifunctional electrocatalysts for hydrogen evolution and hydrazine oxidation. *J. Catal.* **2019**, *373*, 180–189. [CrossRef]
12. Nady, H.; Negem, M. Electroplated Zn-Ni nanocrystalline alloys as an efficient electrocatalyst cathode for the generation of hydrogen fuel in acid medium. *Int. J. Hydrogen Energy* **2018**, *43*, 4942–4950. [CrossRef]
13. Kuznetsov, V.V.; Gamburg, Y.D.; Zhulikov, V.V.; Batalov, R.S.; Filatova, E.A. Re–Ni cathodes obtained by electrodeposition as a promising electrode material for hydrogen evolution reaction in alkaline solutions. *Electrochim. Acta* **2019**, *317*, 358–366. [CrossRef]
14. Youn, J.; Jeong, S.; Kang, H.; Kovendhan, M.; Park, C.; Jeon, K. Effect of carbon coating on Cu electrodes for hydrogen production by water splitting. *Int. J. Hydrogen Energy* **2019**, *44*, 20641–20648. [CrossRef]
15. Martins, P.; Lopes, P.; Ticianelli, E.A.; Stamenkovic, V.R.; Markovic, N.M.; Strmcnik, D. Hydrogen evolution reaction on copper: Promoting water dissociation by tuning the surface oxophilicity. *Electrochem. Commun.* **2019**, *100*, 30–33. [CrossRef]
16. Zhang, X.; Guo, T.; Liu, T.; Lv, K.; Wu, Z.; Wang, D. Tungsten phosphide (WP) nanoparticles with tunable crystallinity, W vacancies, and electronic structures for hydrogen production. *Electrochim. Acta* **2019**, *323*, 134798. [CrossRef]
17. Wu, L.; Pu, Z.; Tu, Z.; Amiinu, I.S.; Liu, S.; Wang, P.; Mu, S. Integrated design and construction of WP/W nanorod array electrodes toward efficient hydrogen evolution reaction. *Chem. Eng. J.* **2017**, *327*, 705–712. [CrossRef]
18. Nikitina, E.V.; Karfidov, E.A. Corrosion of construction materials of separator in molten carbonates of alkali metal. *Int. J. Hydrogen Energy* **2021**, *46*, 16925–16931. [CrossRef]
19. Chernyshev, A.A.; Darintseva, A.B.; Ostanina, T.N.; Panashchenko, I.A.; Orlova, A.A.; Novikov, A.E.; Artamonov, A.S. Electro-crystallization of metals on a rotating drum-cathode. *Int. J. Hydrogen Energy* **2021**, *46*, 16848–16856. [CrossRef]
20. Abd El-Hafez, G.M.; Mahmoud, N.H.; Walcarius, A.; Fekry, A.M. Evaluation of the electrocatalytic properties of Tungsten electrode towards hydrogen evolution reaction in acidic solutions. *Int. J. Hydrogen Energy* **2019**, *44*, 16487–16496. [CrossRef]
21. Badawy, W.A.; Abd El-Hafez, G.M.; Nady, H. Electrochemical performance of tungsten electrode as cathode for hydrogen evolution in alkaline solutions. *Int. J. Hydrogen Energy* **2015**, *40*, 6276–6282. [CrossRef]
22. Chen, L.; Fan, J.L.; Gong, H.R. Atomistic simulation of mechanical properties of tungsten-hydrogen system and hydrogen diffusion in tungsten. *Solid State Commun.* **2020**, *306*, 113772. [CrossRef]
23. Zhang, J.; Li, C.M. Nanoporous metals: Fabrication strategies and advanced electrochemical applications in catalysis, sensing and energy systems. *Chem. Soc. Rev.* **2012**, *41*, 7016–7031. [CrossRef] [PubMed]
24. Tappan, B.C.; Steiner, S.A.; Luther, E.P. Nanoporous Metal Foams. *Angew. Chem. Int. Ed.* **2010**, *49*, 4544–4565. [CrossRef]
25. Walcarius, A. Mesoporous materials and electrochemistry. *Chem. Soc. Rev.* **2013**, *42*, 4098–4140. [CrossRef]
26. Zhang, Q.; Zhang, Z. On the electrochemical dealloying of Al-based alloys in a NaCl aqueous solution. *Phys. Chem. Chem. Phys.* **2010**, *12*, 1453–1472. [CrossRef]
27. Graf, M.; Roschning, B.; Weissmüller, J. Nanoporous Gold by Alloy Corrosion: Method-Structure-Property Relationships. *J. Electrochem. Soc.* **2017**, *164*, C194. [CrossRef]
28. Hakamada, M.; Mabuchi, M. Preparation of Nanoporous Palladium by Dealloying: Anodic Polarization Behaviors of Pd-M (M=Fe, Co, Ni) Alloys. *Mater. Trans.* **2009**, *50*, 431–435. [CrossRef]

29. Nikitina, E.V.; Karfidov, E.A.; Kazakovtseva, N.A. Anodic selective dissolution of copper alloys in chloride and carbonate melts. *J. Alloy Compd.* **2020**, *845*, 156238. [CrossRef]
30. Ye, L.; Ouyang, Z.; Chen, Y.; Liu, S. Recovery of rhenium from tungsten-rhenium wire by alkali fusion in KOH-K$_2$CO$_3$ binary molten salt. *Int. J. Refract. Met. Hard Mater.* **2020**, *87*, 105148. [CrossRef]
31. Kuznetsova, O.G.; Levin, A.M.; Sevostyanov, M.A.; Bolshih, A.O. The improvement of rhenium recovery technology from W-Re alloys. *J. Phys. Conf. Ser.* **2018**, *1134*, 012032. [CrossRef]
32. Entezari-Zarandi, A.; Azizi, D.; Nikolaychuk, P.A.; Larachi, F.; Pasquier, L.C. Selective Recovery of Molybdenum over Rhenium from Molybdenite Flue Dust Leaching Solution Using PC88A Extractant. *Metals* **2020**, *10*, 1423. [CrossRef]
33. Andreev, Y.Y. *Electrochemistry of Metals and Alloys*; High Education and Science: Moskow, Russia, 2016; 320p.

Article

Ab Initio Study of the Mechanism of Proton Migration in Perovskite LaScO$_3$

Alexander Y. Galashev [1,2,*], Dmitriy S. Pavlov [1], Yuri P. Zaikov [1,2] and Oksana R. Rakhmanova [1,2]

[1] Institute of High-Temperature Electrochemistry Ural Branch of the Russian Academy of Sciences, 620990 Yekaterinburg, Russia; dima343@mail.ru (D.S.P.); dir.zaikov@mail.ru (Y.P.Z.); oksana_rahmanova@mail.ru (O.R.R.)

[2] Institute of Chemical Engineering, Ural Federal University Named after the First President of Russia B.N. Yeltsin, 620002 Yekaterinburg, Russia

* Correspondence: galashev@ihte.uran.ru

Citation: Galashev, A.Y.; Pavlov, D.S.; Zaikov, Y.P.; Rakhmanova, O.R. *Ab Initio* Study of the Mechanism of Proton Migration in Perovskite LaScO$_3$. *Appl. Sci.* **2022**, *12*, 5302. https://doi.org/10.3390/app12115302

Academic Editor: Sang Hyuk Im

Received: 22 April 2022
Accepted: 21 May 2022
Published: 24 May 2022

Publisher's Note: MDPI stays neutral with regard to jurisdictional claims in published maps and institutional affiliations.

Copyright: © 2022 by the authors. Licensee MDPI, Basel, Switzerland. This article is an open access article distributed under the terms and conditions of the Creative Commons Attribution (CC BY) license (https:// creativecommons.org/licenses/by/ 4.0/).

Abstract: The mechanism of proton motion in a LaScO$_3$ perovskite crystal was studied by ab initio molecular dynamics. The calculations were performed at different temperatures, locations, and initial velocity of the proton. Different magnitudes and directions of the external electric field were also considered. It is shown that initial location and interaction between proton and its nearest environment are of great importance to the character of the proton movement, while the magnitude and direction of the initial velocity and electric field strength are secondary factors characterizing its movement through the LaScO$_3$ crystal. Four types of proton-jumping between oxygen atoms are determined and the probability of each of them is established. Energy barriers and characteristic times of these jumps are determined. The probable distances from a proton to other types of atoms present in perovskite are calculated. It is shown that the temperature determines, to a greater extent, the nature of the motion of a proton in a perovskite crystal than the magnitude of the external electric field. The distortion of the crystal lattice and its polarization provoke the formation of a potential well, which determines the path for the proton to move and its mobility in the perovskite crystal.

Keywords: electric field; molecular dynamics; perovskite; polaron; proton

1. Introduction

High-temperature solid oxide fuel cells (SOFCs) are made entirely of solid materials. As a solid electrolyte, they use metal oxides which, unlike carbonate electrolytes, do not corrode the anode and cathode. The metal-oxide material used as a solid electrolyte should be as dense and thin as possible in order to reduce the ohmic resistance. On the one hand, the high operating temperature of SOFCs has the advantage that these devices can run on a variety of fuels, including natural gas. Moreover, their operation does not require catalysts based on noble metals (for example, Pt). When reusing waste heat in SOFC, its efficiency can be brought up to 70–75%. However, the high operating temperatures of SOFCs create certain difficulties. Manufacturing SOFCs requires high thermal stability of materials, which increases the cost of the system. There is also difficulty in connecting different components in SOFC at high temperatures. The use of proton conductors exhibits to achieve the higher ionic conductivity at lower temperatures (~500–600 °C). The possibility of creating a solid ionic conductor with a lower operating temperature based on LaScO$_3$ perovskite was studied. At high pressure, a solid solution with a perovskite type structure was formed [1]. For the synthesized material of the composition (Li$_{0.4}$Ce$_{0.15}$La$_{0.67}$)ScO$_3$, an ionic conductivity of 1.1×10^{-3} S·cm^{-1} at 623 K was achieved. A number of works have been devoted to the development of protonic electrolytes [2–7]. Each of them is dedicated to a particular system of materials or manufacturing technology. Sufficiently high conductivity (~10^{-3} S·cm^{-1}) was observed for LaNbO$_4$ doped with calcium in a humid atmosphere at 800 °C [8]. Importantly, this result was achieved using materials that do not contain Ba or Sr, which are not suitable for fuel cells operating in a CO_2 atmosphere.

$LaScO_3$ perovskite is a promising proton conductor [9–12]. It exhibits a fairly high structural stability. Even at high impurity levels, $LaScO_3$ is characterized by orthorhombic symmetry in the range from room temperature to about 1993 K [13]. The structural skeleton of this compound is made up of rigid ScO_6 octahedra. The formation of oxygen vacancies, which increase the proton conductivity, occurs when La is replaced by Sr [14]. The proton conductivity (6×10^{-3} S·cm^{-1}) obtained upon reaching the ratio of Sr to La x = 0.2, i.e., for the $La_{0.8}Sr_{0.2}ScO_{3-\delta}$ compound, still remains one of the highest [15]. This doped perovskite is a promising proton-conducting electrolyte and can be used in proton–ceramic fuel cells. However, a targeted enhancement of proton conductivity requires a fundamental study of the mechanism of proton migration in the corresponding solid electrolyte. Among the most reliable methods for studying the mechanism of conduction performed at the atomic level are atomistic calculations in combination with neutron diffraction and nuclear magnetic resonance (NMR). The possible implementation of the proton dynamics in a crystal with a perovskite structure has not yet been fully investigated [16].

We used an ab initio molecular dynamics approach to comprehensively investigate the proton dynamics in a defect-free perovskite $LaScO_3$ crystal. In this new study, we investigated the influence of initial proton velocities and locations, as well as the influence of electric field and temperature on proton behavior in $LaScO_3$ perovskite. Carrying out these studies can confirm or refute the hypothesis about the polaron character of proton movement along the perovskite crystal lattice. Doped ceramics $LaScO_3$ is also promising material for the use as a solid electrolyte in SOFCs.

The aim of this work is to investigate the mechanism of proton movement in defect-free $LaScO_3$ perovskite as a function of temperature, magnitude of the applied electric field, initial velocity of the proton, and its primary location.

2. Materials and Methods

All calculations were performed using the SIESTA software package. These ab initio molecular dynamics calculations are based on the density functional theory and are implemented within the plane wave model. For all considered atoms, only valence electrons were taken into account in the calculations. The calculation of the exchange–correlation functional was based on the Perdew–Burke–Ernzerhof (PBE) formalism [17]. In our calculations, the periodic supercell consisted of 16 La atoms, 16 Sc atoms, and 48 oxygen atoms. The initial configuration of the perfect $LaScO_3$ system is shown in Figure 1. The resulting parameters of the initial lattice were: a = 11.6003997 Å, b = 8.107899684 Å, c = 11.328470354 Å, $\alpha = \beta = \gamma = 90°$, and the volume of the created system was 1068.87 Å^3. Projecting the plane-wave onto specific orbitals allowed us to compute local states and populations. The Monkhorst–Pack algorithm [18] was used to generate all $10 \times 10 \times 1$ k-points. The cutoff energy of plane waves was 400 Ry. This was found to be sufficient to achieve convergence in the context of calculating relative energies. The integration of the equals of motion in realization of ab initio molecular dynamics was carried out by the Verlet method with a time step of 1 fs.

The calculations were performed according to the block diagram shown in Figure 2. The input data are the charges of the nuclei, the number of electrons, and the coordinates of the atoms. Plane waves were used as basic functions. The calculation procedure was completed in accordance with the specified criterion for the accuracy of determining the density.

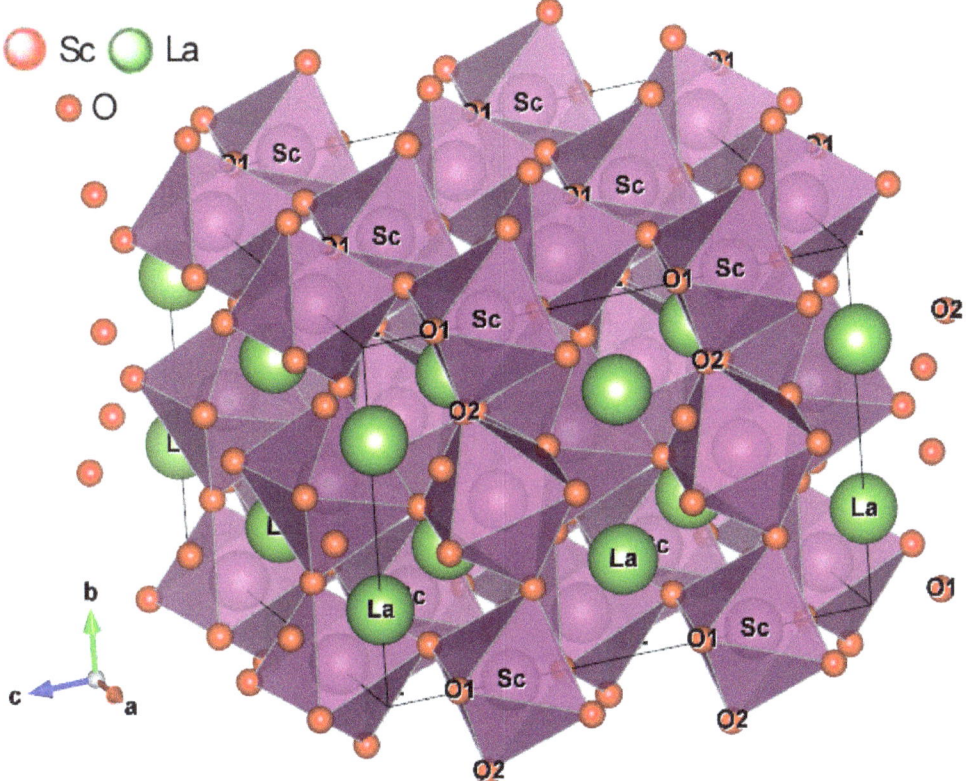

Figure 1. Initial crystal structure of LaScO$_3$ in the absence of a proton: large circles denote the A-site cations, the regions bounded by octahedra shows the localization of the B-site cations, and small circles denote oxygen atoms.

Before each ab initio molecular dynamics calculation, geometric optimization was performed. The reference value for the change in the total energy of the system during the dynamic relaxation of atoms was 0.001 eV. As a result of this preliminary preparation, a structure relaxed to a stable state was obtained, the final stress (pressure) in which was equal to 0.0007 Ry/Bohr^3. Next, a proton was placed in the resulting system, which occupied different positions:

(1) in the Sc octahedron plane between the oxygen atoms at point 1, which has average coordinates (0.44, 0.75, 0.23),
(2) in the Sc octahedron plane near the oxygen atom at point 2, which has average coordinates (0.56, 0.75, 0.23),
(3) in the plane of finding La atoms at localization near the oxygen atom at point 3, which has average coordinates (0.33, 0.75, 0.94).

In each of these three cases of a specific initial position of the proton, five coordinates were used, differing in x and z coordinates. Accordingly, for each case (1-3), five different molecular dynamics calculations were performed. The average values of the initial coordinates of the proton (mentioned above) are also the initial coordinates of the proton for one of the runs in each series.

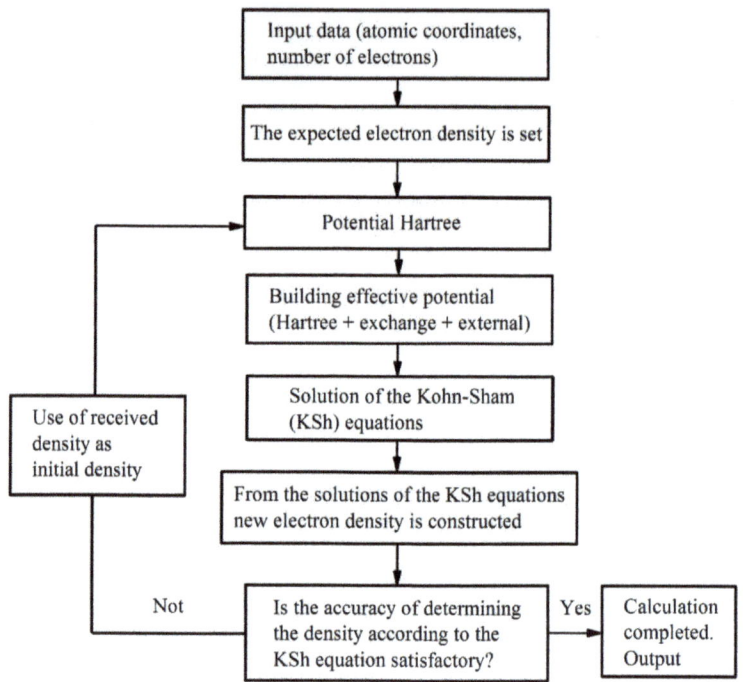

Figure 2. Block diagram of the calculation algorithm of the density functional theory method.

First-principle molecular dynamics calculations were performed for systems containing a proton. The initial velocity of the proton varied from 0 to 0.37418×10^5 m/s (from 0 to 0.5 Bohr/fs). An electric field acted on the proton, and the intensity vector was in the xz plane. The directions of this vector varied, and the magnitude of the tension varied from 0 to 4.95×10^{10} V/m (from 0 to 4.95 V/Å). Note that the maximum value of the electric field strength specified here approaches the value of the atomic electric field, which appears as a result of the interaction of atoms in perovskite oxide $SrTiO_3$ (4×10^{11} near the oxygen atom and 10^{12} V/m near the Ti and Sr atoms) [19]. However, it is significantly higher than the field strength (10^3–10^4 V/m) used to perform the cycle of intercalation/deintercalation of lithium in the molecular dynamics simulation of the silicene anode functioning [20].

Nine series of five runs each with ~1000 time steps were performed in the temperature range from 153 K to 1100 K. The last temperature value is close to the operating temperature of the solid-state fuel cells, using $LaScO_3$-based compounds as the electrolyte. Data on the temperature, applied electric field, initial velocity of the proton, and its primary location are presented in Table 1. To determine the moments of time, at which the transition of a proton from one oxygen atom to another took place, an appropriate algorithm was developed, the code of which was implemented in the Python 3.8 language. The block diagram of the program for determining the energy barrier and the time of the proton transition between oxygen atoms is shown in Figure 3. This algorithm is based on tracking the change in the potential energy of the proton. The peaks in this dependence characterize the change in the average virtual location of the proton. The time interval between these peaks determines the time of its transition from one oxygen atom to another.

Table 1. Characteristics * of the initial data and the calculated temperatures for the series of ab initio molecular dynamics calculations.

Run Number	1	2	3	4	5	6	7	8	9
E_x	0.5	0.5	0	1.5	1.5	1.5	3.5	3.5	3.5
E_y	0	0	0	0	0	0	0	0	0
E_z	0.5	0.5	0	1.5	1.5	1.5	3.5	3.5	3.5
E	0.7071	0.7071	0	2.1213	2.1213	2.1213	4.9497	4.9497	4.9497
v_{0x}	0.1058	0.2645	0.1058	0.2645	0.1058	0.2645	0	0	0
v_{0y}	0	0	0	0	0	0	0	0	0
v_{0z}	0	0.2645	0	0.2645	0.1058	0.2645	0	0	0
v_0	0.1058	0.3741	0.1058	0.3741	0.1496	0.3741	0	0	0
The initial location of the proton	1 (0.44, 0.75, 0.23)	1 (0.44, 0.75, 0.23)	1 (0.44, 0.75, 0.23)	2 (0.56, 0.75, 0.23)	3 (0.33, 0.75, 0.94)	3 (0.33, 0.75, 0.94))	3 (0.33, 0.75, 0.94)	3 (0.33, 0.75, 0.94)	3 (0.33, 0.75, 0.94)
T, K	685	848	682	904	135	255	153	669	1099

* Electric field strength and its x, y, z-components are presented in 10^{10} V/m; the initial velocity of the proton and its x, y, z-components are expressed in 10^5 m/s; the average initial locations of the proton are defined in Å.

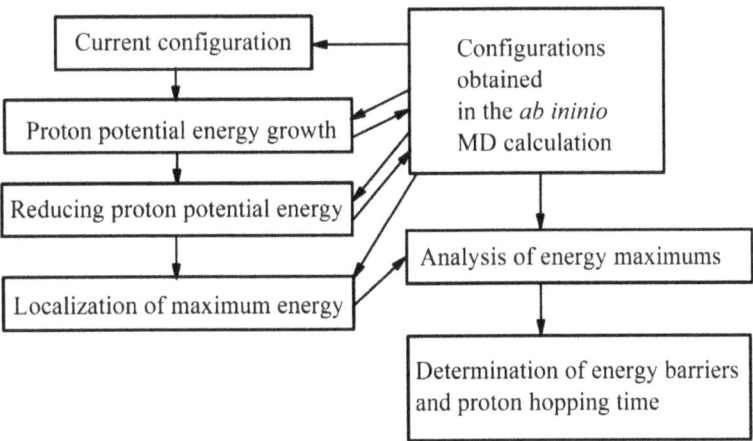

Figure 3. The block diagram of searching for proton transitions.

We used the method of constructing Voronoi polyhedrons to trace the mutual arrangement of the proton and the atoms of the LaScO$_3$ compound closest to it. Polyhedra with the proton in the center were built at each time step. The polyhedra were constructed for three separate subsystems: oxygen, scandium, and lanthanum. Thus, the geometric neighbors closest to the proton were established for O, Sc, and La for the entire time of its presence in the system. Due to the small size of the system and the short residence time of the proton in it, we do not consider the changes occurring in its complete environment, since they are small. That is, we will not present the distribution of polyhedra by the number of faces and the distribution of faces by the number of sides that characterize the rotational symmetry of the subsystems under consideration. Instead, we will consider the angular placement of the proton relative to different kinds of atoms. The angles under consideration are formed by a proton and a pair of atoms included in the immediate environment of the proton. The environment is defined using the Voronoi polyhedron. There is a proton at the vertex of the angle under consideration, the sides of the angle are formed by segments, which extreme points are the proton and one of the atoms of the subsystem under consideration. We have previously performed a similar sounding to the structure by moving a charged particle (in this case, a proton) [21].

3. Results

Let us trace the trajectories of a proton in cases, where its initial placement corresponds to its average initial coordinates obtained in five calculations of the same type. Horizontal projections of the configurations of the LaScO$_3$ crystal corresponding to the time instant of 1 ps are shown in Figure 4.

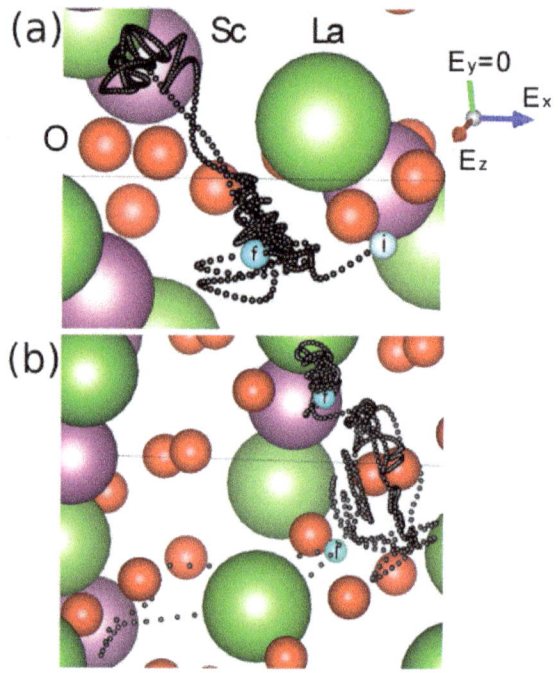

Figure 4. Horizontal projections of the configurations of the LaScO$_3$ crystal, corresponding to the time instant of 1 ps, (**a**) under the electric field with an intensity of $E = 0.7071 \times 10^{10}$ V/m and (**b**) without the electric field; the figure also shows the trajectories of the proton for the specified time interval, and the symbols i and f denote the beginning and the end of the trajectory, respectively.

The upper figures reflect the above system obtained in series 1, and the lower figures show a similar system resulting from the series 3. In series 1, a constant electric field with an intensity $E = 0.7071 \times 10^{10}$ V/m acted on the proton, the series 3 was performed in the absence of an electric field. The trajectory of a proton is enclosed in a more limited space under the effect of the electric field. It is noteworthy that a large part of the path of the proton does not pass along the direction determined by the strength of the electric field (line in the figure), but in the direction forming a certain angle with the direction of the vector E. In either case, the trajectory of the ion turns out to be rather entangled. It can get into the vicinity of another (different from the original) oxygen atom and stay there for some time. It can also approach adjacent La and Sc atoms, which are not initially adjacent to it. In addition, the proton can not only move away from its initial location, nor approach it again, moving in the direction opposite to the direction of the passed path. Thus, the electric field does not affect the choice of the preferred location for the proton in the LaScO$_3$ crystal. Most likely, the place of its virtual location is determined by the attraction of a proton to any closest oxygen atom.

The temperature dependence of the average potential energy of perovskite LaScO$_3$ is not monotonic (Figure 5). Sharp changes in the energy U are observed at temperatures $T < 153$ K and in the temperature range $669 \leq T \leq 685$ K. In this temperature range, the highest energy U corresponds to the state with the highest value of the electric field and the least energy appears when the field is missing. The largest value of the energy U is also observed at $T = 1099$ K at the maximum value of the electric field strength. The behavior of a proton in a system essentially depends on the temperature of the perovskite. We can say with confidence that no proton jumps are observed at temperatures below 400 K; the proton vibrates around the oxygen atom near which it was originally located. In the temperature range of 600–800 K, random proton jumps are observed between oxygen atoms. Most of the proton jumps are observed at the temperatures above 800 K, when the proton has a sufficient speed to overcome the potential barrier. A visual representation of the proton motion in the LaScO$_3$ perovskite is provided in Supplementary data. This film representation shows the behavior of the proton in the calculation 1, when the temperature in the model was 685 K.

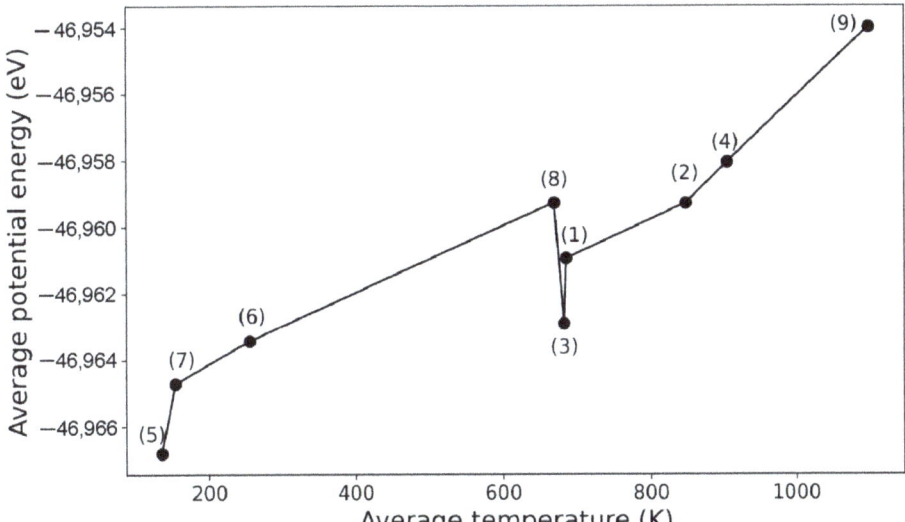

Figure 5. Dependence of the average potential energy on the average temperature for the calculated states of the LaScO$_3$ perovskite. The numbers indicate the series numbers from Table 1.

The behavior of a proton in a perovskite system in the temperature range of 680–850 K makes it possible to distinguish the time interval, during which the proton jumps to a neighboring oxygen atom, or to another place of temporal localization. In the time sweep of the potential energy, such an interval is always included between the local extrema of the function $U(t)$. Figure 4 shows the $U(t)$ functions obtained in the series 1–3, defined by Table 1. The time intervals of the proton jump to a new sedentary position are highlighted in Figure 6 with red stripes. In all cases, the duration of the interval for the jump did not exceed 100 fs, i.e., 100 time steps. Obviously, during such a time interval, the proton completely "forgets" the initial direction of its velocity. The further speed of the proton is determined by the interatomic interaction, temperature, and, in part, the direction of the electric field strength. The electric field does not completely determine the direction of motion of the proton, but it affects the instantaneous values of the potential energy: the greater electric field results in the more significant fluctuations of the $U(t)$ function. Comparison of Figure 6a,b testifies in favor of the fact that the frequency of the proton jumps can be influenced by the direction and magnitude of its initial velocity. In particular, at a higher value of the velocity, the time intervals between the proton jumps increase. At the same time, a comparison of Figure 4a,c shows that the number of proton jumps in the presence of an electric field is likely to decrease.

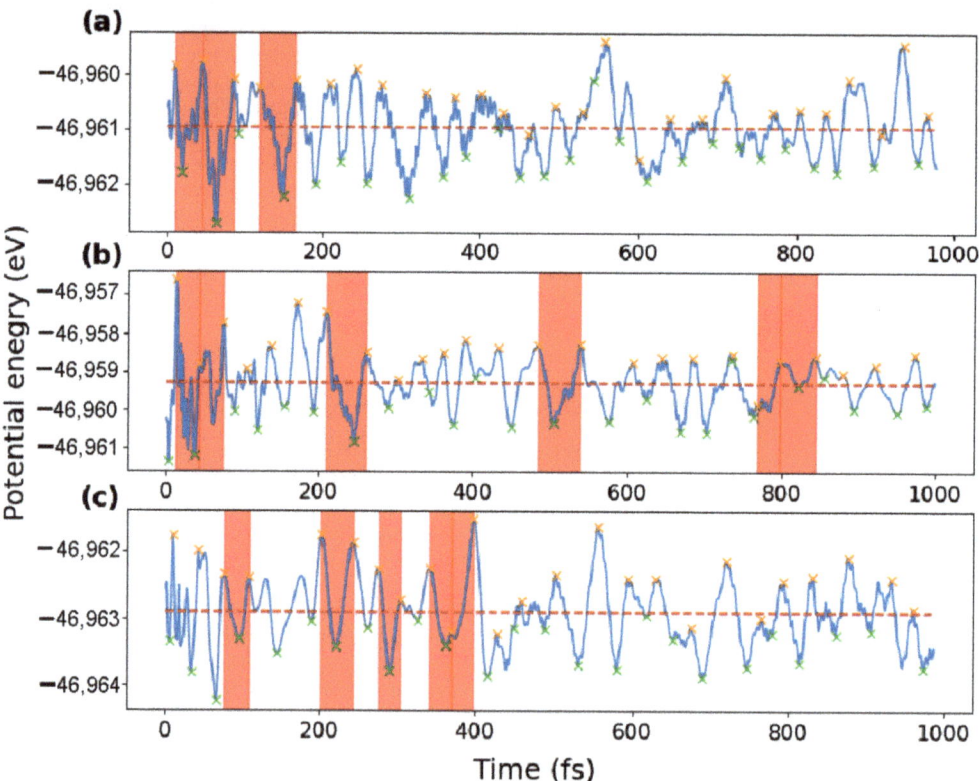

Figure 6. Change in the potential energy of the perovskite system over time in calculations (**a**) 1, (**b**) 2, and (**c**) 3; red stripes show the proton jumps among oxygen atoms; the average potential energy is shown by the dashed line; the initial placement of a proton corresponds to its average initial coordinates for each selected series of five runs.

The types of possible jumps made by a proton in a perovskite are shown in Figure 7. The directions of the proton movement are shown by light segments connecting the light circles (proton designations). In case I, the proton changes its position, passing from the oxygen atom belonging to one octahedron to the oxygen atom assigned to another octahedron, and the transition occurs between La atoms. In case II, the proton moves along only one La atom and passes from one oxygen atom to another along the edge of the octahedron, with both oxygen atoms belonging to the same octahedron. In cases III and IV, the proton also passes one La atom, but the oxygen atoms are separated by an octahedron, i.e., together they belong to three adjacent octahedra. The forward and backward paths of the proton in these cases are not equivalent due to the difference in the atomic environment. Therefore, we consider them separately, assuming that case III corresponds to a direct transition, i.e., the proton hits the plane of the octahedron, and case IV corresponds to a reverse transition.

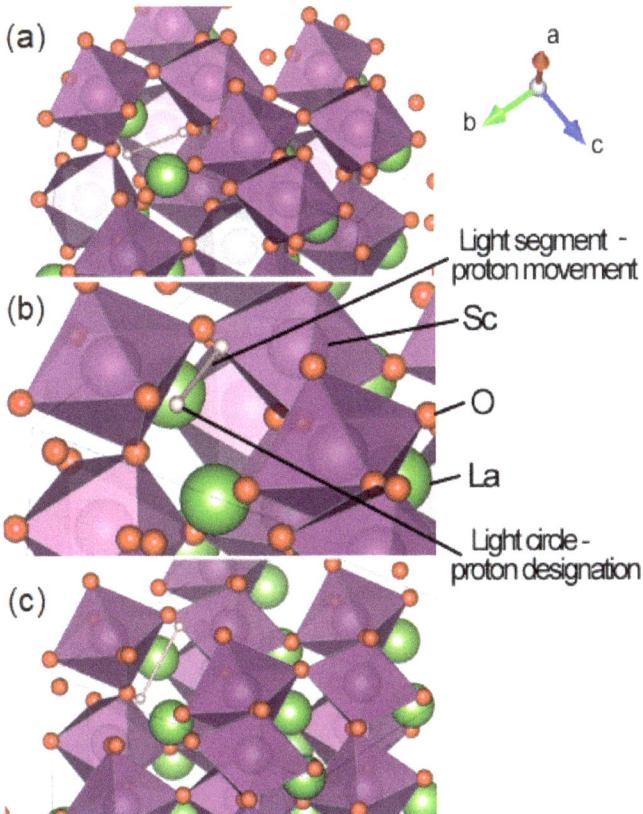

Figure 7. Types of proton jumps in perovskite (**a**) I, (**b**) II, (**c**) both, as III as IV.

Figure 7 shows the characteristics of proton hopping classified according to the four types defined above. The jumps of the first type can be characterized as the migration of a proton in the lanthanum plane (the plane between the octahedra). The second type of jumps takes place on the plane of scandium (in the plane of the octahedrons). The jumps of the third type include the movement of a proton from the plane of lanthanum to the plane of scandium (transition to the plane of the octahedron). The fourth type of transition expresses the exit of the proton from the scandium plane and its return to the lanthanum plane.

The characteristics of jumps of different types are shown in Figure 8. Jumps of the first type are characterized by the maximum energy barrier height, the shortest jump time (since in this case the proton acquires the highest kinetic energy), and the minimum number of jumps of this type. To make a jump of the second type, the proton also needs to acquire a sufficiently large kinetic energy to overcome the second highest energy barrier. In this case, the jump time is the longest. The number of such jumps is a quarter of all jumps performed. The barrier for the jumps of the third type is slightly less than that for the jumps of the second type, and the time needed for such jumps is noticeably shorter than that for the jumps of the second type, but it is slightly longer than for the jumps of the first type. The proportion of the jumps of the third type turns out to be the largest. The jumps of the fourth type are characterized by the lowest energy barrier and the jumps time is a little longer than that for the jumps of the third type. The number of such jumps, like jumps of the second type, is a quarter of all jumps.

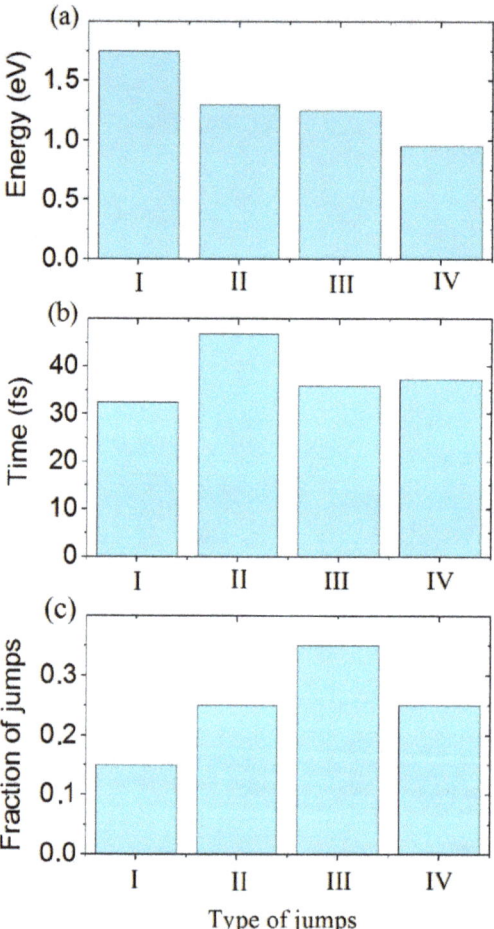

Figure 8. Parameters for performing different types of proton jumps: (**a**) the average height of the energy barrier, (**b**) the average time of the jump, and (**c**) the proportion of four types of jumps.

The migration time of a proton between the planes of lanthanum and scandium is practically independent of the direction, while the fraction of transitions into the plane of scandium is much greater than from the plane. Consequently, this jump is more beneficial for the proton than the motion in the lanthanum plane, so it prefers to move in the direction of the scandium plane. On the other hand, the movement of the proton in the scandium plane and exit from it are equally probable (the fractions of jumps of these types are the same). However, to move in the plane, the proton needs to overcome a more significant potential barrier than to exit from it. Therefore, the jump time in the scandium plane is longer.

Figure 9 shows the time variation of the hydrogen run in the Cartesian coordinates from the series 3, in which 4 proton jumps were observed by the time of 400 fs. Attention should be paid to the behavior of the y coordinate, according to which it is best to diagnose jumping by a proton. A strong change in the y coordinate ends by the time of 400 fs, and further changes in this coordinate are insignificant. Small fluctuations in the y coordinate near a constant value indicate the absence of proton jumps.

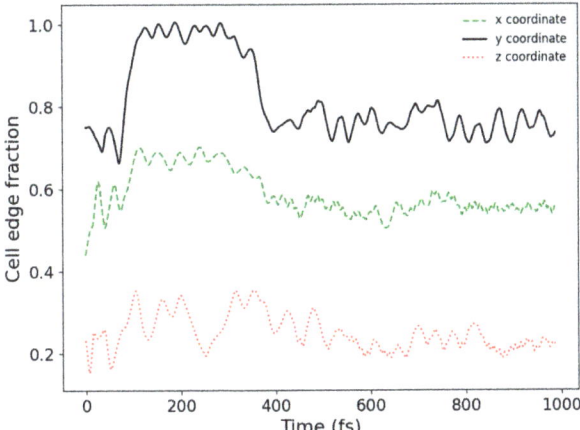

Figure 9. Time dependence of the proton Cartesian coordinates related to series 3; the initial placement of a proton corresponds to its average initial coordinates for the series.

The diffusion coefficient D was determined in terms of the mean square of the proton displacement. According to the Arrhenius law, the dependence of the diffusion coefficient on the reciprocal temperature can be represented as a linear dependence $\ln D(1/T)$. To calculate the average diffusion activation energy, the data presented in Figure 8a,c were used, i.e., the average activation energy was determined, taking into account the contribution of each type of proton jump. The calculated dependence $\ln D(1000/T)$, showing the estimated temperature dependence of the proton diffusion coefficient in perovskite $LaScO_3$, is shown in Figure 10. A similar dependence was determined for protons diffusing in yttrium-doped crystalline perovskite $BaCeO_3$ in a narrow temperature range (from 725 to 770 K) by the neutron scattering method [22]. These values fit satisfactorily into the dependence $\ln D(1000/T)$ obtained by us for $LaScO_3$. Ceramics $BaCe_{0.8}Y_{0.2}O_{3-\delta}$ is one of the most highly conductive proton conductors. The perovskite $LaScO_3$ studied by us can also be attributed to this class of solid proton conductors.

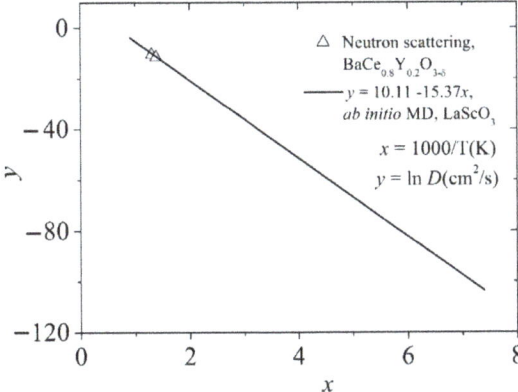

Figure 10. Dependence of $\ln D$ on the parameter $1000/T$ at a constant weighted average activation energy $E_a = 1.255$ eV for the movement of a proton along defect-free perovskite $LaScO_3$; also presented here are the values of a similar characteristic for $BaCeO_3$ doped with yttrium obtained in the experiment on neutron scattering [22].

Figure 11 shows the distributions of the distances between the proton and different types of atoms in the LaScO$_3$ perovskite system. The first peak of the "proton–lanthanum" distance distribution is split into two distinct subpeaks. This testifies the fact that, with a high probability, in addition to the nearest La atom, there are always other La atoms located at a very close distance from the proton. The highest (second) peak in this distribution shows that the majority of other La atoms are located at distances from 4 Å to 6 Å from the proton. The first peak in the "proton–scandium" distribution is not split and is well-resolved. The most probable distance from the proton to the Sc atom is 2.2 Å. The next two most pronounced peaks of this distribution are at 4.5 Å and 6 Å. It is in the vicinity of these distances that the majority of the following Sc atoms adjacent to the proton are located. The "proton–oxygen" distribution has a clearly pronounced peak at 1.0 Å, which shows that hydrogen forms a bond only with one oxygen atom, located in its field. In this case, the spectrum of distances between the proton and the rest of the oxygen atoms blurs as this distance increases. This feature of the "oxygen" spectrum is associated with a large number of O atoms in the system compared to other atoms and with their high mobility.

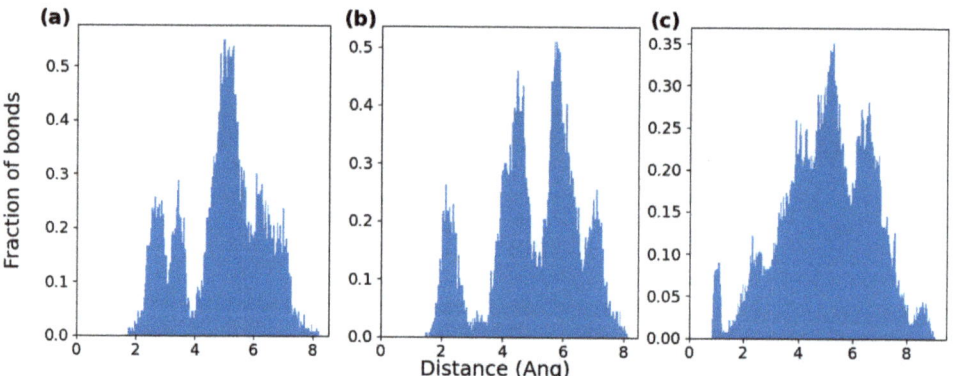

Figure 11. Distribution of distances between the proton and other atoms of the perovskite system: (a) lanthanum, (b) scandium, and (c) oxygen, obtained over a time interval of 1 ns in series 1.

Figure 12 indirectly reflects the location of the proton by the distribution of angles that a proton forms with a pair of any atoms of the same type (i.e., O, Sc, and La). The top row of figures shows the angular distributions obtained in the absence of the electric field, and the bottom row of figures shows the same distributions when the field strength $E = 0.7 \times 10^{10}$ V/m acts in the [10] direction in the systems under consideration. The temperature of the system in the absence and in the presence of an electric field is approximately the same ($T = 683 \pm 2$ K). It is seen that the corresponding angular distributions (when the angles are formed by atoms of the same type) are in good agreement with each other. The significant identity of the corresponding angular spectra indicates that the electric field has no significant effect on the proton movement in the LaScO$_3$ crystal.

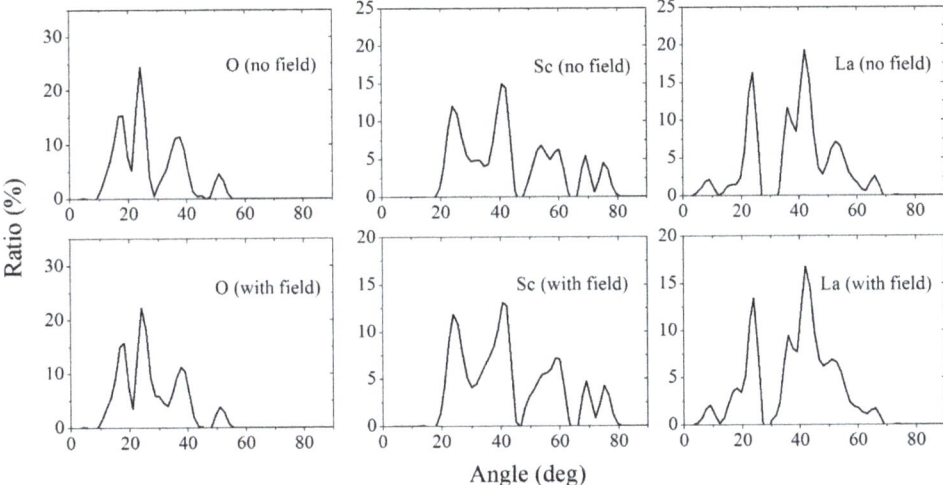

Figure 12. Angular distributions of the nearest geometrical neighbors of the proton, obtained during its migration in the LaScO$_3$ crystal lattice in the absence of the electric field (series 3) and in the presence of the field $E = 0.7 \times 10^{10}$ V/m (series 1) at $T = 683 \pm 2$ K; the averaged initial position of the proton is classified as position 1 (see Table 1) and its initial velocity is 0.1058×10^5 m/s; the captions within the figure show the type of neighbors and the absence or presence of the electric field (in brackets).

Due to the limited space for the proton to move along the LaScO$_3$ crystal, which is associated with the rigid structure of this compound, an increase in the angle, at the top of which the proton is located, when the sides pass through the centers of two atoms of the same type (O, Sc, and La), means an increase in the average distance between the proton and these atoms. Due to the high mobility of the proton, it more likely interacts with the closer atoms. Figure 12 shows that the smallest angular range is characterized by the interaction between the proton and oxygen atoms. It is this interaction that is decisive for the proton. However, the structural skeleton of a compound is not determined by oxygen, but by the metals forming this compound, i.e., Sc and La. The angular range of interaction of the proton with La is less than that with Sc. Therefore, it is easier to regulate the physical properties of the LaScO$_3$ compound by replacing La (rather than Sr) atoms with other atoms.

As can be seen from Figure 13, the situation with the distribution of preferential interactions does not change, as the temperature of the system under consideration increases from 153 K to 1099 K. Even at the highest temperature, as before, the smallest angular range is observed for the proton interactions with oxygen atoms, and the largest one is for that with Sc. In all cases considered, the shape of the angular spectra does not undergo too large changes as the temperature of the system increases. Nevertheless, for the oxygen and lanthanum subsystems, the disappearance or the appearance of separate subpeaks in the angular distribution is observed at the temperature variation. Such changes are less pronounced for the scandium subsystem. Since the shape of the angular distribution characterizes indirectly the interaction between the proton and the atoms selected in the compound, the resulting response of the system to an increase in temperature means that the replacement of La by other atoms should create a corresponding effect at high temperatures.

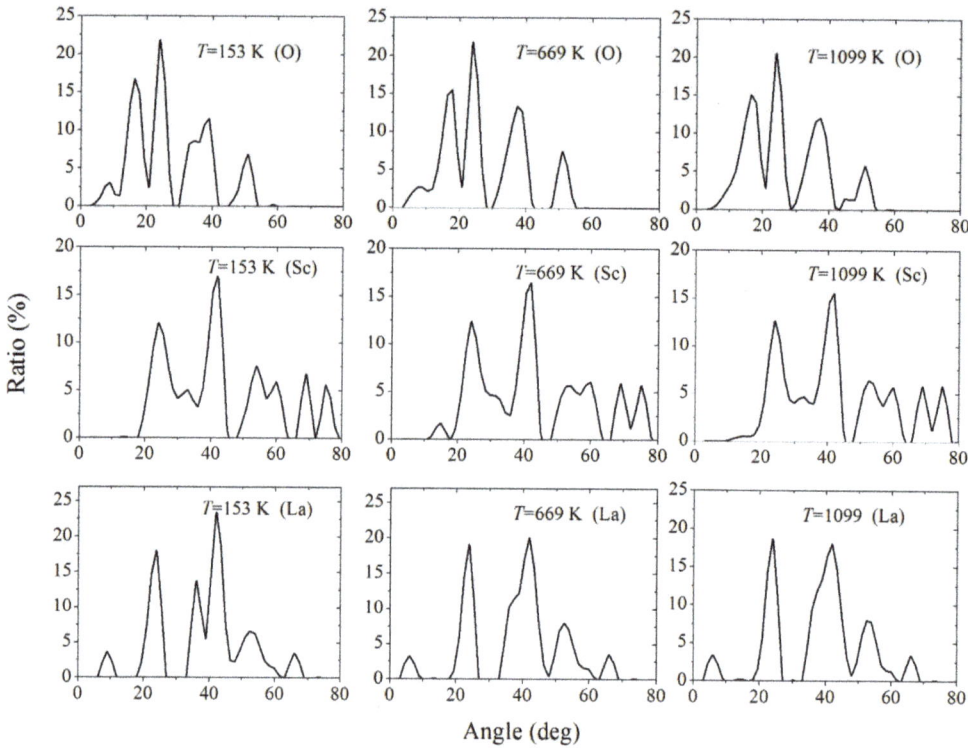

Figure 13. Angular distributions of the nearest geometrical neighbors of the proton, obtained during its migration in the LaScO$_3$ crystal lattice in the presence of an electric field of strength $E = 4.9 \times 10^{10}$ V/m (series 7–9) at different temperatures; the initial position of the proton corresponds to position 3 (as classified in Table 1) and its initial velocity is 0; the captions within of the figure reflect the temperature and the type of neighbors (in brackets).

4. Discussion

The proton conduction is an Arrhenius-type thermal activated process, phenomenologically reminiscent of a polaron-type electronic conductivity. The proton jump times can be accurately represented by the proton polaron model [16]. Conceptually, a proton satisfies the definition of a polaron, since it can be represented as a trapped charge in an elastic crystal field. The polaron nature of proton conductivity was potentially proposed in [23]. However, such a concept as a "proton polaron" has not yet become widespread in the study of the conductivity in perovskites, including the structures of the ABO$_3$ type.

The mechanism for the appearance of a polaron in an ionic crystal with an implanted proton is obvious. Longitudinal optical oscillations in such a system occur when positive and negative ions move in opposite directions. As a result of such oscillations, regions of excess (relative to the equilibrium value) charge of different signs alternating in space arise in the crystal. In the area where the proton is located, an excess of negative charge will most likely appear. It is clear that in such a system, there must be an interaction between the proton and longitudinal optical phonons. Here, the notion of a "polarization cloud" that moves along with the proton becomes appropriate. This is how a new quasi-particle, the polaron, appears. Consequently, the problem of the polaron, and not the "bare" proton, comes to the fore. Moreover, the residual interaction between the polaron and phonons is weak. A polaron drags a region of lattice distortion behind it, so its effective mass is greater than the mass of a "bare" proton, and the energy, on the contrary, is less than that of a "bare" proton. If a proton creates a deformation, which linear dimensions exceed the lattice constant, one can speak of a large-radius polaron. Otherwise, we are dealing with a small-radius polaron.

In this paper, we consider the proton–phonon interaction in the structure of the ABO_3 perovskite in the form of a polaron. In addition to the aforementioned facts, we analyze the problem arising with the control of the proton conductivity in the $LaScO_3$ perovskite using an external electric field. The path of the proton migration in the perovskite structure is determined not by the direction of the field strength, but by the elastic deformations of the crystal lattice of this compound. The mass of the La atom is 3.1 times the mass of the Sc atom. Therefore, the La-O valence phonon mode has a lower energy compared to the Sc-O mode. Consequently, in the $LaScO_3$ perovskite, the thermally activated La-O mode acts as the main driving force for the interaction of the low-energy phonon mode and the diffusion mode of protons. Polarons appear due to the deformation of the crystal lattice. Polaron states are characterized by distinct absorption peaks in the visible (VIS) and near-IR (NIR) ranges of the optical absorption spectrum. Dielectric spectroscopy is an important tool in the study of polaron states. In particular, the electrical transition between polaron states largely determines the temperature-dependence of the relaxation frequency for the DC electrical conductivity.

When the crystal lattice contracts, the protons slow down. As the temperature rises, protons resist pressure more and more and retain a higher diffusion coefficient. With the increasing pressure on the sample, the activation barrier for proton diffusion increases. The activation energy for proton diffusion depends on the degree of the lattice deformation, which in turn is determined by the elastic properties [24]. Proton transfer is facilitated by cooperative processes such as lattice vibrations. In perovskites of the ABO_3 type, the level of proton conductivity depends on the nature of the atoms in this compound. Typically, the proton conductivity increases as the elements A and B become less negative. The proton conductivity in perovskite can be increased by increasing the number of oxygen vacancies, i.e., by doping. In particular, the formation of oxygen vacancies is facilitated by the substitution of Sr for La in the $LaScO_3$ perovskite. However, this way of increasing the proton conductivity is associated with fundamental limitations. First, the formation of an oxygen vacancy is associated with the breaking of the oxygen–metal (O–M) bond, which requires a certain amount of energy. Second, the formation of an electrically neutral vacancy is associated with the reduction of neighboring cations [25], and, therefore, with changes in the chemical energy. Third, at an excessive number of oxygen vacancies, the stability of the compound containing them can be violated [26]. Thus, only a limited number of oxygen vacancies can be created in a perovskite of the ABO_3 type.

The perovskites based on rare earth elements demonstrate nontrivial electrical properties, such as metal-insulator transition, high-temperature superconductivity, etc. [27]. The existence of the tunable bandwidth of the materials is explained by the variations in localized and delocalized electron patterns.

The technology of high-performance solar cells production requires uniform large-area perovskite films obtained by the crystallization process [28]. It should be noted that the main problems of high-efficiency solar cells are the high production cost and decrease in their efficiency with an increase in the size of the solar module [29,30].

Polymer conductors have their own criteria for proton conductivity. Acidic hydrophilic groups often provide high proton conductivity of a compound (proton polymer conductor) with a supramolecular network [31]. For the effective work of the proton polymer conductors, it is necessary to achieve their significant acidity, i.e., high degree of protonation of hydrophilic groups. The acidity index is the main factor in the synthesis of the polymer conductor with good proton-conducting ability.

5. Conclusions

Although oxygen vacancies are currently playing a decisive role in characterizing devices based on ABO_3 perovskites, we investigated the proton conductivity in defect-free perovskites based on metal oxides. Understanding the mechanism of proton conductivity in a defect-free material, which occupies at least 80% of even perovskite doped and enriched with oxygen vacancies, is important for the further improvement of environmentally friendly energy devices. We investigated the behavior of the proton in the perovskite LaSrO$_3$ crystal depending on its initial location and velocity, external electric field, and temperature. The discontinuous nature of proton diffusion in perovskite was established and four types of jumps determined by the structural features of the perovskite were identified. Average execution time for all types of proton jumps is between 32 and 47 fs. The activation diffusion energy varies in the range of $0.95 \leq E_a \leq 1.75$ eV. The difference in the frequencies of different types of jumps can be as high as 2.3. It is found that proton migration in solid defect-free perovskite LaScO$_3$ is characterized by the equation

$$\ln D = 10.11 - 15.37(1000/T),$$

where the diffusion coefficient D is in cm^2/s and the temperature T is in K.

The scale of the proton's displacements in the presence of an electric field and without it was established by the visual observation on the proton trajectory and the partition of the space occupied by atoms, using Voronoi polyhedra. Regardless of the presence of an electric field in the entire temperature range under study, the angles formed by a proton with oxygen ions form a smaller number of peaks (from 4 to 5) than the angles created by a proton with lanthanum ions (5–6) or scandium ions (6–7). This is due to the stronger attraction of the proton to oxygen. The set of ab initio molecular dynamic tests is evidence in favor of the polaron character of proton migration in the LaSrO$_3$ perovskite. This is confirmed by the fact that the immediate environment of the proton and the temperature of the system determine the trajectory of its motion to a much greater extent than the external electric field.

Further improvement of semiconductor electrolytes should be based on taking into account the mechanism of proton migration along a perfect perovskite crystal. This, along with investigating the effect of oxygen vacancies, could significantly improve the performance of the clean energy devices.

Supplementary Materials: The following supporting information can be downloaded at: https://www.mdpi.com/article/10.3390/app12115302/s1.

Author Contributions: Conceptualization, A.Y.G.; methodology, A.Y.G. and D.S.P.; software, D.S.P.; validation, A.Y.G. and D.S.P.; formal analysis, A.Y.G. and D.S.P.; investigation, A.Y.G. and D.S.P.; resources, D.S.P.; data curation, D.S.P.; writing—original draft preparation, A.Y.G.; writing—review and editing, A.Y.G. and O.R.R.; visualization, D.S.P.; supervision, A.Y.G. and Y.P.Z.; project administration, Y.P.Z. and O.R.R.; funding acquisition, Y.P.Z. and O.R.R. All authors have read and agreed to the published version of the manuscript.

Funding: This work is partly supported by Government of Russian Federation, the State Assignment number FEUZ-2020-0037, registration number 075-03-2022-011; and is partly executed in the frame of the scientific theme of Institute of high-temperature electrochemistry UB RAS, number FUME-2022-0005, registration number 122020100205-5.

Institutional Review Board Statement: Not applicable.

Informed Consent Statement: Not applicable.

Data Availability Statement: All the details regarding the reported results can be obtained upon private request from the authors.

Acknowledgments: These calculations were performed on a hybrid cluster type computer "URAN" at the N.N. Krasovskii Institute of Mathematics and Mechanics of the Ural Branch of the Russian Academy of Sciences with a peak performance of 216 Tflop/s and 1864 CPU.

Conflicts of Interest: The authors declare no conflict of interest.

References

1. Zhao, G.; Suzuki, K.; Hirayama, M.; Kanno, R. Syntheses and characterization of novel perovskite-type LaScO$_3$-based lithium ionic conductors. *Molecules* **2021**, *26*, 299. [CrossRef] [PubMed]
2. Hossain, S.; Abdalla, A.M.; Jamain, S.N.B.; Zaini, J.H.; Azad, A.K. A review on proton conducting electrolytes for clean energy and intermediate temperature-solid oxide fuel cells. *Renew Sust. Energ. Rev.* **2017**, *79*, 750–764. [CrossRef]
3. Dai, H.; Kou, H.; Wang, H.; Bi, L. Electrochemical performance of protonic ceramic fuel cells with stable Ba-ZrO3-based electrolyte: A mini-review. *Electrochem. Commun.* **2018**, *96*, 11–15. [CrossRef]
4. Kim, J.; Sengodan, S.; Kim, S.; Kwon, O.; Bu, Y.; Kim, G. Proton conducting oxides: A review of materials and applications for renewable energy conversion and storage. *Renew. Sust. Energ. Rev.* **2019**, *109*, 606–618. [CrossRef]
5. Loureiro, F.J.A.; Nasani, N.; Reddy, G.S.; Munirathnam, N.R.; Fagg, D.P. A review on sintering technology of proton conducting BaCeO$_3$-BaZrO$_3$ perovskite oxide materials for Protonic Ceramic Fuel Cells. *J. Power Sources* **2019**, *438*, 226991. [CrossRef]
6. Rashid, N.L.R.M.; Samat, A.A.; Jais, A.A.; Somalu, M.R.; Muchtar, A.; Baharuddin, N.A.; Wan Isahak, W.N.R. Review on zirconate-cerate-based electrolytes for proton-conducting solid oxide fuel cell. *Ceram. Int.* **2019**, *45*, 6605–6615. [CrossRef]
7. Li, J.; Wang, C.; Wang, X.; Bi, L. Sintering aids for proton-conducting oxides–a double-edged sword? A mini review. *Electrochem Commun.* **2020**, *112*, 106672. [CrossRef]
8. Haugsrud, R.; Norby, T. Proton conduction in rare-earth ortho-niobates and ortho-tantalates. *Nat. Mater.* **2006**, *5*, 193–196. [CrossRef]
9. Lesnichyova, A.S.; Belyakov, S.A.; Stroeva, A.Y.; Kuzmin, A.V. Proton conductivity and mobility in Sr-doped LaScO$_3$ perovskites. *Ceram. Int.* **2021**, *47*, 6105–6113. [CrossRef]
10. Kumar, S.; Kaswan, J.; Satpati, B.; Shukla, A.K.; Gahtori, B.; Pulikkotil, J.J.; Dogra, A. LaS-cO$_3$/SrTiO$_3$: A conducting polar heterointerface of two 3d band insulating perovskites. *Appl. Phys. Lett.* **2020**, *116*, 051603. [CrossRef]
11. Sun, L.; Devakumar, B.; Guo, H.; Liang, J.; Li, B.; Wang, S.; Sun, Q.; Huang, X. Synthesis, structure, and luminescence characteristics of far-red emitting Mn^{4+}-activated LaScO$_3$ perovskite phosphors for plant growth. *RSC Adv.* **2018**, *8*, 33035–33041. [CrossRef] [PubMed]
12. Zhang, W.; Hu, Y.H. Progress in proton-conducting oxides as electrolytes for low-temperature solid oxide fuel cells: From materials to devices. *Energy Sci. Eng.* **2021**, *9*, 984–1011. [CrossRef]
13. Kim, Y.; Kim, Y.M.; Shin, J.; Chara, K. LaInO$_3$/BaSnO$_3$ polar interface on MgO substrates. *APL Mater.* **2018**, *6*, 096104. [CrossRef]
14. Fan, L.; Rautama, E.L.; Lindén, J.; Sainio, J.; Jiang, H.; Sorsa, O.; Han, N.; Flox, C.; Zhao, Y.; Li, Y.; et al. Two orders of magnitude enhancement in oxygen evolution reactivity of La$_{0.7}$Sr$_{0.3}$Fe$_{1-x}$Ni$_x$O$_{3-\delta}$ by improving the electrical conductivity. *Nano Energy* **2022**, *93*, 106794. [CrossRef]
15. Nomura, K.; Takeuchi, T.; Kamo, S.-I.; Kageyama, H.; Miyazaki, Y. Proton conduction in doped LaScO$_3$ perovskites. *Solid State Ion.* **2004**, *175*, 553–555. [CrossRef]
16. Slodczyk, A.; Colomban, P.; Willemin, S.; Lacroix, O.; Sala, B. Indirect Raman identification of the proton insertion in the high-temperature [Ba/Sr][Zr/Ti]O$_3$-modified perovskite protonic conductors. *J. Raman Spectrosc.* **2009**, *40*, 513–521. [CrossRef]
17. Perdew, J.P.; Burke, K.; Ernzerhof, M. Generalized gradient approximation made simple. *Phys. Rev. Lett.* **1996**, *77*, 3865–3868. [CrossRef]
18. Monkhorst, H.J.; Pack, J.D. Special points for Brillouin-zone integrations. *Phys. Rev. B* **1976**, *13*, 5188–5192. [CrossRef]
19. Muller, K.; Krause, F.F.; Beche, A.; Schowalter, M.; Galioit, V.; Loffler, S.; Verbeeck, J.; Zweck, J.; Schattschneider, P.; Rosenauer, A. Atomic electric fields revealed by a quantum mechanical approach to electron picodiffraction. *Nat. Commun.* **2014**, *5*, 5653. [CrossRef]
20. Galashev, A.Y. Computational investigation of silicene/nickel anode for lithium-ion battery. *Solid State Ion.* **2020**, *357*, 115463. [CrossRef]
21. Galashev, A.E.; Ivanichkina, K.A. Computer study of silicene channel structure based on the transport of Li$^+$. *Rus. J. Phys. Chem. A* **2021**, *95*, 724–729. [CrossRef]
22. Braun, A.; Chen, Q. Experimental neutron scattering evidence for proton polaron in hydrated metal oxide proton conductors. *Nat. Commun.* **2017**, *8*, 15830. [CrossRef] [PubMed]
23. Krasnoholovets, V.V.; Tomchuk, P.A.; Lukyanets, S.P. Proton transfer and coherent phenomena in molecular structures with hydrogen bonds. *Adv. Chem. Phys.* **2003**, *125*, 351–548. [CrossRef]
24. Chen, Q.; Huang, T.-W.; Baldini, M.; Hushur, A.; Pomjakushin, V.; Clark, S.; Mao, W.L.; Manghnani, M.H.; Braun, A.; Graule, T. Effect of compressive of compressive strain on the Raman modes of the dry and hydrated BaCe$_{0.8}$Y$_{0.2}$O$_3$ proton conductor. *J. Phys. Chem. C* **2011**, *115*, 24021–24027. [CrossRef]
25. Pavone, M.; Muñoz-García, A.B.; Ritzmann, A.M.; Carter, E, A, First-principles study of lanthanum strontium manganite: Insights into electronic structure and oxygen vacancy formation. *J. Phys. Chem. C* **2014**, *118*, 13346–13355. [CrossRef]
26. Park, M.H.; Lee, D.H.; Yang, K.; Park, J.-Y.; Yu, G.T.; Park, H.W.; Materano, M.; Mittmann, T.; Lomenzo, P.D.; Mikolajick, T.; et al. Review of defect chemistry in fluorite-structure ferroelectrics for future electronic devices. *J. Mater. Chem. C* **2020**, *8*, 10526–10550. [CrossRef]
27. Rudra, M.; Halder, S.; Saha, S.; Dutta, A.; Sinha, T.P. Temperature dependent conductivity mechanisms observed in Pr$_2$NiTiO$_6$. *Mater. Chem. Phys.* **2019**, *230*, 277–286. [CrossRef]

28. Green, M.; Dunlop, E.; Ebinger, J.H.; Yoshita, M.; Kopidakis, N. Solar cell efficiency tables (version 57). *Prog. Photovolt.* **2021**, *29*, 3–15. [CrossRef]
29. Deng, Y.; van Brackle, C.H.; Dai, X.; Zhao, J.; Chen, B. Tailoring solvent coordination for high-speed, room-temperature blading of perovskite photovoltaic films. *Sci. Adv.* **2019**, *5*, eaax7537. [CrossRef]
30. Rolston, N.; Scheideler, W.J.; Flick, A.C.; Chen, J.P.; Elmaraghi, H. Rapid open-air fabrication of perovskite solar modules. *Joule* **2020**, *4*, 2675–2692. [CrossRef]
31. Liu, H.; Li, R.; Lu, J.; Liu, Z.; Wang, S.; Tian, H. Proton conduction studies on four porous and nonporous coordination polymers with different acidities and water uptake. *Cryst. Eng. Comm.* **2020**, *22*, 6935–6946. [CrossRef]

Article

Mechanical and Thermal Properties of Aluminum Matrix Composites Reinforced by In Situ Al₂O₃ Nanoparticles Fabricated via Direct Chemical Reaction in Molten Salts

Liudmila A. Yolshina [1,*], Aleksander G. Kvashnichev [1], Dmitrii I. Vichuzhanin [2] and Evgeniya O. Smirnova [2]

1. Institute of High-Temperature Electrochemistry, Ural Branch of RAS, 620990 Ekaterinburg, Russia
2. Institute of Engineering Science, Ural Branch of RAS, 620049 Ekaterinburg, Russia
* Correspondence: yolshina06@rambler.ru; Tel.: +7-3433745044

Citation: Yolshina, L.A.; Kvashnichev, A.G.; Vichuzhanin, D.I.; Smirnova, E.O. Mechanical and Thermal Properties of Aluminum Matrix Composites Reinforced by In Situ Al₂O₃ Nanoparticles Fabricated via Direct Chemical Reaction in Molten Salts. *Appl. Sci.* 2022, *12*, 8907. https://doi.org/10.3390/app12178907

Academic Editors: Suzdaltsev Andrey and Oksana Rakhmanova

Received: 10 August 2022
Accepted: 2 September 2022
Published: 5 September 2022

Publisher's Note: MDPI stays neutral with regard to jurisdictional claims in published maps and institutional affiliations.

Copyright: © 2022 by the authors. Licensee MDPI, Basel, Switzerland. This article is an open access article distributed under the terms and conditions of the Creative Commons Attribution (CC BY) license (https://creativecommons.org/licenses/by/4.0/).

Abstract: The development of novel methods for industrial production of metal-matrix composites with improved properties is extremely important. An aluminum matrix reinforced by "in situ" α-Al₂O₃ nanoparticles was fabricated via direct chemical reaction between molten aluminum and rutile TiO₂ nanopowder under the layer of molten salts at 700–800 °C in air atmosphere. Morphology, size, and distribution of the in situ particles, as well as the microstructure and mechanical properties of the composites were investigated by XRD, SEM, Raman spectra, and hardness and tensile tests. Synthesized aluminum–alumina composites with Al₂O₃ concentration up to 19 wt.% had a characteristic metallic luster, their surfaces were smooth without any cracks and porosity. The obtained results indicate that the "in situ" particles were mainly cube-shaped on the nanometer scale and uniform matrix distribution. The concentration of Al₂O₃ nanoparticles depended on the exposure time and initial precursor concentration, rather than on the synthesis temperature. The influence of the structure of the studied materials on their ultimate strength, yield strength, and plasticity under static loads was established. It is shown that under static uniaxial tension, the cast aluminum composites containing aluminum oxide nanoparticles demonstrated significantly increased tensile strength, yield strength, and ductility. The microhardness and tensile strength of the composite material were by 20–30% higher than those of the metallic aluminum. The related elongation increased three times after the addition of nano-α Al₂O₃ into the aluminum matrix. Composite materials of the Al-Al₂O₃ system could be easily rolled into thin and ductile foils and wires. They could be re-melted for the repeated application.

Keywords: metal-matrix composite; aluminum; nano-alumina; molten salt synthesis; mechanical properties

1. Introduction

Metal-matrix composites based on aluminum and its alloy have been the subject of great scientific and practical interest over the last decades [1,2]. The reason is the possibility to obtain lightweight metallic materials with unique properties (enhanced hardness, ductility, elastics modulus, and many others) by combining the advantages of constructional metals with the merits of filler materials. Metal-matrix composites based on aluminum and aluminum-metal-matrix composite alloys (AMMC) are promising for aerospace, military, automobile, and electronic areas because of their low density, high specific stiffness, strong wear resistance, reduced thermal expansion coefficient and high thermal conductivity [1,2]. These materials sustain their stability characteristics over a wide temperature range, have high electrical and thermal conductivities, and low sensitivity to surface defects [3–5].

AMMCs are fabricated both by ex situ synthesis (e.g., liquid ingot casting and powder metallurgy), where reinforcement powder particles are added to the metal melt, and by in situ synthesis (e.g., exothermic dispersion, reactive hot pressing, reactive infiltration, and direct melt reaction), where particles are directly synthesized in the metal melt [6–12]. The reinforcement in AMMCs can have a form of continuous/discontinuous fibers, whiskers, or particulates. The most frequently applied fillers are ceramic ones such as alumina, silicon carbides or nitride, carbon fiber materials, or graphene [13,14].

At present, researchers focus on the possibility of introducing the above-mentioned ceramic additives in the form of nanoparticles into the aluminum matrix. When moving to nanoscale hardeners, we have the opportunity to obtain an entirely new class of lightweight materials with increased specific strength, plasticity, and crack resistance. The mechanisms of deformation and fracture in aluminum–alumina composite materials reinforced by nanosized alumina particles can be fundamentally different from those in classical aluminum alloys.

There are some problems arising in the synthesis of Al-Al_2O_3 composites: inhomogeneity of the distribution of oxide nanocrystals in the volume of the metal matrix; the tendency of nano-objects to agglomeration inside the matrix (1–3 μm); low wettability of oxide particles by molten aluminum (lattice consistency of the reinforcing phase and aluminum), and degradation of oxide materials (chemical, mechanical, etc.).

The mechanical addition of alumina particles into the metal melt by means of powder metallurgy [15–17] results in the interface pollution, poor wetting ability, and mechanical properties. The composites synthesized via the ex situ methods have high porosity and a lot of cracks. They are characterized by the appearance of "soft" and "hard" phases [17] because of the difference in aluminum particles' concentration and agglomeration [15]. Hence, the mechanical properties of composites produced by powder metallurgy methods can vary greatly.

Selective laser melting can be used for the creation of advanced ball-milled Al-Al_2O_3 nanocomposites with a two-fold increase in Vickers micro-hardness compared to the initial Al [18]. The mechanical properties of Al-nano Al_2O_3 produced by ball-milling and hot-pressing under uniaxial pressure of 50 MPa were improved due to the utilization of nano-sized reinforcement particles [19]. To date, the compaction of several layers of aluminum foil and aluminum oxide powder, followed by rolling from 2 to 20 times, is another promising method of composite materials' fabrication [20–22]. As a result, layered laminated Al-Al_2O_3 composite structures are formed. The uneven distribution of aluminum oxide particles in the aluminum matrix, as well as the significant agglomeration of oxide particles [21], are the main problems of the laminated composite production that cause the uneven properties of the composite. Such defects are usually eliminated by increasing the number of rolling cycles. In addition, the non-isomorphism of the composites' properties is also a problem.

Sometimes, both powder metallurgy and rolling are combined for the synthesis of composites, which makes it possible to obtain Al-Al_2O_3 composites with a better distribution of dispersing particles [23].

Direct reaction synthesis (DRS) refers to the process in which reactant powders or compacts of reactant powders are directly added to molten metal, and ceramic reinforcing particles are formed in situ through the exothermic reaction between reactants or between the reactant and the component of the melt. During the in situ process, the reinforcement phases formed in the matrix metal by direct chemical reactions are thermodynamically stable, free of surface contamination, and disperse more uniformly. These factors lead to the stronger particle–matrix bonding at the "in-situ" process.

DRS in situ synthesis can be applied as a commercial method because of its simplicity, low cost, and near-net-shape-forming capability. The biggest problem of the traditional DRS method is the necessity of a long mixing time at a high temperature, which causes the agglomeration of small initial particles [24,25]. The introduction of alumina to the molten aluminum matrix by the "in situ" process was proved by different independent physical–chemical methods such as XRD, Raman spectroscopy, and others [26].

Nowadays, synthesizing Al-Al_2O_3 nanocomposites is an important practical and fundamental task in the field of inorganic chemistry. There is a very high need to create a high-tech method for the synthesis of non-porous aluminum–alumina nanocomposites with improved mechanical properties.

The novelty of the present work is that the aluminum matrix was reinforced by "in situ" perfectly wetted and well-distributed Al_2O_3 nanoparticles, formed via a direct chemical reaction between molten aluminum and rutile TiO_2 nanopowder under a molten salt layer at 700–750 °C in air atmosphere. The microstructure and mechanical properties of the composites were investigated by XRD, Raman spectroscopy, SEM, and tensile testing.

This research aims to develop a method for the industrial large-scale "in situ" synthesis of aluminum-based matrix composites, reinforced with ceramic nanoparticle inclusions. These composites are formed via the reaction between molten aluminum and the salt electrolyte flux, which contains rutile TiO_2 nanopowder as the precursor. The effect of nano-alumina inclusions content on mechanical and thermal properties of the Al-nano-Al_2O_3 composite is analyzed.

2. Experimental

Titanium dioxide nanopowders were produced by electrochemical oxidation of titanium foil of 99.9 wt.% purity ("VSMPO-Avisma", Ekaterinburg, Russia) in the chloride-nitrate melt according to the method described in detail elsewhere [27,28]. The interaction between molten aluminum and nano- and micro-oxide powders of titanium dioxide was performed in molten alkali halide media in an alumina crucible containing the mixture of alkali chlorides with a weighed portion of oxide additive. Salt mixtures of alkali chlorides with melting points lower than 700 °C, with small additions of some fluorides, were used as a base electrolyte with the concentration of titanium dioxide ranging from 0.2 to 2 wt.%. We used commercial "chemically pure" salts provided by "Lanhit" Ltd., Moscow, Russia.

The aluminum sample A95 of 99.5 wt.% purity ("Rusal", Russia) consisted of a 1 cm high disc with a diameter strictly equal to the crucible diameter. The disc was loaded into a layer of the dry oxide–salt mixture at the bottom of the crucible. The primary admixture of technical aluminum was iron—about 0.5 wt.%. The aluminum disc was covered by a layer of dry alkali chlorides. The alumina crucible was exposed inside the vertical heating furnace at the temperatures range of 700–800 °C for 3 to 5 h. After the exposure, the liquid aluminum globule was poured into the new cold alumina crucible. After the solidification, the salt mixture was dissolved in distilled water. The aluminum composite ingot was separated and dried. Cooled to the ambient temperature, aluminum–alumina composite ingots were divided into four parts. One part was used for the determination of Al_2O_3 content in the composite, the second part was used for XRD analysis, the third one was poured with current-carrying gum, polished by six different silicon solutions using a Struers disc-finishing machine (Copenhagen, Denmark), and examined by means of scanning electron microscopy (SEM, Raman spectroscopy (Wotton-under-Edge, England, UK), microhardness testing). The fourth biggest part was remelted to form cylindrical samples for a servo-hydraulic testing machine.

The processing of the studied composites is schematically shown in Figure 1.

The structures of titanium dioxide and aluminum–alumina composite were studied by means of SEM, XRD, and Raman spectroscopy. In order to determine the material composition and the composite structure, the purity of the oxide phase was analyzed using Raman spectra by a Renishaw U 1000 micro-Raman spectrometer (UK) connected to a Leica DML microscope equipped with 50× and 100× magnification lenses. An Ar^+ laser (Cobalt model) with a wavelength of 532 nm and power output of 20 mW was used as the excitation source. The diameter of the laser spot was about 1 μm; the acquisition time was set to 20 s. The XRD analysis was performed on a «RIGAKU» DNAX 2200PC diffractometer at ambient temperature with Cu Kα radiation. Subsequent analysis of the vibrational spectra was carried out using the embedded Wire 3.0 software (Wotton-under-Edge, England, UK); the spectral lines were approximated using the Fityk 0.9.8 curve fitting and data analysis application. The typical average spectra obtained are presented for each sample.

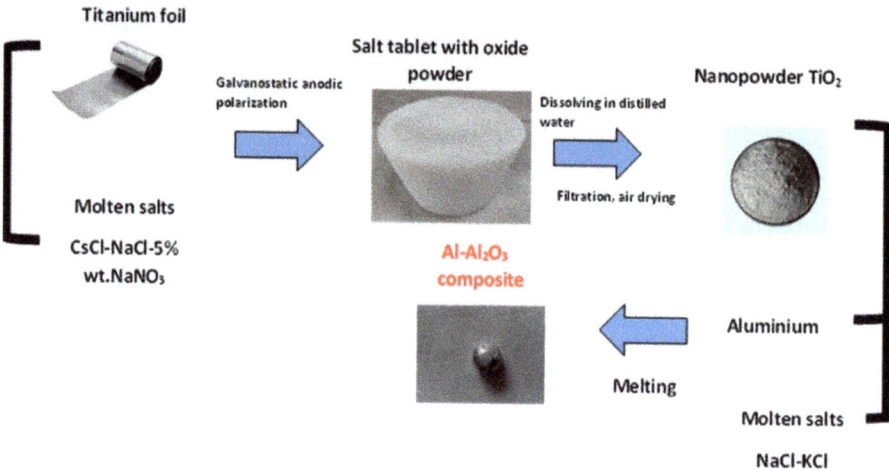

Figure 1. Processing schematic of Al/Al$_2$O$_3$ nanocomposite.

The volumetric method determined the quantitative content of aluminum oxide in the produced AMMC. Volumetric analysis was carried out by dissolving a composite sample in 30% hydrochloric acid, followed by recalculating the released hydrogen per mass of pure aluminum in the sample according to reaction (1):

$$2Al + 6HCl \rightarrow 2AlCl_3 + 3H_2(g) \uparrow \qquad (1)$$

The sample was then examined as follows: scanning electron microscopy (SEM) was performed using JEOL 5900LV and TESCAN MIRA3 microscopes; the sample surface was characterized by a Veeko Wyko NT 1100 optic profilometer-profilograph in the Vertical Scanning Interferometry (VSI) regime; the fatigue strength was examined by a servo-hydraulic testing machine INSTRON 8801, the sample microhardness was tested a WIN–HCU FISCHERSCOPE HM 2000 XYm system, which is designed for Martens hardness testing with the WIN–HCU software at a maximum load P from 1000 mN to 2000 mN, a loading time of 20 s, a time of exposure under loading of 15 s, and an unloading time of 20 s according to the ISO 14577 standard [29].

The characteristics measured by indentation were used for the calculation of the following parameters: the ratio of the indentation hardness to the contact elasticity modulus HIT/E* [30], the elastic recovery $R_e = \frac{h_{max}-h_p}{h_{max}} \times 100\%$ [31], the power ratio [32], and the plasticity index $\delta_A = 1 - \frac{W_e}{W_t}$ [29] that characterizes the ability of a material to resist elastoplastic strain. Ten measurements of indentation characteristics at each load were performed. The error of measurements was determined with a confidence probability $p = 0.99$.

3. Results and Discussion

3.1. Formation of Nano-Titanium Dioxide as a Precursor

The formation of titanium nano-oxide powders as precursors of composite synthesis was carried out by electrochemical oxidation of pure titanium in a chloride-nitrate melt in an argon atmosphere [27]. The main function of the nitrate melt is corrosion protection against the aggressive environment due to the formation of dense protective oxide layers on metals. Chloride-nitrate melts allow synthesizing both dense protective coatings and nanocrystalline powders of metal oxides such as Al$_2$O$_3$, TiO$_2$, ZrO$_2$, and PbO [33,34] varying the nitrate concentration in the flux. Due to the alkali nitrates' thermal instability, the temperature range of their application is narrow, while amorphous oxides are usually formed at low temperatures. It is necessary to synthesize oxides using salt electrolytes with higher melting points to increase the temperature of the crystalline metal oxide synthesis.

The metal oxidation in the chloride-nitrate salt melt bulk at the current densities ranging from 3 to 5 mA/cm^2 lead to the accumulation of significant amounts of oxide nanopowder in the bulk salt electrolyte. The chemical composition and modification of the formed oxides depended mainly on the oxidation temperature.

Despite the nanometer size of titanium dioxide crystallites (Figure 2a), the X-ray diffraction pattern contained main rutile titanium dioxide lines of 2θ angle at 27.5°, 36.16°, 39.26°, 41. 32°, 44.14°, 54.42°, 56.72°, 63.44°, and 69.11° with diffraction plans (110), (101), (200), (111), (210), (211), (220), (002), and (301), respectively (JCPDS no. #88-1175) [35] (Figure 2c). The broadening of lines or a halo was not observed, which may be explained by the fact that the formed oxide particles were well-faceted cubic crystals, and not spheroids, as is often the case, which happens in the synthesis of nano-oxides by the sol-gel method. Titanium dioxide nanopowders synthesized in a molten chloride-nitrate electrolyte at the current density of 3.5 mA/cm^2 were characterized by a high degree of crystallinity and took the form of titanium dioxide cubes with a size up to 30 nm (Figure 2a).

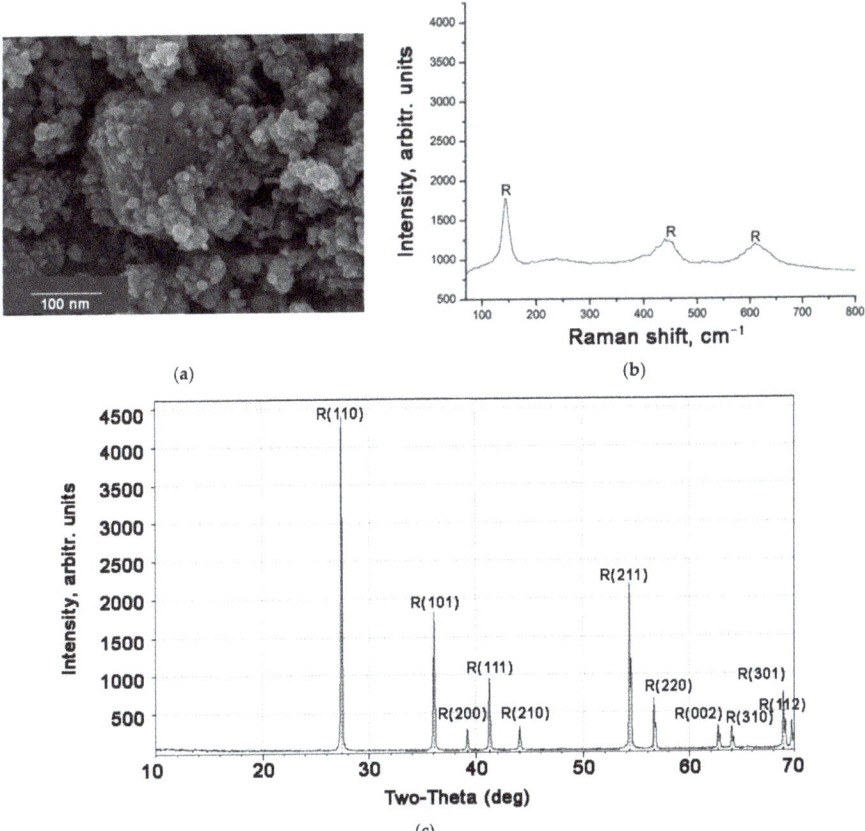

Figure 2. SEM image (a), Raman spectrum (c), and XRD pattern (b) of TiO$_2$ nanopowder in rutile modification synthesized in NaCl-CsCl-5% NaNO$_3$ at 700 °C.

According to the results of the Raman spectroscopy of the sample obtained at 700 °C, peaks characteristic for the modification of rutile TiO$_2$ were determined in the Raman bands: 144 cm^{-1}, 448 cm^{-1}, 612 cm^{-1} (Figure 2b).

Further, rutile titanium dioxide nanopowder formed under the above conditions was used as a precursor.

Various industrially sol-gel methods for nano- and micro-powders production in the form of rutile and anatase were previously studied as precursors for the synthesis of the nanocomposite. However, all these nanopowders were sedimented to the bottom of the crucible and reaction (3) (Table 1) did not proceed. The titanium dioxide nanopowder in the rutile modification synthesized by the electrochemical oxidation of titanium in a chloride–nitrate melt was the only precursor with which the above reaction was successful.

Table 1. Gibbs energies of some possible reactions in the Al-TiO$_2$-NaCl-O$_2$ system.

Reaction		$\Delta G_{700\,°C}$	$\Delta G_{750\,°C}$	$\Delta G_{800\,°C}$
$13TiO_2 + 4NaCl = 2Na_2Ti_6O_{13} + TiCl_4(g)$	(2)	333.9	326.0	318.0
$3TiO_2 + 4Al = 3Ti + 2Al_2O_3$	(3)	−437.6	−430.85	−424.0
$13Ti + 4NaCl + 13O_2 = 2Na_2Ti_6O_{13} + TiCl_4(g)$	(4)	−9642.6	−9535.7	−9429.0
$3TiO_2 + 7Al = 3AlTi + 2Al_2O_3$	(5)	−643.7	−643.2	−624.7
$3TiO_2 + 13Al = 3Al_3Ti + 2Al_2O_3$	(6)	−804.8	−788.9	−773.0

3.2. Interaction between Molten Aluminum and Titanium Nanodioxide under the Layer of Molten Halides and Analysis of Interaction Products

The AMMC material dispersion reinforcement is typically achieved by casting or powder metallurgy, with the ceramic particles being introduced ex situ into a solid or liquid matrix. A new AMMC production method has been developed on the basis of the controlled "in-situ" chemical reactions; reinforcing fillers are formed during the chemical interaction between the matrix components and reactive additives [9,12]. Such AMMCs demonstrate excellent mechanical and physical properties, because coherent (i.e., having a standard atomic layer at the interface) or partially coherent interfaces are formed between the matrix and new phases that arise in "in-situ" reactions.

The uniform distribution of fine particles of alumina in the aluminum matrix was achieved by the modified method of molten salt synthesis. Molten halides (alkali or alkali-earth chlorides and/or fluorides) are the optimal reactive media for the formation of aluminum–alumina composites, as was shown in our previous papers [26,36].

The synthesis of alumina particles inside aluminum is carried out by a one-step chemical reaction of titanium dioxide with molten aluminum under the layer of molten halides in air atmosphere by reaction (3); therefore, newly formed alumina nanoparticles are perfectly wetted by the liquid metal.

Obviously, these reactions, except for the direct interaction between TiO$_2$ and molten sodium chloride, are thermodynamically feasible at sufficiently low temperatures of 700–800 °C. The presented process is a special case of aluminothermy.

The interaction between liquid aluminum and fine titanium dioxide particles by reactions (5) and (6) can proceed up to the formation of the intermetallic AlTi and Al$_3$Ti titanium aluminide compounds. The Gibbs energies of reactions (5) and (6) are more negative than that of reaction (3). It makes reactions (5) and (6) more favorable.

Titanium aluminides were not detected in the reaction product by X-ray phase analysis, Raman spectroscopy, EDS spectra of composites, and mapping sample surfaces in X-rays. Intermetallic compounds of aluminum and titanium or titanium species were not detected by means of an inductively coupled plasma optical emission spectrometer OPTIMA 4300 DV (Perkin Elmer, USA) after dissolving the resulting composite in hydrochloric acid. Thus, it can be argued that the interaction occurred according to the mechanism described above by reaction (3).

The reason for such a result is the appearance of a side reaction (4) in the system. The interaction between titanium atoms, sodium chloride, and gaseous oxygen according to reaction (4) resulted in the formation of Na$_2$Ti$_6$O$_{13}$ nanowires and nanorods, which was experimentally proved [37]. The value of Gibbs energy of this reaction is 15 times greater that of all other reactions.

The composites obtained in this way were non-porous with a characteristic metallic luster (Figure 3). They exhibited high electrical conductivity and are excellently suited for mechanical processing. It was experimentally proved that the aluminum oxide content did not change from the surface of the aluminum drop to its center. The density of the obtained composite averaged 2.6977 g/cm^3, which is very close to that of pure aluminum (2.6989 g/cm^3), and indicates the absence of porosity in the composite, which is typical for such samples obtained by traditional methods.

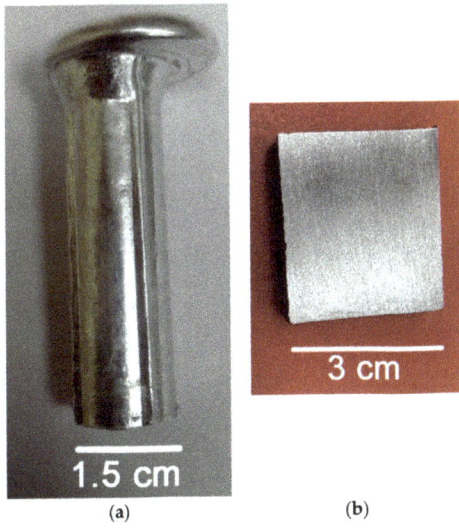

Figure 3. Optic photo of the ingot (**a**) and cross-section (**b**) of Al-Al$_2$O$_3$.

According to the X-ray diffraction data, the thus obtained material was composed of Al and α-Al$_2$O$_3$ [36].

Raman spectra of the oxide inclusions were collected from several crystals inside the aluminum matrix. The average spectrum of the oxide inclusion is presented in Figure 4. It is obvious that this spectrum was a standard spectrum of α-Al$_2$O$_3$ with the characteristic bonds at 378 cm^{-1}, 417 cm^{-1}, 430 cm^{-1}, 448 cm^{-1}, 576 cm^{-1}, 644 cm^{-1}, and 749.5 cm^{-1} [38]. It confirmed the XRD data about the synthesis of α-Al$_2$O$_3$ inside the aluminum matrix [36].

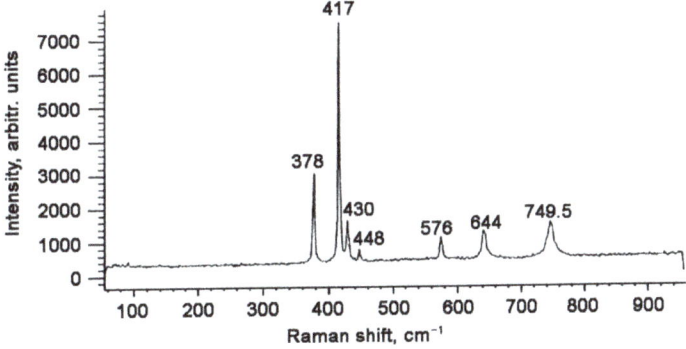

Figure 4. Raman spectrum of the oxide inclusion inside the Al matrix.

Micro- and nanosized inclusions of aluminum oxide were also clearly visible in the micrographs in the optic microscope (Figure 5a) and in backscattered electrons in a scanning electron microscope (Figure 5b,c). The EDS spectrum of inclusion showed the content of aluminum and oxygen in the stoichiometric ratio of Al_2O_3; inclusions of aluminum oxide had sizes close to 100 nm.

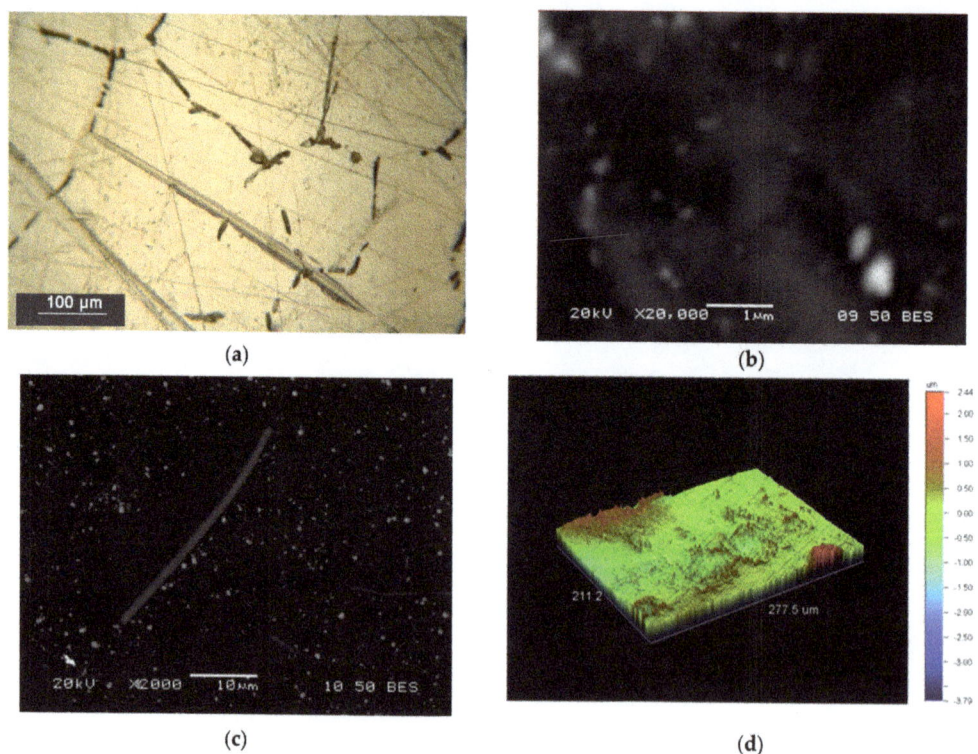

Figure 5. Optic photo of the cross-section of Al-α-Al_2O_3 composite (**a**) (×40); BES images of Al-α-Al_2O_3 composite (**b**,**c**); 3D image of Al-α-Al_2O_3 composite (**d**).

A very uniform distribution of oxide inclusions inside aluminum grains of the cross-section of Al-Al_2O_3 composite is presented in Figure 6.

Inclusions in the form of long rods were crystals of iron aluminide Al_3Fe (Figure 6), which were present in the starting aluminum as an admixture. According to the phase measurements of the cross-section of this sample, the Al area is 97.5%, Al_3Fe-1.4%, and Al_2O_3-1.2%. The 3D image of the cross-section of AMMC (Figure 5d) also shows a uniform distribution of alumina nanoparticles inside the grain of the aluminum matrix.

To prove the possibility of finding nanoparticles mainly inside the grain rather than at the grain boundary, as would be expected from classical considerations, we calculated the value of the critical nucleus (r_c) during the crystallization of aluminum using well-known Formula (7) [39]:

$$r_C = \frac{2\sigma T_{12} M}{Q_{12} \Delta T \rho}, \qquad (7)$$

where σ is the surface tension of the metal melt; T_{12} is the melting point of the metal; M is the molecular weight; Q_{12} is the heat of crystallization (heat of fusion); ΔT is the supercooling of the melt, at which the nucleation of the crystallization center occurred; ρ is the density of the metal.

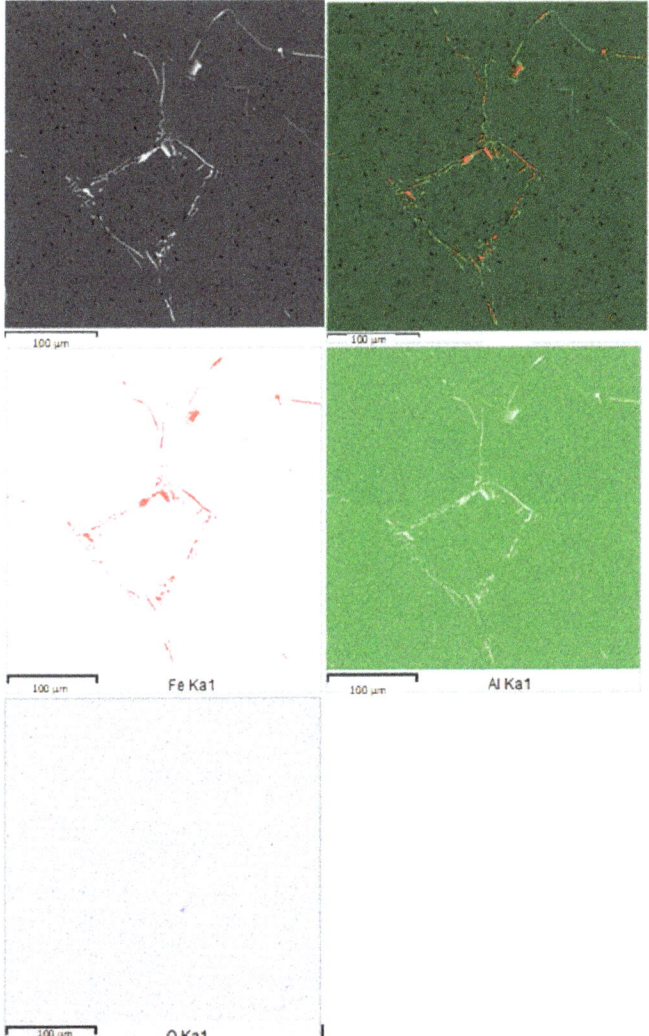

Figure 6. Elemental distribution map of the Al-1.2% Al_2O_3 composite cross-section.

As a result, the value of the radius of the critical nucleus of Al_2O_3 was r_c = 108.33 nm.

Thus, the size of the critical nucleus was three times larger than the initial size of the nanoparticles (30 nm), and the calculated value of synthesized α- alumina in the aluminum matrix was very close to the experimental one, which varied from 100 to 150 nm.

Thus, it can be assumed that nano-α aluminum oxide formed as a result of the chemical interaction between fused aluminum and nanopowder of titanium dioxide and was not grouped along the grain boundaries but was uniformly distributed in the volume of the aluminum grain (Figures 5 and 6), which is a favorable factor in assessing the mechanical properties of the composite.

The composite samples were dissolved in hydrochloric acid HCl solution to determine the quantitative content of aluminum oxide in the produced AMMCs using the volumetric method. The concentration of Al_2O_3 in the composite was determined by the difference in hydrogen evolution.

The data in Table 2 elucidates that an increase in the exposure time and concentration of titanium dioxide in the melt leads to an increase in the mass fraction of aluminum oxide in the composite. However, when the titanium dioxide concentration increased significantly above 1% and exposure time was up to 9 h, it was not possible to obtain samples with a reinforcing phase concentration of more than 19%. A slight decrease in the mass fraction of aluminum oxide in the composite caused by an increase in temperature to 750 °C could be associated with the interaction between atomic titanium and a salt flux, which resulted in the formation of sodium titanate.

Table 2. The influence of the concentration of TiO_2 precursor, temperature, and exposure time on the concentration of Al_2O_3 inside the obtained Al-Al_2O_3 composite.

% TiO_2 Added	T Exposure, h	T Interaction, °C	% Al_2O_3 in the Ingot
0.5	3	700	7.8 ± 0.2
0.5	5	700	13.5 ± 0.3
0.5	3	750	4.6 ± 0.2
0.5	5	750	13.0 ± 0.3
1.0	3	700	8.2 ± 0.2
1.0	5	700	18.4 ± 0.2
1.0	3	750	8.7 ± 0.3
1.0	5	750	10.1 ± 0.3

As can be seen from Table 2, we varied the concentration of aluminum oxide of AMMC not by introducing a certain amount of aluminum oxide into the aluminum matrix, but by changing the synthesis time and temperature. The content of aluminum oxide formed by the reaction (3) did not depend on the titanium dioxide weight. Titanium dioxide was a necessary seed for the formation of aluminum oxide in the composite; when interacting in a salt melt, it formed sodium hexatitanate nanofibers in stoichiometric amounts [37].

3.3. Thermal Analysis of Composite Al-Nano-Al_2O_3

Thermal analysis of Al-Al_2O_3 composites was measured by a thermal analyzer STA 449C Jupiter, NETZSCH at a heating rate of 10 K/min, the temperature determination error was <1.5 K. Synchronous analysis of samples of aluminum–nano-alumina composites was carried out by differential scanning calorimetry (DSC) and thermogravimetry as a function of temperature in air atmosphere. In the first case, the sample was heated to temperatures up to 723 °C, cooled and, without removing the sample from the crucible, heated repeatedly to the same temperatures, and then the sample was heated to the temperature of 800 °C. The samples of the initial aluminum and studied composites exhibited congruent melting, since the melting peaks were reproduced during repeated cooling and heating cycles, while the lines were expressed in one DSC anomaly with the same areas during each repeated cooling/heating cycle. The onset points changed no more than 0.3 °C, hence the maximum temperatures, indicating the termination of the melting process of the samples, differed slightly and fluctuated in the range of 669.9–671.4 °C (Table 3).

Table 3. Onset points, melting temperatures, and Gibbs energies of melting of Al-Al_2O_3 samples.

№	Al_2O_3, wt.%	T_{ons}, °C	T_m, °C	ΔG, J/g
1	0	660.3	669.9	−10.63
2	7.8	660.5	671.4	−10.55
3	18.4	660.6	669.4	−10.8
4	4.6	660.2	668.4	−10.91

This phenomenon illustrates the property of peak reproducibility (Figure 7), which is a reflection of the reversibility of the phase transition during cooling and subsequent heating. The closeness of the value for the obtained composites to the parameters of the original metal can be primarily due to the indicators of the wettability of the dispersed oxide nanophase by aluminum. The data obtained by the thermogravimetry simultaneously with the differential scanning calorimetry also revealed that the thermal stability of the obtained composite materials was close to pure aluminum. The recorded change in the mass of the aluminum was equal to 0.74% and that in the composite samples during three heating and cooling cycles did not exceed 0.37%. Such an insignificant change in the mass of the sample during measurements, accompanied by the appearance of lines attributed to endothermic anomalies during DSC, directly indicated the presence of the process, which is a first-order phase transition with a high reproducibility index.

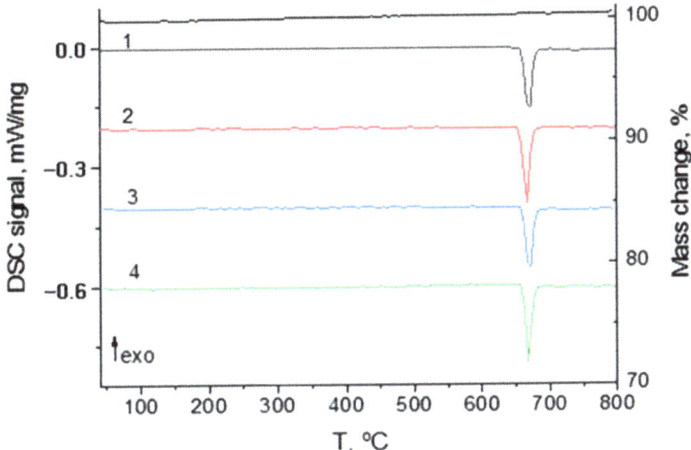

Figure 7. DSC curves (**left**) of initial aluminum (**1**), AMMCs with 7.79% (**2**), 18,4% (**3**), and 4.61% (**4**) of Al_2O_3.

3.4. Measurement of Mechanical Properties of Composite Al-Nano-Al_2O_3

Experimental methods to study mechanics of deformable "Al-nano-α-alumina" composites included hardness studies and tensile testing of the material with data registration in the form of "stress-strain" curves. For the in-depth mechanical studies, samples with an aluminum oxide concentration of 10 and 14 wt.% were selected.

The instrumented indentation data on the specific features of Al and Al-Al_2O_3 behavior under elastoplastic strain are given in Table 4. It is evident that the "in situ" implementation of uniformly distributed nano-α Al_2O_3 (10 and 14 wt.%) into the aluminum matrix decreased in the maximum and residual indentation depth h_{max} and h_p and increased in the Martens hardness H_M and the indentation hardness at all loads HIT. Such simultaneous change in the characteristics h_{max}, h_p, H_M, H_{IT}, W_e, and W_t measured in the indentation process is commonly caused by the material hardening and mainly is observed in aluminum composites with uniformly distributed alumina nanopowders [40–42]. In particular, for the samples "Al- nano-alumina ", an average increase in H_{IT} was nearly 20–25%. Contact elasticity modulus E * decreased for Al-14% Al_2O_3 composite at the load 1000 mN and did not change for all other composites.

The AMMC was characterized by higher resistance to elastoplastic strain as evidenced by an increase in the calculated parameters R_e by factor 1.22–1.52, H_{IT}/E^* by factor 1.2–1.5, and H^3_{IT}/E^{*2} by the factor 2.1–2.7 (Table 5). The plasticity index δ_A did not change at the increased hardness and remained relatively high in the range of 0.972–0.982.

Table 4. Results of microindentation to the surface of Al and Al-Al$_2$O$_3$ samples at different maximum indentor loads P.

	P, mN	h$_{max}$, μm	h$_p$, μm	HM, MPa	H$_{IT}$, MPa	E*, GPa	W$_t$, μJ	W$_e$, μJ
Al in	1000	11.20 ± 0.28	11.03 ± 0.85	310.7 ± 2.8	332.6 ± 0.3	91.1 ± 2.7	3.9 ± 0.2	0.1 ± 0.005
Al in	2000	16.05 ± 0.35	15.83 ± 0.91	301.7 ± 1.4	323.2 ± 4.5	97.4 ± 3.1	11.0 ± 0.6	0.2 ± 0.005
Al-10% Al$_2$O$_3$	1000	10.25 ± 0.7	10.07 ± 0.7	372.8 ± 3.0	398.4 ± 3.2	91.7 ± 5.5	3.7 ± 0.4	0.1 ± 0.005
Al-10% Al$_2$O$_3$	2000	14.48 ± 0.5	14.22 ± 0.6	374.2 ± 4.4	399.9 ± 7.4	92.7 ± 3.1	9.7 ± 0.6	0.3 ± 0.005
Al-14% Al$_2$O$_3$	1000	10.18 ± 0.04	9.95 ± 0.003	381.7 ± 1.1	407.0 ± 7.7	74.8 ± 1.1	3.5 ± 0.1	0.1 ± 0.005
Al-14% Al$_2$O$_3$	2000	15.07 ± 0.24	14.85 ± 0.24	343.7 ± 1.0	367.7 ± 1.1	94.5 ± 2.2	10.6 ± 0.1	0.3 ± 0.005

Table 5. Elastic recovery R$_e$, ratios H$_{IT}$/E* and H$_{IT}^3$/E*^{2}, and plasticity δ$_A$ at different maximum indentor loads P for the surface of Al and Al-Al$_2$O$_3$ samples.

	P, mN	R$_e$, %	H$_{IT}$/E*	H$_{IT}^3$/E*^{2}	δ$_A$
Al in	1000	1.47	0.00365	0.00000443	0.974
Al in	2000	1.34	0.00332	0.00000356	0.982
Al-10% Al$_2$O$_3$	1000	1.80	0.00434	0.00000752	0.973
Al-10% Al$_2$O$_3$	2000	1.80	0.00434	0.00000744	0.969
Al-14% Al$_2$O$_3$	1000	2.24	0.00544	0.0000120	0.972
Al-14% Al$_2$O$_3$	2000	1.58	0.00389	0.00000557	0.972

Mechanical tensile tests of aluminum–aluminum oxide composite samples were carried out in order to determine the strength and plastic properties. The tests were carried out in accordance with RF GOST 1497-84 on IV type samples (Figure 8).

Figure 8. Typical Al-Al$_2$O$_3$ sample for stress–strain tests.

Typical room temperature stress–strain curves of the aluminum–alumina composites reinforced with varying alumina content are shown in Figure 9.

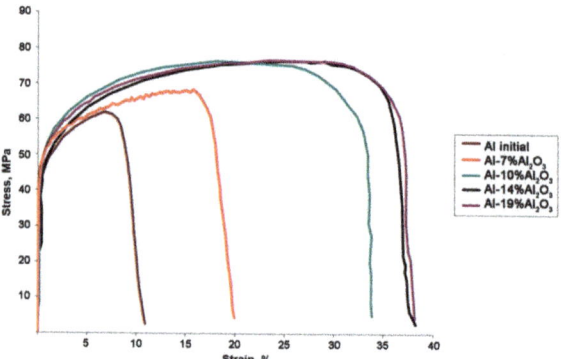

Figure 9. Typical stress–strain curves of the aluminum–alumina composites reinforced with varying alumina content.

It is evident that the yield strength did not increase with the increase of nano-α-alumina content up to 14 wt.% in the composite (Figure 10). This is in opposition to conclusions drawn in [43] that the increase in tensile strength was observed only up to α-Al_2O_3 concentrations of about 5 wt.% and the tensile strength decreased at other concentrations. The composites with an addition of 7–14 wt.% of α-Al_2O_3 possessed tensile strengths of 67.56 and 78.16 MPa, which were 1.10 and 1.27 times greater than that of the aluminum matrix sample (61.38 MPa). As can be seen from Figure 8, a further increase in the content of aluminum oxide in the composite to 19 wt.% neither improved, nor worsened, the mechanical properties of the composite. Thus, it can be considered that the aluminum oxide concentration of 14 wt.% was optimal from the point of view of the combination of all properties.

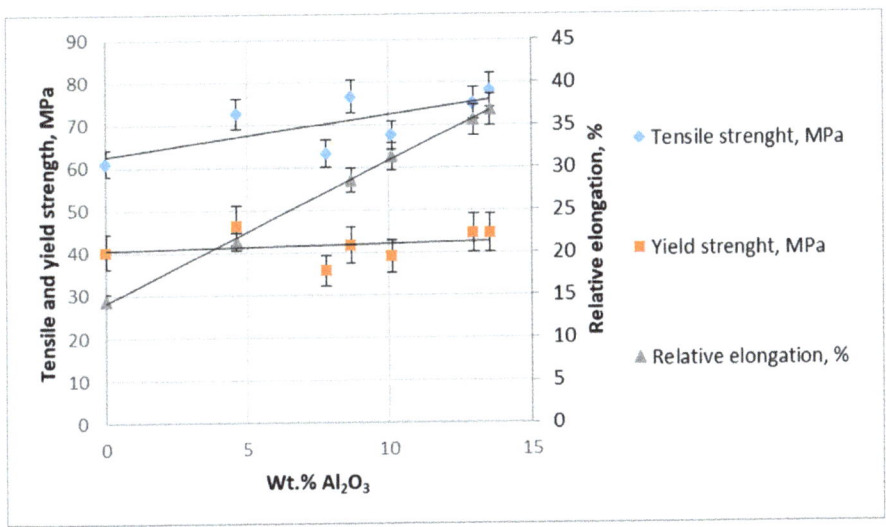

Figure 10. Dependence of tensile and yield strengths of AMMCs on the alumina concentration.

A great simultaneous increase in the sample elongation accompanied this strong improvement of tensile strength. At the same time, significant hardening was accompanied by the elongation of the samples with the addition of 7–14 wt.% of α-Al_2O_3, which increased from 1.64 to 3.06 times with the greater addition of alumina and reached values of 21.32% and 39.77%, compared with 13% for pure aluminum. This means that the hardness, elasticity, and ductility of the aluminum-nano-α-Al_2O_3 composite improved simultaneously according to the oxide concentration with uniformly distributed α-Al_2O_3 nanoparticles inside the aluminum grains. Commonly, improving the hardness of materials leads to brittle deformation, however, the addition of up to 14 wt.% of nano-α-alumina into the aluminum matrix allows increasing simultaneously the strength, hardness, ductility, and elasticity of the resultant aluminum-nano-α-Al_2O_3 composite.

Figure 11 shows SEM images of the tensile fracture surfaces of Al-Al_2O_3 nanocomposites. The fracture mode of the initial matrix Al is predominantly not ductile, with several dimples all over the surface. As shown in Figure 11b, 14 wt.% nano Al_2O_3-reinforced composite also displayed dimples similar to the Al matrix, except that the length and depth scale of the dimples was very different. No cracks on the fracture image of Al- nano-α Al_2O_3 nanocomposite were observed at tensile strength tests.

When the Al_2O_3 content in the composite was increased, the number and depth of the dimples decreases significantly, and the dimples' size increased at least three times, leading to a very ductile type of failure. This three-fold increase in the length of the cracks was in excellent agreement with the three-times-increased elongation of AMMC under room temperature stress–strain tests. As seen in Figure 11b, nanoparticles of α-alumina did not agglomerate in the aluminum matrix and were uniformly distributed even after tensile strength tests. Cracks or porosity were not observed because of the very uniform distribution of α-Al_2O_3 nanoparticles in the aluminum matrix.

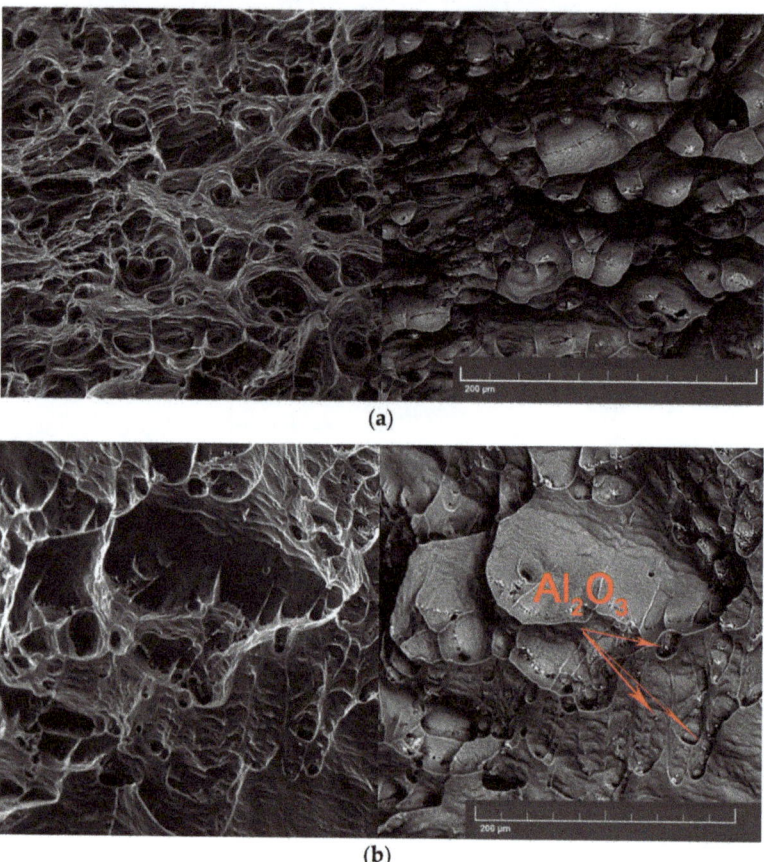

Figure 11. Fracture morphologies of Al (**a**) and AMMC with 14 wt.% of Al_2O_3 (**b**) after room temperature stress–strain tests.

3.5. Strengthening Mechanism of Al-Al_2O_3 Nanocomposites

The strengthening mechanism of Al-Al_2O_3 nanocomposites can be described by the strengthening of fine nanoparticles. Fine Al_2O_3 particles can act throughout the matrix as barriers to the dislocations and thus the elevated mechanical properties may significantly improve the hardness and ductile properties of the composites compared to the base alloy. The contributions of different strengthening mechanisms were estimated according to well-known formulas [43] (Table 6).

The ultra-hard α-Al_2O_3 nanoparticles in the Al matrix acted as barriers to the motion of dislocations generated in the matrix. The higher particle density caused the Orowan strengthening, which is the main influence according to Table 6. It is obvious that the experimental values were much closer to the sum of the load transfer ($\Delta\sigma_{load}$) and Orowan strengthening ($\Delta\sigma_{Orowan}$). The yield strength of the composites increased mostly because of the Orowan strengthening.

Table 6. Contributions of different strengthening mechanisms to the yield stress of Al- nano-Al$_2$O$_3$ composites.

Al$_2$O$_3$, wt.%	$\Delta\sigma_{load}$, MPa	$\Delta\sigma_{Orowan}$, MPa	$\Delta\sigma_{CTE}$, MPa	$\Delta\sigma_{theor}$, MPa	$\Delta\sigma_{load} + \Delta\sigma_{Orowan}$, MPa	$\Delta\sigma_{exp}$, MPa
7.0	0.9763	35.7268	4.4312	41.1343	36.70314	35.76
8.7	1.1948	36.4985	5.76531	43.4585	37.69324	41.69
10.1	1.1908	37.4905	3.4322	42.1135	38.68129	39.08
13.0	2.0617	44.6348	4.6656	51.3621	46.69649	44.58
14.0	2.1555	45.4353	5.6504	53.2413	47.59082	44.56

However, an increased addition of α-Al$_2$O$_3$ nanoparticles up to 14 wt.% did not lead to the agglomeration of the reinforcement, and we did not observe any subsequent degradation of the hardness values as opposed to Srivastana et al. and Su et al. [1,7]. Further increase in the concentration of highly dispersed α-Al$_2$O$_3$ nanoparticles up to 19 wt.% did not result in any hardness improvements as compared to the composite containing 14 wt.% of α-Al$_2$O$_3$, but we did not observe any deterioration in the hardness. The concentrations above 19 wt.% Al$_2$O$_3$ in the composite were not produced by our method. The strengthening from fine nanoparticles can describe the strengthening mechanism of Al-Al$_2$O$_3$ nanocomposites. Fine Al$_2$O$_3$ particles spread throughout the matrix can act as barriers to the dislocations, and thus mechanical properties of composites (hardness and ductile properties) are improved as compared to the base alloy.

The increase in hardness and strength (approximately 1.2–1.3 times) of the new composite material compared to the original aluminum is explained by the inclusion of α-Al$_2$O$_3$ particles into the aluminum matrix. The α-Al$_2$O$_3$ particles are in their strongest allotropic corundum modification, which has a hardness close to that of a diamond. An unusual increase in the relative elongation of the composite by a factor of three makes it possible to roll up to 15 μm thick foil on conventional rolling equipment with a deformation of about 500%, as well as to draw 0.2 mm thick wire from Al-Al$_2$O$_3$ nanocomposites (Figure 12).

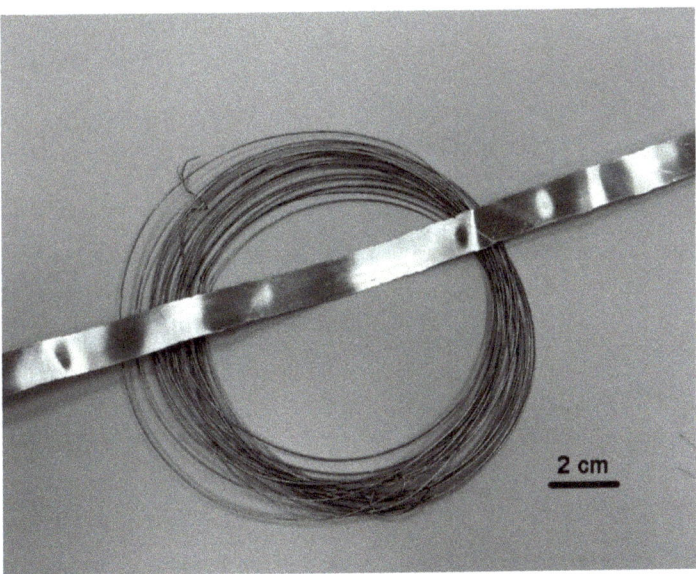

Figure 12. Optic images of 15-μm-thick foil and 0.6-mm-diameter wire of Al-Al$_2$O$_3$ nanocomposite.

4. Conclusions

Nanocomposite Al-α-nano-Al$_2$O$_3$ materials were synthesized via the chemical interaction between a salt melt containing rutile TiO$_2$ nanopowder and molten aluminum under the layer of molten halides in air atmosphere at the temperatures of 700–800 °C. Aluminum oxide nanoparticles were wholly wetted by aluminum and uniformly distributed over the "in situ" metal volume. The concentration of 100 nm Al$_2$O$_3$ particles formed inside the aluminum matrix depended entirely on the exposure time and concentration of titanium dioxide in the melt. The mass fraction of aluminum oxide in the composite may have been increased up to 19 wt.%.

Melting temperatures of composites with 4–19 wt.% of nano-α- Al$_2$O$_3$ did not differ from the aluminum melting point because of the uniform distribution of nanometer alumina inclusions in the composites.

It was revealed that the tensile strength, yield strength, hardness, and ductility of cast Al-α-nano-Al$_2$O$_3$ composites in uniaxial tension increased up to 20–30% in the presence of α-Al$_2$O$_3$ nanoparticles in the structure of aluminum. The tensile strength reached maximum for the composites with 14 wt.% of α-nano-Al$_2$O$_3$, and it was the same for those with 19 wt.% of α-nano-Al$_2$O$_3$. This remarkable improvement of tensile strength was accompanied by a significant simultaneous increase in the elongation of nanocomposite by three times as opposed to the initial aluminum. The reason for such a combination of mechanical properties was the formation of very small nanoparticles of α-Al$_2$O$_3$ that were fully wetted and uniformly distributed in the aluminum matrix.

Composite materials of the Al-Al$_2$O$_3$ system could be easily rolled into thin and ductile foils and wires.

Thus, the formed metal composite materials of the Al-Al$_2$O$_3$ system combined the advantages of a low-melting metal matrix and the properties of a ceramic dispersive filler α-Al$_2$O$_3$, which makes this system an extremely attractive structural material, that is very light, strong, hard, and elastic.

A possible mechanism for the simultaneous increase in mechanical properties and elongation in tension when true aluminum oxide nanoparticles are introduced into the Al-Al$_2$O$_3$ composite will be the subject of our further study.

Author Contributions: Data curation, L.A.Y.; Formal analysis, L.A.Y.; Investigation, A.G.K., D.I.V. and E.O.S.; Methodology, L.A.Y.; Writing—original draft, L.A.Y. All authors have read and agreed to the published version of the manuscript.

Funding: The work was funding by the Research Program № 122020100210-9 (IHTE UB RAS), Russian Academy of Sciences, Ural Branch, Russia.

Institutional Review Board Statement: Not applicable.

Informed Consent Statement: Not applicable.

Data Availability Statement: Not applicable.

Acknowledgments: This work was carried out using equipment at the Shared Access Center for the Composition of Compounds, Institute of High-Temperature Electrochemistry, Ural Branch, RAS (Ekaterinburg, Russia) for analytical support and the Shared Access Centre "Plastometry", Institute of Engineering Science, Ural Branch, RAS (Ekaterinburg, Russia) for mechanical testing.

Conflicts of Interest: The authors declare that they have no known competing financial interest or personal relationships that could have appeared to influence the work reported in this paper.

References

1. Srivastava, N.; Chaudhari, G.P. Strengthening in Al alloy nanocomposites fabricated by ultrasound assisted solidification technique. *Mater. Sci. Eng. A* **2016**, *651*, 241–247. [CrossRef]
2. Yolshina, L.A.; Muradymov, R.V.; Vichuzhanin, D.I.; Smirnova, E.O. Enhancement of the mechanical properties of aluminum-graphene composites. *AIP Conf. Proc.* **2016**, *1785*, 040093. [CrossRef]
3. Gaitonde, V.N.; Karnik, S.R.; Jayaprakash, M.S. Some Studies on Wear and Corrosion Properties of Al5083/Al$_2$O$_3$/Graphite Hybrid Composites. *J. Miner. Mater. Charact. Eng.* **2012**, *11*, 695–703.

4. Schultz, B.F.; Ferguson, J.B.; Rohatgi, P.K. Microstructure and hardness of Al_2O_3 nanoparticle reinforced Al–Mg composites fabricated by reactive wetting and stir mixing. *Mater. Sci. Eng. A* **2011**, *530*, 87–97. [CrossRef]
5. Chikova, O.A.; Finkelstein, A.B.; Schaefer, A. Microstructures, mechanical properties ingot AlSi7Fe1 after blowing oxygen through melt. *Acta Metall. Slovaca. B* **2017**, *23*, 4–11.
6. Karnesky, R.A.; Meng, L.; Dunand, D.C. Strengthening mechanisms in aluminum containing coherent Al_3Sc precipitates and incoherent Al_2O_3 dispersoids. *Acta Mater.* **2007**, *55*, 1299–1308. [CrossRef]
7. Su, H.; Gao, W.; Feng, Z.; Lu, Z. Processing, microstructure and tensile properties of nano-sized Al_2O_3 particle reinforced aluminum matrix composites. *Mater. Des.* **2012**, *36*, 590–596. [CrossRef]
8. Luan, B.F.; Hansen, N.; Godfrey, A.; Wu, G.H.; Liu, Q. High strength $Al–Al_2O_3p$ composites: Optimization of extrusion parameters. *Mater. Des.* **2011**, *32*, 3810–3817. [CrossRef]
9. Hamid, A.A.; Ghosh, P.K.; Jain, S.C.; Ray, S. The influence of porosity and particles content on dry sliding wear of cast in situ $Al(Ti)–Al_2O_3(TiO_2)$ composite. *Wear* **2008**, *265*, 14–26. [CrossRef]
10. Shafiei-Zarghani, A.; Kashani-Bozorg, S.F.; Zarei-Hanzaki, A. Microstructures and mechanical properties of Al/Al_2O_3 surface nano-composite layer produced by friction stir processing. *Mater. Sci. Eng. A* **2009**, *500*, 84–91. [CrossRef]
11. Kok, M.; Ozdin, K. Wear resistance of aluminium alloy and its composites reinforced by Al_2O_3 particles. *J. Mater. Processing Technol.* **2007**, *183*, 301–309. [CrossRef]
12. Azarniya, A.; Hosseini, H.R.M.; Jafari, M.; Bagheri, N. Thermal decomposition of nanostructured Aluminum Titanate in an active Al matrix: A novel approach to fabrication of in situ $Al/Al_2O_3–Al_3Ti$ composites. *Mater. Des.* **2015**, *88*, 932–941. [CrossRef]
13. Tong, X.C.; Fang, H.S. Al-TiC Composites In Situ-Processed by Ingot Metallurgy and Rapid Solidification Technology: Part I. Microstructural Evolution. *Metall. Mater. Trans. A* **1998**, *29*, 875–891. [CrossRef]
14. Yolshina, L.A.; Muradymov, R.V.; Korsun, I.V.; Yakovlev, G.A.; Smirnov, S.V. Novel aluminum-graphene and aluminum-graphite metallic composite materials: Synthesis and properties. *J. Alloys Compd.* **2016**, *663*, 449–459. [CrossRef]
15. Yadav, M.; Kumaraswamidhas, L.A.; Singh, S.K. Investigation of solid particle erosion behavior of $Al-Al_2O_3$ and $Al-ZrO_2$ metal matrix composites fabricated through powder metallurgy technique. *Tribol. Int.* **2022**, *172*, 107636. [CrossRef]
16. Zhang, Z.; Fan, G.; Tan, Z.; Zhao, H.; Xu, Y.; Xiong, D.; Li, Z. Towards the strength-ductility synergy of Al_2O_3/Al composite through the design of roughened interface. *Compos. Part B* **2021**, *224*, 109251. [CrossRef]
17. Liu, J.; Cao, G.; Zhu, X.; Zhao, K.; An, L. Optimization of the microstructure and mechanical properties of heterogeneous $Al-Al_2O_3$ nanocomposites. *Mater. Today Commun.* **2020**, *25*, 101199. [CrossRef]
18. Han, Q.; Setchi, R.; Evans, S.L. Synthesis and characterisation of advanced ball-milled $Al-Al_2O_3$ nanocomposites for selective laser melting. *Powder Technol.* **2016**, *297*, 183–192. [CrossRef]
19. Liu, J.; Huang, X.; Zhao, K.; Zhu, Z.; Zhu, X.; An, L. Effect of reinforcement particle size on quasistatic and dynamic mechanical properties of $Al-Al_2O_3$ composites. *J. Alloys Compd.* **2019**, *797*, 1367–1371. [CrossRef]
20. Vogel, T.; Ma, S.; Liu, Y.; Guo, Q.; Zhang, D. Impact of alumina content and morphology on the mechanical properties of bulk nanolaminated Al_2O_3-Al composites. *Compos. Commun.* **2020**, *22*, 100462. [CrossRef]
21. Liu, W.; Ke, Y.; Sugio, K.; Liu, X.; Guo, Y.; Sasaki, G. Microstructure and mechanical properties of Al_2O_3-particle-reinforced Al-matrix composite sheets produced by accumulative roll bonding (ARB). *Mater. Sci. Eng. A* **2022**, *850*, 143574. [CrossRef]
22. Sadeghi, B.; Cavaliere, P.; Balog, M.; Pruncu, C.I.; Shabani, A. Microstructure dependent dislocation density evolution in micro-macro rolled Al_2O_3/Al laminated composite. *Mater. Sci. Eng. A* **2022**, *830*, 142317. [CrossRef]
23. Gao, T.; Liu, L.; Liu, G.; Liu, S.; Li, C.; Li, M.; Zhao, K.; Han, M.; Wu, Y.; Liu, X. In–situ synthesis of an Al–based composite reinforced with nanometric $γ–Al_2O_3$ and submicron AlB_2 particles. *J. Alloys Compd.* **2022**, *920*, 165985. [CrossRef]
24. Tjong, S.C.; Ma, Z.Y. Microstructural and mechanical characteristics of in situ metal matrix composites. *Mater. Sci. Eng.* **2000**, *29*, 49–113. [CrossRef]
25. Zhao, Y.; Zhang, S.; Chen, G. Aluminum matrix composites reinforced by in situ Al_2O_3 and Al_3Zr particles fabricated via magnetochemistry reaction. *Trans. Nonferrous Met. Soc. China* **2010**, *20*, 2129–2133. [CrossRef]
26. Elshina, L.A.; Muradymov, R.V.; Kvashnichev, A.G.; Vichuzhanin, D.I.; Molchanova, N.G.; Pankratov, A.A. Synthesis of new metal-matrix $Al–Al_2O_3$–graphene composite materials. *Russ. Metall. (Met.)* **2017**, *2017*, 631–641. [CrossRef]
27. Elshina, L.A.; Kudyakov, V.Y.; Malkov, V.B.; Elshin, A.N. High-Temperature Electrochemical Synthesis of Oxide Thin Films and Nanopowders of Some Metal Oxides. *Glass Phys. Chem.* **2008**, *34*, 617–622. [CrossRef]
28. Elshina, L.A.; Kvashnichev, A.G.; Pelegov, D.V. Electrochemical Synthesis of Titanium Oxide Nanopowders in a Molten Mixture of Alkali Chlorides and Nitrates. *Russ. Metall. (Met.)* **2021**, *2021*, 1029–1035. [CrossRef]
29. Savrai, R.A.; Skorynina, P.A.; Makarov, A.V.; Osintseva, A.L. Effect of Liquid Carburizing at Lowered Temperature on the Micromechanical Characteristics of Metastable Austenitic Steel. *Phys. Met. Metallogr.* **2020**, *121*, 1015–1020. [CrossRef]
30. Cheng, Y.T.; Cheng, C.M. Relationships between hardness, elastic modulus and the work of indentation. *Appl. Phys. Lett.* **1998**, *73*, 614–618. [CrossRef]
31. Petrzhik, M.I.; Levashov, E.A. Modern methods for investigating functional surfaces of advanced materials by mechanical contact testing. *Crystallogr. Rep.* **2007**, *52*, 966–974. [CrossRef]
32. Mayrhofer, P.H.; Mitterer, C.; Musil, J. Structure-property relationships in single- and dual-phase nanocrystalline hard coatings. *Surf. Coat. Technol.* **2003**, *174–175*, 725–731. [CrossRef]

33. Yolshina, L.A.; Kudyakov, V.Y.; Malkov, V.B.; Molchanova, N.G. Corrosion and electrochemical behavior of aluminium treated with high-temperature pulsed plasma in CsCl-NaCl-NaNO$_3$ melt. *Corros. Sci.* **2011**, *53*, 2015–2026. [CrossRef]
34. Yolshina, L.A.; Malkov, V.B.; Yolshin, A.N. The influence of formation conditions on the electrochemical behavior of lead oxide in sulfuric acid solution. *J. Power Sources* **2009**, *191*, 36–41. [CrossRef]
35. Ijadpanah-Saravy, H.; Safari, M.; Khodadadi-Darban, A.; Rezaei, A. Synthesis of titanium dioxide nanoparticles for photocatalytic degradation of cyanide. *Anal. Lett.* **2014**, *47*, 1772–1782. [CrossRef]
36. Yolshina, L.A.; Kvashinchev, A.G. Chemical Interaction of Liquid Aluminum with Metal Oxides in Molten Salts. *Mater. Des.* **2016**, *105*, 124–132. [CrossRef]
37. Yolshina, L.A.; Kvashnichev, A.G.; Pelegov, D.V.; Pryakhina, V.I. Molten salt synthesis and characterization of 1D sodium hexatitanate nanowires. *Colloids Interface Sci. Commun.* **2021**, *42*, 100398. [CrossRef]
38. Elshina, L.A.; Elshina, V.A. Synthesis of a Nanocrystalline α-Al$_2$O$_3$ Powder in Molten Halides in the Temperature Range 700–800 °C. *Russ. Metall. (Met.)* **2020**, *2020*, 138–141. [CrossRef]
39. Batyshev, A.I. *Crystallization Metals and Alloys Under Pressure*; Kristallizatsia Metallov i Splavov Pod Davleniem; Metallurgy: Moscow, Russia, 1977; 151p.
40. Hossein-Zadeh, M.; Razavi, M.; Mirzaee, O.; Ghaderi, R. Characterization of properties of Al–Al$_2$O nano-composite synthesized via milling and subsequent casting. *J. King Saud Univ. Eng. Sci.* **2013**, *25*, 75–80. [CrossRef]
41. Vorozhtsov, S.A.; Zhukov, I.A.; Vorozhtsov, A.B.; Zhukov, A.S.; Eskin, D.G.; Kvetinskaya, A.V. Synthesis of Micro- and Nanoparticles of Metal Oxides and Their Application for Reinforcement of Al-Based Alloys. *Adv. Mater. Sci. Eng.* **2015**, *2015*, 718207. [CrossRef]
42. Dang, X.; Zhang, B.; Zhang, Z.; Hao, P.; Xu, Y.; Xie, Y. Microstructural evolutions and mechanical properties of multilayered 1060Al/Al-Al$_2$O$_3$ composites fabricated by cold spraying and accumulative roll bonding. *J. Mater. Res. Technol.* **2021**, *15*, 3895–3907. [CrossRef]
43. Ma, P.; Jia, Y.; Gokuldoss, P.K.; Yu, Z.; Yang, S.; Zhao, J.; Li, C. Effect of Al$_2$O$_3$ Nanoparticles as Reinforcement on the Tensile Behavior of Al-12Si Composites. *Metals* **2017**, *7*, 359. [CrossRef]

Article

Effect of Zinc Aluminum Magnesium Coating on Spot-Welding Joint Properties of HC340LAD + ZM Steel

Xiang Chen [1], Xinjie Peng [1], Xinjian Yuan [1,*], Ziliu Xiong [2], Yue Lu [2], Shenghai Lu [2] and Jian Peng [1]

1. State Key Laboratory of Mechanical Transmission, College of Materials Science and Engineering, Chongqing University, Chongqing 400044, China
2. HBIS Group Technology Research Institute, Shijiazhuang 050023, China
* Correspondence: xinjianyuan@cqu.edu.cn; Tel.: +86-136-1825-5866

Abstract: Zn-Al-Mg (zinc, aluminum and magnesium)-coated steel is gradually replacing traditional hot-dip galvanized steel due to its excellent corrosion resistance, self-healing properties and good surface hardness. However, the effect of Zn-Al-Mg coating on the resistance spot-welding joint properties of HC340LAD + ZM steel plates is not clear, and there are few systematic studies on it. In this paper, L16 (4^3) orthogonal experiments were designed on Zn-Al-Mg-coated steel HC340LAD + ZM (thickness = 1 mm). In addition, taking the tensile shear force as the main evaluation standard, the optimal spot-welding process properties could be achieved when the welding current, the welding time and the electrode loading were 10 kA, 14 cycles and 2.6 kN, respectively. On this basis, the formation mechanism, microstructure and corrosion properties of two plates of steels, with or without zinc, aluminum and magnesium coating under different welding times, were studied. The presence of Zn-Al-Mg coating slightly affected the mechanical properties of welding joints. However, the corrosion current of the body material containing Zn-Al-Mg plating was 7.17 times that of the uncoated plate.

Keywords: zinc; aluminum and magnesium coating; HC340LAD + ZM; spot welding; corrosion

Citation: Chen, X.; Peng, X.; Yuan, X.; Xiong, Z.; Lu, Y.; Lu, S.; Peng, J. Effect of Zinc Aluminum Magnesium Coating on Spot-Welding Joint Properties of HC340LAD + ZM Steel. Appl. Sci. 2022, 12, 9072. https://doi.org/10.3390/app12189072

Academic Editor: Suzdaltsev Andrey

Received: 2 August 2022
Accepted: 5 September 2022
Published: 9 September 2022

Publisher's Note: MDPI stays neutral with regard to jurisdictional claims in published maps and institutional affiliations.

Copyright: © 2022 by the authors. Licensee MDPI, Basel, Switzerland. This article is an open access article distributed under the terms and conditions of the Creative Commons Attribution (CC BY) license (https://creativecommons.org/licenses/by/4.0/).

1. Introduction

Hot-dip galvanized steel is widely used in automobiles, construction, home appliances and other industries due to its good welding performance and excellent corrosion resistance [1–3]. In addition, the corrosion resistance of hot-dip galvanized steel is mainly through the physical barrier protection of the surface coating and the electrochemical protection of zinc as an anode for steel under the action of corrosive media [4–6]. With technological development and industrial upgrading, when appropriate amounts of Mg and Al (usually 1–4 wt%) are added to zinc coating, a Zn-Al-Mg coating is formed, which can greatly improve the corrosion resistance of hot-galvanized steel [7–12]. Thus, the Zn-Al-Mg-coated steel comes into being.

Zn-Al-Mg-coated steel, as an upgrade product of hot-dip galvanized steel, not only has a much lower corrosion rate than hot-dip galvanized steel in the same environment but also has a good bending strength ratio, excellent surface hardness, low cost and other advantages. Therefore, Zn-Al-Mg-coated steel is widely used in automotive outer plates, brackets and internal components [13–15].

As we all know, resistance spot-welding has become the main welding process of automobile steel plate links due to its fast welding speed, excellent welding quality, high degree of automation and low cost [16–21]. The coating of Zn-Al-Mg-plated steel is melted first in the process of spot-welding due to the existence of plating, and the Zn element in the coating is the first to evaporate, reducing the temperature of the steel [9,15,22]. However, a part of the Al-Mg elements will remain in the weld, the formation process of the melting core is more complex than that of traditional galvanized steel.

Therefore, improving the quality of the spot-welding joint of Zn-Al-Mg-coated steel and its welding process parameters are key problems. In addition, the main factors affecting the mechanical properties of resistance spot-welding joints are welding current, welding time and electrode loading [23–25]. Welding current mainly provides heat for the welding process. When the temperature exceeds the melting temperature at the workpiece–workpiece interface, the welding core begins to produce. Not only is the size of the welding core related to the size of the welding current, but also the welding time and electrode loading have a great impact on it [26,27].

In order to explore the influence of Zn-Al-Mg coating on the resistance spot-welded joint properties of HC340LAD + ZM steel, the optimal welding process parameters of HC340LAD + ZM steel were explored through an orthogonal experiment design. Then, based on the optimal parameters, HC340LAD + ZM with Zn-Al-Mg coating and HC340LA without coating were designed as experimental welding parameters to explore the nucleation process of the spot-welding joint and the influence of the coating on the properties of the spot-welding joint and, further, to study the influence of the Zn-Al-Mg coating on the corrosion resistance of the spot-welding joint.

2. Materials and Methods

In order to study the effect of the coating on HC340LAD + ZM steel resistance spot welding, a cold-rolled, low-alloy, high-strength HC340LA steel (sheet thickness = 1.4 mm) was prepared as the experimental material. The other one was an HC340LAD + ZM steel, which was developed from HC340LA steel after adding zinc, aluminum and magnesium coating; the amount of coating was 180 g/m^2. The chemical composition and mechanical properties of the matrix are shown in Table 1.

Table 1. Chemical composition and mechanical properties of HC340LAD + ZM steel.

Chemical Composition/wt%							Mechanical Property		
C	Mn	Si	P	S	Nb	Ti	Yield Strength/MPa	Tensile Strength/MPa	Elongation/%
≤0.1	≤0.05	≤1.3	≤0.03	≤0.045	≤0.03	≤0.03	385	475	28

The matrix micromorphology and the coating cross-section morphology are also shown in Figure 1. The mechanical properties and matrix microstructure of the two kinds of steel plates are almost consistent. In addition, the matrix microstructure of both is composed of ferrite and a small amount of pearlite. The differences only lie in the presence of Zn-Al-Mg coating.

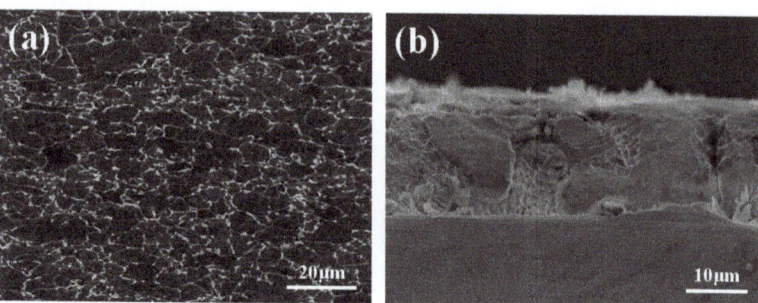

Figure 1. (a) Micromorphology of matrix and (b) coating cross-section morphology.

The two experimental steel plates were cut and processed into several 125 mm × 40 mm plate samples, and the processed samples were cleaned and dried with ethanol (99.5%) to avoid the metallurgical contamination of the spot-welding experiment by oil pollution and other factors. The cleaned samples were tested for spot-welding experiments under the lap state, as shown in Figure 2, and the tensile samples and metallographic samples were prepared. In addition, the metallographic samples were welded twice, and a spot-welding experiment was carried out on the OBARA SIV21 small original spot-welding machine.

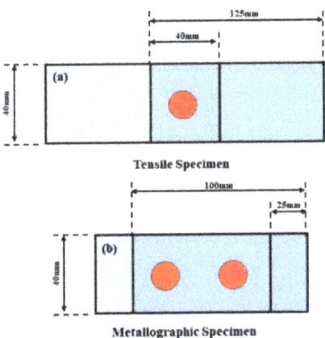

Figure 2. Size diagram of tensile sample and metallographic sample.

The tensile shear force of the spot-welding samples was measured by the SANS microcomputer servo-controlled hydraulic universal testing machine with a tensile speed of 2 mm/min. The rectangular samples (15 mm × 7.5 mm) of the metallographic and scanning samples were prepared from the spot-welding joints. The corrosion solution was 4% nitrate alcohol solution, and the corresponding corrosion time was 14 s. Then, the Leica DM 2000 microscope and TESCAN VEGA3 tungsten filament scanning electron microscope were used to observe the metallographic and scanning section of the welding joint. The micro-hardness value was measured with the HXS-1000AKY hardness meter (300 g, 10 s), and its micro-hardness distribution was measured along the horizontal and vertical direction of the welding joint.

Firstly, welding current, welding time and electrode loading were chosen as variables to design the orthogonal experiment. Then, mechanical properties and spot-welding joint parameters (pressure down rate, welding joint diameter, welding joint thickness) were measured systematically. Furthermore, the splash degree was assessed according to the vast amount of splash distance generated in the spot-welding process, and the splash size was graded in a section of 50 cm. The A/B/C/D splashes were, respectively, represented by splash distances of 150–100 cm; 100–50 cm; 50–0 cm; and no splash. The value range of welding current, welding time and electrode loading was preliminarily selected according to the thickness, strength and galvanizing layer of the plate and combined with previous tests and related documents (the current = 8–11 kA, the time = 12–18 cycles (1 cycle = 0.02 s), the electrode loading = 1.8–3.0 kN). The optimal welding parameters of HC340LAD + ZM were obtained through the orthogonal experiment, then the welding current and electrode loading of two kinds of steel, HC340LAD + ZM with Zn-Al-Mg coating and HC340LA without coating, were fixed under the optimal parameters. Taking welding time as the main variable, the nucleation process of the spot-welding joint of two kinds of steels and the effect of coating on the properties of the spot-welding joint were explored, and the change in corrosion resistance before and after welding was studied. The welding joint and the unwelded body material with the optimal welding parameters of 10 kA, 14 cycles and 2.6 kN for HC340LAD + ZM steel with Zn-Al-Mg coating and HC340LA without coating were selected, and three sets of 15 mm × 15 mm samples were prepared with wire cutting. Then, the polarization curve was tested after alcohol cleaning, with a scanning speed of 2 mV/s.

3. Experimental Results

3.1. Orthogonal Experiments

Then, the orthogonal tests were designed, and the specific test process parameters are shown in Table 2.

Table 2. Welding results of HC340LAD + ZM orthogonal test.

Number	Welding Current (kA)	Welding Time (Cycles)	Electrode Loading (kN)	Pull Shear (kN)	Splash Situation	Melting Core Diameter (mm)	Welding Point Thickness (mm)	Down Rate (%) < 30%
1	8	12	1.8	6.58	CCCD	5.73	2.70	3.57
2	8	14	2.2	9.41	CCDD	6.16	2.69	4.11
3	8	16	2.6	9.51	CDDD	6.17	2.72	3.04
4	8	18	3	7.74	CDDD	6.41	2.55	8.93
5	9	12	2.2	8.75	CCCC	6.44	2.61	6.79
6	9	14	1.8	7.47	CDCC	6.41	2.63	5.98
7	9	16	3	9.54	BBBC	6.44	2.47	11.88
8	9	18	2.6	8.74	BBBC	6.67	2.35	16.25
9	10	12	2.6	10.67	BBBD	6.44	2.66	5.18
10	10	14	3	10.21	BBBB	6.54	2.59	7.50
11	10	16	1.8	8.55	BBBC	6.67	2.45	12.41
12	10	18	2.2	9.85	BBBB	6.83	2.38	15.18
13	11	12	3	9.07	BBBB	6.72	2.38	15.18
14	11	14	2.6	9.34	BBBB	6.44	2.24	20.09
15	11	16	2.2	10.21	BBBB	6.67	2.30	17.86
16	11	18	1.8	10.55	BBBB	6.86	2.38	15.00

Note: Splash "A", "B", "C" and "D" mean "A splash", "B splash", "C splash" and "no splash", respectively.

With tensile shear as the main parameter, the mechanical properties of joints under each group of parameters were inserted into SPSS for multi-factor variance and extreme difference analysis, and the optimal process parameters were as follows: welding current, welding time and electrode loading were 10 kA, 14 cycles and 2 kN, respectively; the order of the influence of the three factors on tensile shear was current > pressure > time.

3.2. Macro- and Micro-Morphology

According to the orthogonal experiment results, the optimal welding process parameters were as follows: welding current, welding time and electrode loading were 10 kA, 14 cycles and 2 kN, respectively. Then, the two kinds of steel with coating and without coating were welded separately on the basis of the optimal welding process parameters by fixing the welding current and electrode loading and by changing the welding time to explore the dynamic nucleation process of the spot-welding process and the influence of the coating on the nucleation quality of the welding joint. The macroscopic morphology of the welds under different welding times is shown in Figure 3. With the increase in welding time, the diameter of the welds of the two plates of steel also increased. In addition, with the increase in welding time, the HC340LA steel without coating showed a better quality welding spot surface without splash, while HC340LAD + ZM steel with coating had rough joints on the welding spot surface with fluctuations. Furthermore, the splash phenomenon appeared after 10 cycles, and the center of the welding point became yellow.

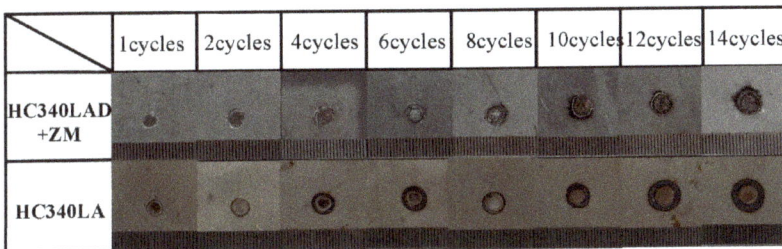

Figure 3. Macroscopic morphology under different welding times (10 kA, 2.6 kN).

Figure 4 shows the change in the welding joint's diameter, thickness, pressure down rate and tensile shear force with welding time. It can be clearly seen from Figure 4 that the joule heat and the welding joint diameter of HC340LAD + ZM and HC340LA increased with the increase in welding time. When the welding time was one cycle, the welding joint diameter of HC340LA without coating was 0.83 mm larger than that of the HC340LAD + ZM steel with coating. With the increase in welding time, the gap in the welding joint diameter became smaller. When the welding time was 10 cycles, the welding joint diameter was almost the same, and then the increase in the welding joint diameter tended to be flat, while the thickness of the welding points was the opposite, and the thickness of both plates of steel showed a downward trend. When the welding time exceeded 14 cycles, the welding point thickness of the two plates of steel tended to coincide. The welding joint thickness of the 16 cycles was 2.56 mm and 2.5 mm, respectively. The corresponding downpressure rate was 14.88% and 10.71%.

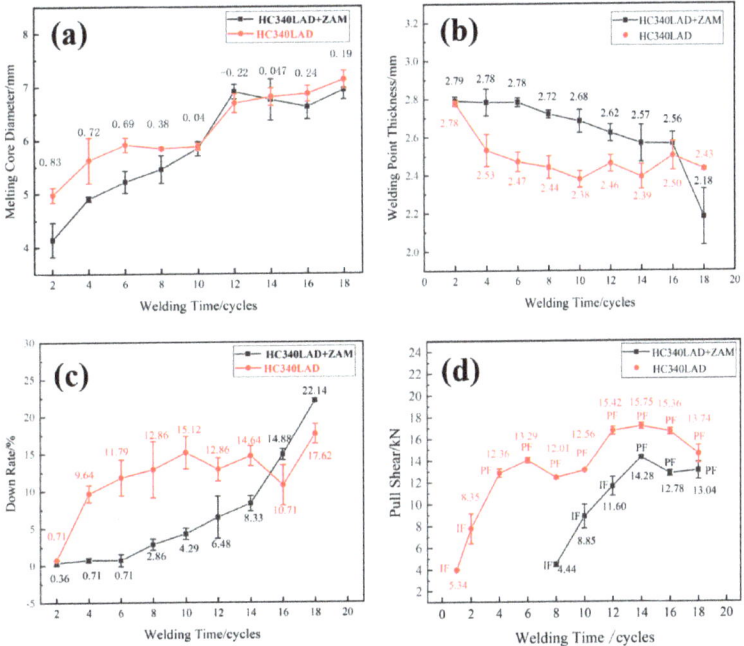

Figure 4. The change in (**a**) welding diameter, (**b**) thickness, (**c**) down−rate and (**d**) tensile shear with welding time.

Figure 4d shows the tensile shear force of HC340LA steel without coating and HC340LAD + ZM steel with Zn-Al-Mg coating. As the welding time increased, the tensile shear force rose first and then decreased, reaching its maximum after 14 cycles with the tensile shear forces of 14.28 kN and 15.75 kN. However, the tensile shear force of HC340LA steel without coating was higher than that of HC340LAD + ZM steel with Zn-Al-Mg coating under the same welding parameters. When the welding time of HC340LA without coating was one cycle, the welding core had already appeared, and the tensile shear force at this time was 5.34 kN. HC340LAD + ZM with coating did not nucleate until the welding time was eight cycles, and the corresponding tensile shear force was 4.44 kN. Combined with the diameter and the thickness of the welding point, the existence of the Zn-Al-Mg coating reduced the weldability. Although the diameter of the welding point was large enough after six cycles, an effective welding core was still not formed at this time. Because of the existence of Zn-Al-Mg coating, the contact resistance between the two plates was smaller, and there was less heat input from the welding center area, thus failing to effectively form the welding core. Therefore, with the increase in the welding time, the impact of the coating on the welding time became smaller, and HC340LAD + ZM with coating did not nucleate until the welding time was eight cycles. Meanwhile, the zinc with a low melting point evaporated and gasified the heat during welding, which reduced the effective heat input generated during welding, resulting in more difficult nucleation.

Figures 5 and 6 show the cross-section morphologies of the HC340LA without coating and the HC340LAD + ZM with coating at different welding times. Figure 5 shows the cross-section morphology of HC340LA steel without coating when the welding time was 1~18 cycles. As shown in the figure, the uncoated HC340LA steel was very sensitive to the change in welding time, and the width of the molten core gradually increased with the increase in welding time, and the thickness of the welding point also decreased. When the welding time was one cycle, there was no complete nucleation, but a union appeared between the plates. When the welding time was two cycles, the nucleation began to obviously appear, forming a small welding core area. After four cycles, nucleation completed, and the columnar crystal appeared obviously perpendicular to the plastic ring. At this time, a shrinkage hole defect appeared. Because there was so little molten metal, it shrank as it cooled, and holes formed. As the welding time increased, the height of the melting core decreased, and the width of the melting core also gradually increased, which was basically unchanged after reaching 14 cycles. Figure 6 shows the cross-section morphology of the HC340LAD + ZM steel with coating when the welding time was 1~18 cycles. Due to the existence of the coating, the welding of the coated HC340LAD + ZM steel requires a greater welding heat input. On the one hand, the existence of the coating made the contact resistance between the plates smaller and made the heat input generated during welding lower, resulting in difficult nucleation. On the other hand, because of the low melting point of Zn, the main heat generated was first absorbed by the coating, resulting in less heat input obtained by the parent material, Fe, which could not effectively nucleate. Thence, nucleation appeared initially when the welding time was eight cycles; when the welding time reached 10 cycles, the welded joint nucleated completely.

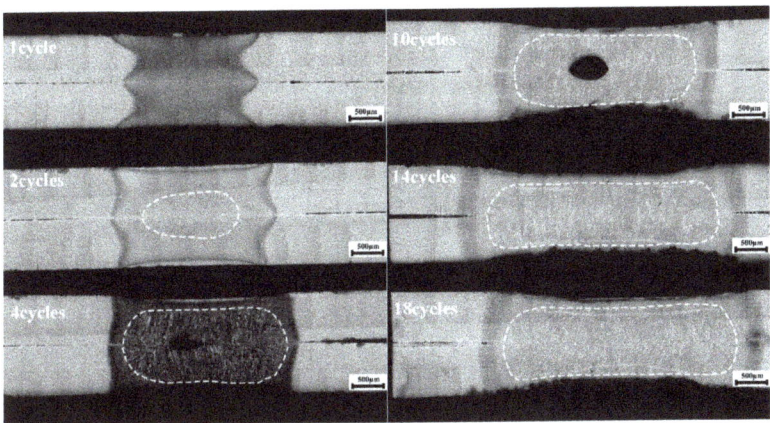

Figure 5. Morphology of HC340LA joint at different welding times (10 kA, 2.6 kN).

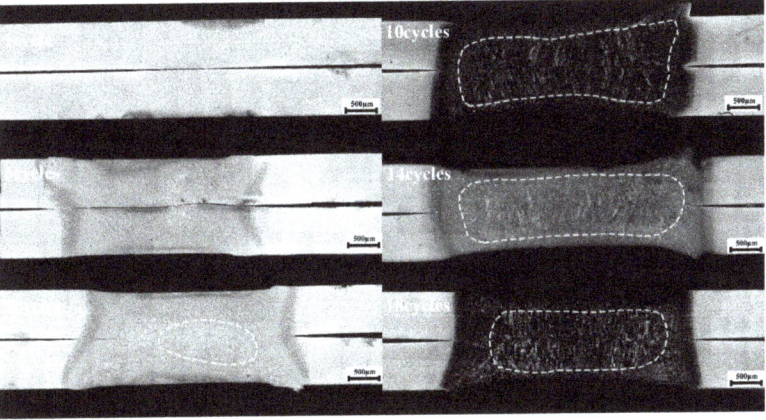

Figure 6. Morphology of HC340LAD + ZM joint at different welding times (10 kA, 2.6 kN).

3.3. Fracture Form and Hardness Distribution

The failure modes of the welding point fracture of HC340LAD + ZM with Zn-Al-Mg coating and HC340LA without coating are mainly divided into two types: interface fracture (IF = interfacial failure) and pull-out fracture (PF = plug failure). Figure 7a,b reveal the morphologies of IF macroscopic tensile shear fracture and IF fracture, respectively. The macroscopic fracture shows that the fracture occurred at the two interfaces of the welding point, while the morphology of the microscopic fracture is an obvious cleavage step. Figure 7c,d show the morphologies of PF macroscopic tensile shear fracture and PF fracture, respectively. The macroscopic fracture illustrates that pull-out tear occurred in the parent material zone around the welding joint; the welding point was well fused, and no fracture occurred, while the morphology of the microscopic fracture had countless axial dimples.

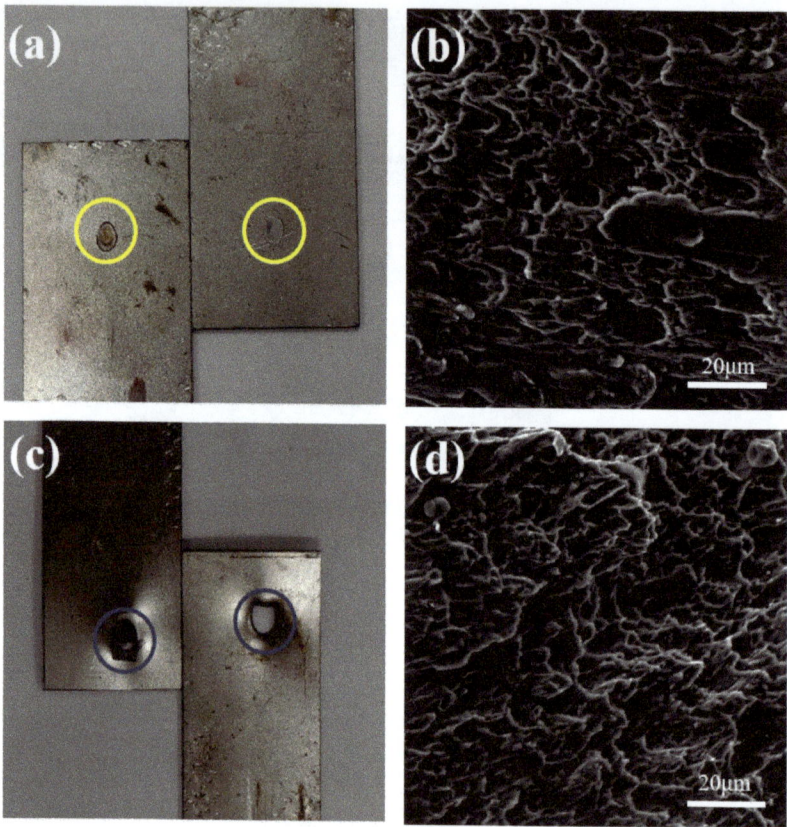

Figure 7. Macroscopic and microscopic morphology of typical failure modes: (**a**) IF, (**b**) SEM of IF, (**c**) PF and (**d**) SEM of PF.

According to Figure 4d, due to the existence of coating, the HC340LAD + ZM steel did not nucleate at a low welding time. Nucleation completed after eight cycles, and the tensile shear force was 4.44 kN. In addition, the quality of the welding joint was poor, and the welding core was not firmly bonded. In this context, an IF-type interface fracture of the welding joint occurred. After 14 cycles, the quality of the welding joint nucleation was better, and the tensile shear force was 14.28 kN. At this time, a PF-type pull-out fracture occurred. The HC340LA steel without coating had good malleability due to the existence of no coating. After one cycle, although the plates did not melt in the process of chemical metallurgy at this time, the two plates combined under the action of electrode loading. At this time, the tensile shear force was 5.34 kN, and an IF-type interface fracture occurred. When the welding time was four cycles, a welding point with good quality formed, and a PF-type pull-out fracture occurred with a pull shear force of 12.36 kN. This was because, when the welding time was short, the two plates in the welding core area were not closely combined, and the shear force between the welding points was less than the tensile strength of the parent material. At this time, the fracture position occurred in the welding core. When the welding time was long enough, the welding heat was enough; the nucleation was complete, and the shear force of the welding core area was greater than the tensile strength of the parent material, thus, the pull-out fracture occurs. Under the same welding parameters, the tensile shear force of the welding joint with a Zn-Al-Mg coating was smaller than that of the welding joint without coating, which may owe to the existence of the coating, which affected the temperature and thermal cycle of the welding joint, resulting in the low tensile shear force.

Figures 8 and 9 show the hardness value distribution of the welding point under different welding times for HC340LA without coating and HC340LAD + ZM with coating. As shown in the figure, the hardness value distribution of the welding points of both steel results in a "π", high on both sides and low in the middle.

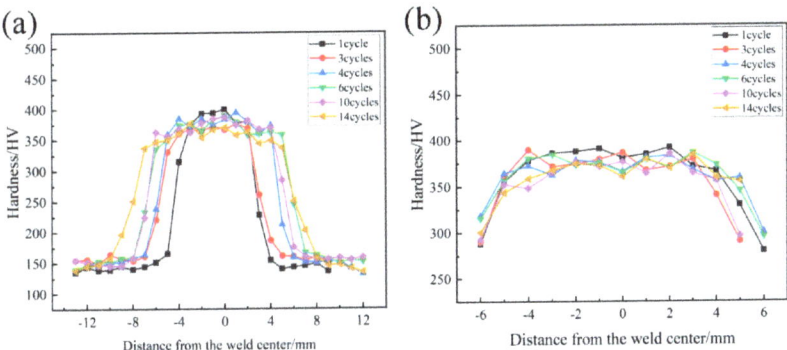

Figure 8. HC340LA nugget test hardness changes at different welding times (**a**) in a horizontal direction and (**b**) a vertical direction.

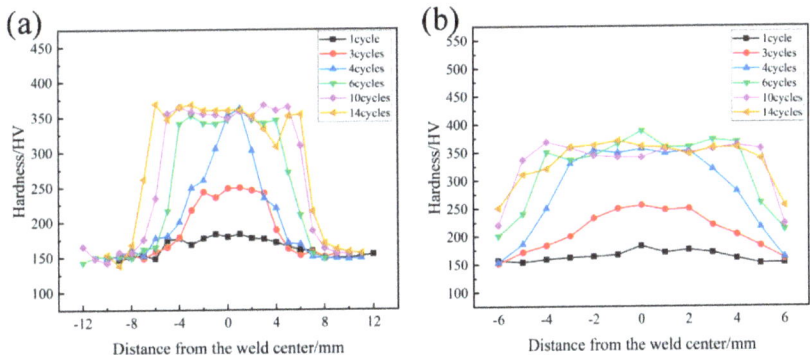

Figure 9. HC340LAD + ZM nugget test hardness changes at different welding times (**a**) in a horizontal direction and (**b**) a vertical direction.

As can be seen in Figure 8, nucleation occurred after one cycle, and even the obvious welding core zone, heat-affected zone and welding core zone appeared. The hardness value of the parent material was 150 HV; the average hardness of the welding core zone was 375 HV, and the hardness value of the heat-affected zone was between the two. In the horizontal direction, it is obvious that the welding core zone widened as the welding time increased. In the vertical direction, it was distributed high on both sides and low in the middle, and the hardness value near the electrode head was 280~300 HV.

As shown in Figure 9, after one cycle and three cycles, the peak temperature of the welding point did not exceed the A_3 line, and the tissue of the welding core zone was mixed with martensite and ferrite. At this time, the hardness value of the central position was low (180 HV). After four cycles, the temperature of some areas near the welding core area exceeded the A_1 line, and a small amount of martensite formed. In this condition, the hardness value of the welding core zone reached 250 HV. After six cycles, on the one hand, due to the influence of the coating, there was no effective connection between the plates, but at this time, the welding core zone was completely austenitic, and a large number of martensite was obtained. In addition, the hardness value rose. On the other hand, due to the existence of the coating, the mechanical properties of the welding joints deteriorated, and the average hardness value of the welding core zone was 355 HV, which was lower than that of the welding joints of HC340LA without coating. In the vertical direction, the hardness value near the two sides of the electrode cap was 240 HV, which was lower than that of the welding joint of the HC340LA without coating (280 HV). This is because, during welding, the surface coating evaporated, splashed and took away the heat; thus, the surface area obtained less heat, eventually forming less martensite, and it had a lower hardness value.

3.4. Corrosion Behaviour Study

The results of the polarization curves for each zone are shown in Figure 10 and Table 2.

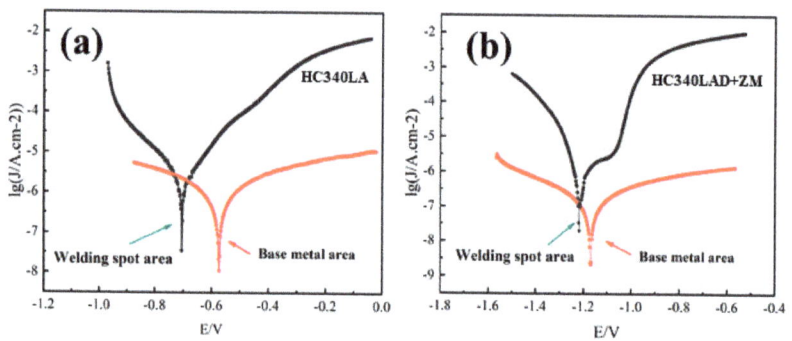

Figure 10. HC340LA (**a**) and Zn−Al−Mg−coated HC340LAD + ZM (**b**) polarization curves.

According to Figure 10 and Table 3, compared with the parent material and the welding point polarization curve of the same material, the corrosion potential of the welding point was lower than that of the corresponding parent material, which illustrates that the corrosion tendency of the welding point was greater than that of the body material, and the change in corrosion current was also consistent. The corrosion currents of the body material and the welding joint of the HC340LA steel without coating were 4.258×10^{-4} mA·cm^{-2} and 3.602×10^{-3} mA·cm^{-2}. The corrosion current decreased by 8.46 times, while the corrosion currents of the parent material and the welding joint of the HC340LAD + ZM with Zn-Al-Mg coating were 5.934×10^{-5} mA·cm^{-2} and 1.323×10^{-4} mA·cm^{-2}. The corrosion current decreased by 2.23 times. This is because the welding heat input increased in the welding process, which led to tissue transformation and the loss of the surface coating of the welding point during the welding process, thus leading to the deterioration of corrosion resistance.

Table 3. Electrochemical parameters of HC340LAD + ZM and HC340LA polarization curves.

Sample	Region	E_{corr}/V	I_{corr}/(mA·cm^{-2})
HC340LA	Base metal	−0.572	4.258×10^{-4}
	Soldered dot	−0.705	3.602×10^{-3}
HC340LAD + ZM	Base metal	−1.167	5.934×10^{-5}
	Soldered dot	−1.218	1.323×10^{-4}

The corrosion potentials of the HC340LAD + ZM base material and the welding joint containing zinc aluminum magnesium coating were (−1.167 V) and (−1.218 V), respectively; the corrosion potentials of the base material and the welding joint of HC340LA without coating were (−0.5721 V) and (−0.705 V), respectively. According to the longitudinal comparison, the HC340LAD + ZM had a lower corrosion potential at both the base material and the welding joint, so its corrosion tendency was greater. The corrosion current was just the opposite. The corrosion currents of the parent material of the zinc–aluminum–magnesium-coated plate and the uncoated plate were 5.934×10^{-5} mA·cm^{-2} and 4.258×10^{-4} mA·cm^{-2}, respectively. The corrosion current of the zinc–aluminum–magnesium plate was 7.17 times that of the uncoated plate. The corrosion currents of the welding joint of the zinc–aluminum–magnesium-coated plate and the uncoated plate were 1.323×10^{-4} mA·cm^{-2} and 3.602×10^{-3} mA·cm^{-2}. The gap in the corrosion current was 27 times greater. The reason why the corrosion potential of the Zn-Al-Mg-coated plate was higher than that of the uncoated plate and the corrosion current was smaller is that the corrosion of the Zn-Al-Mg-coated plate was the first to develop surface corrosion. The coating was mainly zinc, whose corrosion potential is higher (-1 V), and the uncoated plate was the corrosive matrix ferrite (-0.4 V). Due to the presence of primary crystal phase Zn; Al; MgZn$_2$; binary co-crystal tissue Zn/MgZn$_2$; and ternary co-crystal tissue Zn/MgZn$_2$/Al in the zinc–aluminum–magnesium coating—in which MgZn$_2$ and Mg$_2$Zn$_{11}$ had higher microhardness, better compactness and better single-phase corrosion resistance—the zinc and aluminum magnesium plates had higher corrosion potential, but a slower corrosion rate [28–30].

4. Conclusions

(1) The orthogonal tests showed that the optimal parameters of HC340LAD + ZM steel resistance spot-welding are as follows: welding current, welding time and electrode loading were 10 kA, 14 cycles, and 2.6 kN, respectively. The influence of welding current, welding time and electrode loading on the tensile shear force can be successively ranked as current > pressure > time.

(2) When the fixed welding current (10 kA) and the electrode loading (2.6 kN) increased, the welding point diameter of both steel plates also increased as the welding time increased. Under the same welding parameters, the mechanical properties of the spot-welding joint of the HC340LA steel without coating were slightly better than those of the HC340LAD + ZM with coating. The HC340LA steel without coating began to nucleate after one cycle, and the tensile shear force was 5.34 kN. The nucleation completed after four cycles, and the tensile shear force reached 12.36 kN. The HC340LAD + ZM with coating started to nucleate after eight cycles, and the tensile shear force was 4.44 kN.

(3) In the nucleation test, the tensile shear force of the welding joint of the HC340LA steel without coating was greater than that of the welding joint of the HC340LAD + ZM steel with zinc–aluminum–magnesium coating. The tensile shear forces of the zinc-and-magnesium-coated steel and the uncoated steel increased first and then decreased as the welding time increased, and the tensile shear reached the maximum after 14 cycles. At this time, the tensile shear forces of the welding joint of the zinc-and-magnesium-coated steel and the uncoated steel were 14.28 kN and 15.75 kN, respectively.

(4) The fracture modes of the welding joint of the two kinds of steel can be divided into two types: pull-out fracture and interface fracture. When the welding time was short, and the heat input was small, the nucleation of the welding joint was insufficient, and the tensile shear force was small. Then, an interface fracture occurred. When the welding time increased, and the heat input increased, the nucleation of the welding joint was fully complete, and the tensile shear force exceeded 11.6 kN. At this time, the welding joint was pulled and broken at the parent material near the heat-affected zone.

(5) The hardness of the welding point of the uncoated steel and the coated steel was distributed in a "π" shape, which was high on both sides and low in the middle. The HC340LA steel without coating completed nucleation after one cycle, and the corresponding hardness values of the parent material and the welding joint were 150 HV and 375 HV. However, the hardness value for the HC340LAD + ZM steel with coating was very low after one and two cycles, and the hardness values of the welding core zone were 180 HV and 250 HV, respectively. A large number of martensite appeared in the center of the welding point after four cycles, and the hardness value of the welding core zone reached 355 HV.

Compared with the hardness value of the welding joint of the HC340LA steel without coating and the HC340LAD + ZM steel with coating, the hardness value in the welding core zone and near the surface of the welding joint was higher than that of the same area with the coating.

(6) The corrosion resistance of the welding point was lower than that of the parent material. The corrosion current of the HC340LAD + ZM with zinc–aluminum–magnesium coating was 2.23 times that of the parent material, and the corrosion current of the welding joint of the HC340LA without coating was 8.46 times that of the parent material. The corrosion current of the welding joint of the zinc–aluminum–magnesium-coated steel was 27 times that of the welding joint of the uncoated steel, while the corrosion current of the parent material of the zinc–aluminum–magnesium-coated plate was 7.17 times that of the uncoated plate.

Author Contributions: Conceptualization, X.C. and X.Y.; investigation, Y.L.; resources, Y.L.; data curation, X.C.; writing—original draft preparation, X.C. and X.P.; writing—review and editing, X.C. and X.P.; visualization, Z.X., S.L. and J.P.; supervision, J.P. and X.Y.; project administration, X.Y.; funding acquisition, Z.X. and S.L. All authors have read and agreed to the published version of the manuscript.

Funding: This research was funded by HBIS Company Limited. The authors thank Ba Gen, a senior engineer of Tangshan Iron and Steel company, for their technical support and valuable suggestions.

Conflicts of Interest: The authors declare no conflict of interest.

References

1. Oh, M.S.; Kim, S.H.; Kim, J.S.; Lee, J.W.; Shon, J.H.; Jin, Y.S. Surface and Cut-edge Corrosion Behavior of Zn-Mg-Al Alloy-Coated Steel Sheets as a Function of the Alloy Coating Microstructure. *Met. Mater. Int.* **2016**, *22*, 26–33. [CrossRef]
2. Marder, A.R. The metallurgy of zinc-coated steel. *Prog. Mater. Sci.* **2000**, *45*, 191–271. [CrossRef]
3. Qu, Q.; Yan, C.W.; Wan, Y.; Cao, C.N. Effects of NaCl and SO2 on the initial atmospheric corrosion of zinc. *Corros Sci.* **2002**, *44*, 2789–2803. [CrossRef]
4. Al-Negheimish, A.; Hussain, R.R.; Alhozaimy, A.; Singh, D.D.N. Corrosion performance of hot-dip galvanized zinc-aluminum coated steel rebars in comparison to the conventional pure zinc coated rebars in concrete environment. *Constr. Build. Mater.* **2021**, *274*, 121921. [CrossRef]
5. Zhang, J.; Yu, Z.H. Corrosion Performance of Hot Dip Galvanized Steel. *Adv. Manuf. Syst.* **2011**, *201–203*, 2611–2614. [CrossRef]
6. Ramezanzadeh, B.; Attar, M.M.; Farzam, M. Corrosion performance of a hot-dip galvanized steel treated by different kinds of conversion coatings. *Surf. Coat. Technol.* **2010**, *205*, 874–884. [CrossRef]
7. Dutta, M.; Halder, A.K.; Singh, S.B. Morphology and properties of hot dip Zn-Mg and Zn-Mg-Al alloy coatings on steel sheet. *Surf. Coat. Technol.* **2010**, *205*, 2578–2584. [CrossRef]
8. LeBozec, N.; Thierry, D.; Persson, D.; Stoulil, J. Atmospheric Corrosion of Zinc-Aluminum Alloyed Coated Steel in Depleted Carbon Dioxide Environments. *J. Electrochem. Soc.* **2018**, *165*, C343–C353. [CrossRef]
9. Shimizu, T.; Yoshizaki, F.; Miyoshi, Y.; Andoh, A. Corrosion products of hot-dip Zn-6%Al-3%Mg coated steel sheet subjected to atmospheric exposure. *Tetsu-to-Hagane* **2003**, *89*, 166–173. [CrossRef]
10. Zhang, Y.J.; Xu, M.Y.; Ma, R.N.; Du, A.; Fan, Y.Z.; Zhao, X.; Cao, X.M. Effect of Oxide Scale on Hot Dip Zn-Al-Mg Alloy Coating Prepared by Reduction Combined with Induction Heating. *Metals* **2020**, *10*, 1519. [CrossRef]
11. Zhang, Z.Y.; Yue, Z.W.; Li, X.G.; Jiang, B.; Wang, F.F.; Fan, Z.B.; Yan, F.J.; Li, W.J.; Wu, Y.P.; Mi, C.X.; et al. Microstructure and Corrosion Resistance of Zn-Al-Mg Alloy Diffusion Coating developed on Carbon Steel by Hot Dipping. *Int. J. Electrochem. Sci.* **2020**, *15*, 5512–5519. [CrossRef]
12. Chang, B.H.; Xiao, H.; Zeng, J.L.; Yang, S.; Du, D.; Song, J.L.; Han, G.L. Fluid Flow Characteristics and Weld Formation Quality in Gas Tungsten Arc Welding of a Thick-Sheet Aluminum Alloy Structure by Varying Welding Position. *Appl. Sci.* **2018**, *8*, 1215. [CrossRef]
13. Long, W.Y.; Wu, Y.P.; Gao, W.W.; Hong, S. Corrosion Resistance Behavior and Mechanism of Zn-Al-Mg-RE Coating in Seawater with SRB. *Cailiao Gongcheng* **2018**, *46*, 91–97. [CrossRef]
14. Prosek, T.; Nazarov, A.; Goodwin, F.; Serak, J.; Thierry, D. Improving corrosion stability of Zn-Al-Mg by alloying for protection of car bodies. *Surf.Coat. Technol.* **2016**, *306*, 439–447. [CrossRef]
15. Jiang, G.R.; Chen, L.F.; Wang, H.Q.; Liu, G.H. Microstructure and Corrosion Resistance Property of a Zn-Al-Mg Alloy with Different Solidification Processes. *MATEC Web Conf.* **2017**, *109*. [CrossRef]
16. Gomes, G.F.; Vieville, P.; Durrenberger, L. Dynamic behavior investigation of spot welding machines and its influence on weld current range by modal analysis. *J. Braz. Soc. Mech. Sci.* **2017**, *39*, 765–773. [CrossRef]
17. Tarasov, N.M. Resistance Welding with Compression of Peripheral Zone of Weld Spot. *Weld Prod+* **1973**, *20*, 49–51.

18. Oikawa, M.; Atsumi, K.; Otsuka, Y.; Kawada, N. Development of Condition Monitoring System for Electric Resistance Spot Welding Used to Manufacture Railway Car Bodies. *J. Robot. Mechatron.* **2021**, *33*, 421–431. [CrossRef]
19. Ambroziak, A.; Korzeniowski, M. Using resistance spot welding for joining aluminium elements in automotive industry. *Arch. Civ. Mech. Eng.* **2010**, *10*, 5–13. [CrossRef]
20. Jung, W.; Oh, H.; Yun, D.H.; Kim, Y.G.; Youn, J.P.; Park, J.H. Modified Recurrence Plot for Robust Condition Monitoring of Electrode Tips in a Resistance Spot Welding System. *Appl. Sci.* **2020**, *10*, 5860. [CrossRef]
21. Chen, S.J.; Wu, N.; Xiao, J.; Li, T.M.; Lu, Z.Y. Expulsion Identification in Resistance Spot Welding by Electrode Force Sensing Based on Wavelet Decomposition with Multi-Indexes and BP Neural Networks. *Appl. Sci.* **2019**, *9*, 4028. [CrossRef]
22. Cao, R.; Chang, J.H.; Huang, Q.; Zhang, X.B.; Yan, Y.J.; Chen, J.H. Behaviors and effects of Zn coating on welding-brazing process of Al-Steel and Mg-steel dissimilar metals. *J. Manuf. Processes* **2018**, *31*, 674–688. [CrossRef]
23. Pouranvari, M.; Marashi, S.P.H. Factors affecting mechanical properties of resistance spot welds. *Mater. Sci. Technol.* **2013**, *26*, 1137–1144. [CrossRef]
24. Onar, V.; Ozen, F.; Kekik, M.; Aslanlar, Y.S.; Ozderin, Y.; Aslan, H.; Aslanlar, S. Effect of current and welding time on tensile-peel strength of resistance spot welded TWIP 1000 and martensitic steels. *Indian J. Chem. Technol.* **2019**, *26*, 248–251.
25. Kim, T.; Park, H.; Rhee, S. Optimization of welding parameters for resistance spot welding of TRIP steel with response surface methodology. *Int. J. Prod. Res.* **2005**, *43*, 4643–4657. [CrossRef]
26. Podrzaj, P.; Polajnar, I.; Diaci, J.; Kariz, Z. Overview of resistance spot welding control. *Sci. Technol. Weld. Join.* **2008**, *13*, 215–224. [CrossRef]
27. Wang, N.N.; Qiu, R.F.; Peng, W.J.; Shi, H.X. Resistance Spot Welding between Mild Steel and Stainless Steel. *Appl. Mech. Mater.* **2014**, *675–677*, 23–26. [CrossRef]
28. Zhang, H.T.; Song, J.Q. Microstructural evolution of aluminum/magnesium lap joints welded using MIG process with zinc foil as an interlayer. *Mater. Lett.* **2011**, *65*, 3292–3294. [CrossRef]
29. Guo, L.; Zhang, F.; Lu, J.C.; Zeng, R.C.; Li, S.Q.; Song, L.; Zeng, J.M. A comparison of corrosion inhibition of magnesium aluminum and zinc aluminum vanadate intercalated layered double hydroxides on magnesium alloys. *Front. Mater. Sci.* **2018**, *12*, 198–206. [CrossRef]
30. Bolton, R.; Dunlop, T.; Sullivan, J.; Searle, J.; Heinrich, H.; Westerwaal, R.; Boelsma, C.; Williams, G. Studying the Influence of Mg Content on the Microstructure and Associated Localized Corrosion Behavior of Zn-Mg PVD Coatings Using SVET-TLI. *J. Electrochem. Soc.* **2019**, *166*, C3305–C3315. [CrossRef]

Article

Few-Layer Graphene as an Efficient Buffer for GaN/AlN Epitaxy on a SiO₂/Si Substrate: A Joint Experimental and Theoretical Study

Denis Petrovich Borisenko [1,*], Alexander Sergeevich Gusev [1,*], Nikolay Ivanovich Kargin [1], Petr Leonidovich Dobrokhotov [1], Alexey Afanasievich Timofeev [1], Vladimir Arkhipovich Labunov [1,2], Mikhail Mikhailovich Mikhalik [2], Konstantin Petrovich Katin [1,*], Mikhail Mikhailovich Maslov [1], Pavel Sergeevich Dzhumaev [1] and Ivan Vladimirovich Komissarov [2]

1 Nanoengineering in Electronics, Spintronics and Photonics Institute, National Research Nuclear University "MEPhI", Kashirskoe Shosse 31, 115409 Moscow, Russia
2 Laboratory "Integrated Micro- and Nanosystems", Belarusian State University of Informatics and Radioelectronics, 6 P. Browka Street, 220013 Minsk, Belarus
* Correspondence: dpborisenko@mephi.ru (D.P.B.); asgusev@mephi.ru (A.S.G.); kpkatin@mephi.ru (K.P.K.)

Citation: Borisenko, D.P.; Gusev, A.S.; Kargin, N.I.; Dobrokhotov, P.L.; Timofeev, A.A.; Labunov, V.A.; Mikhalik, M.M.; Katin, K.P.; Maslov, M.M.; Dzhumaev, P.S.; et al. Few-Layer Graphene as an Efficient Buffer for GaN/AlN Epitaxy on a SiO₂/Si Substrate: A Joint Experimental and Theoretical Study. *Appl. Sci.* **2022**, *12*, 11516. https://doi.org/10.3390/app122211516

Academic Editors: Suzdaltsev Andrey and Oksana Rakhmanova

Received: 13 October 2022
Accepted: 9 November 2022
Published: 13 November 2022

Publisher's Note: MDPI stays neutral with regard to jurisdictional claims in published maps and institutional affiliations.

Copyright: © 2022 by the authors. Licensee MDPI, Basel, Switzerland. This article is an open access article distributed under the terms and conditions of the Creative Commons Attribution (CC BY) license (https://creativecommons.org/licenses/by/4.0/).

Abstract: Single-layer (SLG)/few-layer (FLG) and multilayer graphene (MLG) (>15 layers) samples were obtained using the CVD method on high-textured Cu foil catalysts. In turn, plasma-assisted molecular beam epitaxy was applied to carry out the GaN graphene-assisted growth. A thin AlN layer was used at the initial stage to promote the nucleation process. The effect of graphene defectiveness and thickness on the quality of the GaN epilayers was studied. The bilayer graphene showed the lowest strain and provided optimal conditions for the growth of GaN/AlN. Theoretical studies based on the density functional theory have shown that the energy of interaction between graphene and AlN is almost the same as between graphite sheets (194 mJ/m²). However, the presence of vacancies and other defects as well as compression-induced ripples and nitrogen doping leads to a significant change in this energy.

Keywords: GaN/AlN; molecular beam epitaxy; CVD graphene; buffer layer; gallium nitride; GaN-on-Si technology

1. Introduction

Heteroepitaxy of III-nitrides is one of the most widely developed techniques for the manufacturing of devices for optoelectronics (light-emitting diodes [1,2], laser diodes [3,4], ultraviolet emitters [5]) and high-frequency power electronics [6,7]. Due to the lack of cheap GaN or AlN wafers of sufficient size, GaN-based structures are usually deposited on various foreign substrates such as sapphire, 4H-SiC (0001) or silicon (111). The main drawback of sapphire is its low thermal conductivity, whereas silicon carbide substrates are still rather expensive. Silicon wafers are more manufacturable and significantly cheaper [8]. They have sufficient size (up to 8 inches) and higher thermal conductivity than Al_2O_3. The ability to manufacture GaN-based devices in fully depreciated 6- or 8-inch wafer silicon fabrication plants provides the cost competitiveness of GaN-on-Si technology [9]. Despite all the successes, obtaining III-nitrides on Si substrates is still problematic due to the differences in lattice constants (18%) and thermal expansion coefficients (46%) of these materials [10]. Thus, GaN-on-Si technology requires a complex buffer design including different transition layers and superlattices [11,12] to reduce the dislocation density and residual stresses.

In some recent works, researchers regarded graphene as a suitable 2D buffer, which facilitated the high-quality epitaxy of III-nitrides layers on a silicon substrate via both the MOCVD [13–16] and MBE techniques [17–19]. Graphene has a hexagonal crystal lattice comparable to the (0001) plane of a GaN (AlN) crystal in the wurtzite phase. In addition, graphene is thermally stable and can be transferred to an arbitrary substrate. The Van der Waals interaction at the III-nitride/graphene interface provides reliable binding despite some lattice constants mismatch. In addition, gallium atoms can migrate on the graphene surface due to the sufficiently low migration barrier (12 meV) [20]. On the other hand, the nucleation on the pristine graphene is very difficult due to the absence of dangling bonds on its surface [21]. Islands of nuclei appear primarily at the graphene wrinkles. This effect leads to poor morphology and high defect density in GaN epilayers. Oxygen [22] or nitrogen [23,24] plasma pretreatment can be applied to promote GaN nucleation and growth on the graphene surface. In addition, the work function, surface potential, surface oxidation and chemical activity of few-layer graphene highly depend on the number of layers. Therefore, number of layers influence on the density and activity of nucleation sites.

Despite the growing interest in the graphene-assisted epitaxy of III-nitrides, there are only a few studies considering the dependence of the nucleation process on the number of graphene layers [21,25–27]. The only work devoted to the growth of GaN by MBE on graphene-coated SiO_2 substrates was conducted by [25]. However, the authors of [25] considered only XRD data. In our previous works [18,28], GaN and AlN graphene-assisted epitaxy using the PA-MBE method on an amorphous substrate was demonstrated. In the present paper, we investigate the epitaxy of GaN/AlN on few-layer and multilayer CVD graphene. The effect of the number of graphene sheets, surface properties and defectiveness of the 2D buffer on the quality of the resulting III-nitrides epitaxial layers are considered in detail. A theoretical study of the AlN–graphene interaction conducted using the density functional theory complements the experimental results.

2. Experimental Section

We deposited graphene on pure copper foil (99.999%, Alfa Aesar) using atmospheric-pressure chemical vapor deposition (AP-CVD). Two gases, methane and ethylene, were used as precursors. The synthesis took place in a cylindrical quartz reactor with a diameter of 14 mm. We used a custom-made CVD setup (Figure S1e). The sample size was equal to 35×45 mm^2. The temperature and duration of the catalyst annealing were equal to 1 h at 1050 °C, respectively. Two gases, nitrogen N_2 and hydrogen H_2, were involved during annealing. Their flow rates were equal to 100 and 150 cm^3 min^{-1}, respectively. Single-layer (SLG)/few-layer (FLG) and multilayer graphene (MLG) (>15 layers) samples were obtained from the reactor at 1054 °C within 20 min. Subsequent cooling of the reactor in the presence of gaseous nitrogen flow was conducted at a rate of 50 °C min^{-1}. For SLG, in order to preserve the integrity of the graphene during wet-chemical transfer, we used a polymer support, which was subsequently removed using special techniques. In the case of multilayer graphene, a wet-chemical transfer was carried out without polymer support. After the process of graphene synthesis, the copper catalyst was completely etched in an aqueous solution of $FeCl_3$. After removal of the catalyst, the graphene films were carefully transferred to the surface of 3-inch silicon wafers with 500 nm-thick thermal Loxide (SiO_2/Si).

The methods of scanning electron microscopy and scanning probe microscopy were used to study the morphology and surface relief of the experimental samples. For all AFM measurements, a Solver Open microscope (NT-MDT) was applied in the semi-contact scanning mode with a NSG01-brand cantilever. The typical tip curvature radius for this cantilever is 10 nm.

The analysis of the crystal structure and crystalline perfection was carried out by the electron backscatter diffraction (EBSD) and high-resolution X-ray diffraction (HRXRD) methods. An Ultima IV (Rigaku) diffractometer provided XRD patterns of the considered samples. We used the wavelength λ corresponding to CuKα (0.15406 nm).

Deformations, doping, defectiveness, thickness, and quality of graphene were estimated via Raman spectroscopy. We used a confocal micro-Raman spectrometer Confotec

NR500. It provides an excitation wavelength, spectral resolution and spectral accuracy of about 532 nm, 1 cm^{-1}, and ~1 cm^{-1}, respectively.

The number of graphene layers was also estimated using the classical light transmission method. Here, optical quartz plates were used as a substrate. Spectral measurements were performed using a UV-2600 Shimadzu spectrometer.

In this work the plasma-assisted molecular beam epitaxy system Veeco GEN 930 was used to carry out the process of III-nitrides graphene-assisted growth. After low-temperature annealing (200 °C) in the load-lock chamber, the wafer with the graphene buffer was attached to a substrate holder in the growth module. This was followed by high-temperature annealing at growth temperature (in the range of 700–720 °C) under ultrahigh vacuum conditions. First, a 35 nm-thick AlN wetting layer was deposited. The temperature of the Al effusion cell was 1060 °C providing nitrogen enriched conditions. Both shutters (for Al and N cells) were opened simultaneously and remained open throughout the entire time of the AlN nucleation growth (480 s). After that, a 500 nm-thick GaN layer was grown. The temperature of the Ga cell was 970 °C providing metal-rich conditions. A Metal Modulated Epitaxy (MME) technique was used which consists in the pulsed gallium supply to prevent its accumulation on the surface in the form of liquid metal droplets, while the flow of active nitrogen remained constant throughout the entire GaN growth process. The pulse duration (open state time of the Ga shutter), monitored with an optical pyrometer, was 8 s, while the pulse repetition period was 22 s. The GaN growth consisted of 300 periods. The nitrogen consumption in both cases (AlN and GaN growth) was 1.6 sccm at an RF source power of 350 W. The power of the substrate heater was maintained at a constant level throughout the growth process.

The experimental samples were also investigated using an in situ reflection high-energy electron diffraction technique (RHEED). RHEED patterns at all growth stages were monitored using a standard 15 kV RHEED (STAIB Instruments) system equipped with a 6-inch phosphor screen flange and a high-resolution CCD camera. The electron beam in RHEED hits the wafer surface at an angle of inclination (~1°), which makes this method extremely valuable for studying the crystal integrity of graphene samples and can be a valuable tool for checking the structure and quality of graphene on the surface of amorphous substrates before manufacturing various devices. It should be noted that RHEED patterns can be obtained in the entire range of azimuthal angles φ from 0° to 360° due to the possibility of wafer holder rotation controlled by a stepper motor. The beam reflection profiles were obtained at different sample rotation angles in increments of 1°.

To supplement the obtained experimental results, we carried out modelling of the AlN–graphene interaction. Calculations were carried out within the density functional theory with dispersion Grimme corrections [29], which ensure correct accounting for non-covalent interlayer interactions, as provided in the Quantum Espresso package [30,31]. Ultrasoft pseudopotentials with the GGA-PBE functional were used for all elements [32,33]. The cutoff energy of the plane wave basis was chosen to be equal to 45 Ry. The number of points of the Monkhorst-Pack grid [34] in k-space was $4 \times 4 \times 1$. The first order Metfessel-Paxton scheme with 0.01 Ry smearing was used [35]. Structural optimization continued until the residual forces became less than 0.1 meV/Å.

3. Results and Discussion

Figure 1a shows the scanning electron microscopy (SEM) image of 25 μm thick Cu foil after graphene growth. Additional SEM and AFM images of the copper foil surface are given in the supporting information (Figure S1a,b). The surface of Cu is smooth, without pits, with an average grain size of 80–150 μm. The dominant orientation of the copper grains was analyzed using the XRD method (Figure 1b,c).

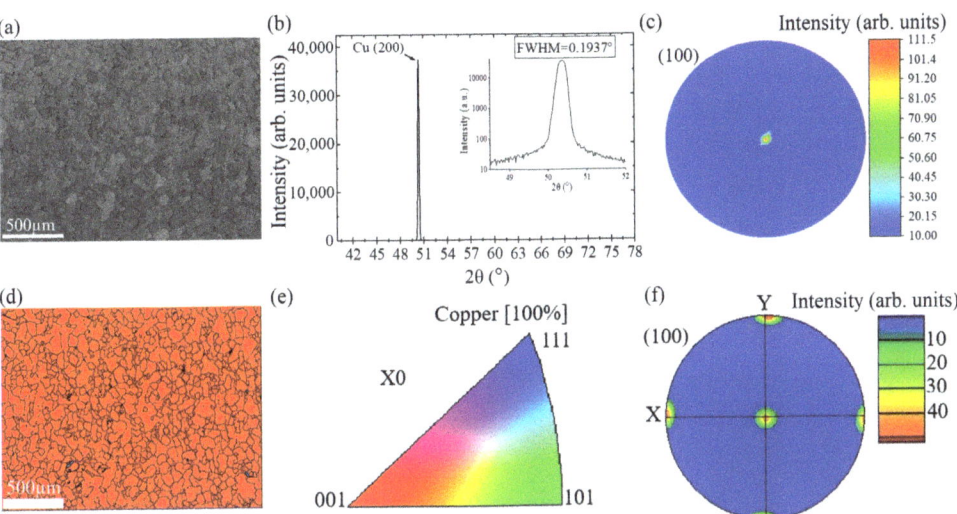

Figure 1. (a) SEM image of the 25 μm Cu foil: (b,c) X-ray diffraction θ-2θ spectrum and pole figure of a Cu (100) sample after graphene growth; the inset shows the rocking curve for the Cu (100) diffraction; (d–f) EBSD analysis of the Cu foil after graphene growth; (d) the crystallographic orientation mapping in direction normal to the sample surface; (e) The color code for this map is given in the stereographic triangle; (f) The corresponding pole figure in [001]-direction (i.e., normal to the sample surface).

As follows from the X-ray diffraction analysis, the copper catalyst used has a predominant (100) orientation (Figure 1b). The measured XRD θ–2θ scan (Figure 1b), shows a Cu (200) peak (2θ = 50.37°). The presence of one diffraction peak indicates a good texture with out-of-plane grain orientation along the <100> direction. The inset to Figure 1b shows the Cu (200) peak profile, with a characteristic small width at half maximum (FWHM) of ~0.19°. The small value of the FWHM suggests that the Cu film has a high crystalline quality. For additional study of the texture, an incomplete X-ray pole figure (PF) (100) was assembled (Figure 1c). The maximum tilt angle at this PF does not exceed 70°, so only the central maxima are visible. To evaluate the domain sizes and microstructure of the Cu (100) foil, an EBSD analysis was performed (Figure 1d–f). Figure 1d shows a crystallographic map of the Cu film orientation using the inverse pole figure (IPF)-Z component, which correlates spatial crystallographic orientations with respect to the normal sample surface. The color uniformity in Figure 1d (corresponding to the stereographic triangle in Figure 1e) suggests that the normal direction of the sample is <100> over the entire map. Figure S1c shows the <100> angular deviation orientation map of copper grains. The dominance of one color demonstrates the mutual direction of all copper grains along the direction with a slight misorientation of several degrees (~2°) (Figure S1d). Figure 1f shows the Cu (100) EBSD PF in the sample normal direction, which is fully consistent with the results of the X-ray diffraction (Figure 1c).

After transfer to an amorphous substrate, the graphene samples were analyzed using the SEM and AFM methods to obtain information on the quality of the surface morphology, the presence of contaminants, wrinkles or various defects. Figure 2a–c shows the quality of the SiO_2/Si surface morphology. The root mean square roughness (RMS) calculated from AFM data is about 0.4 nm.

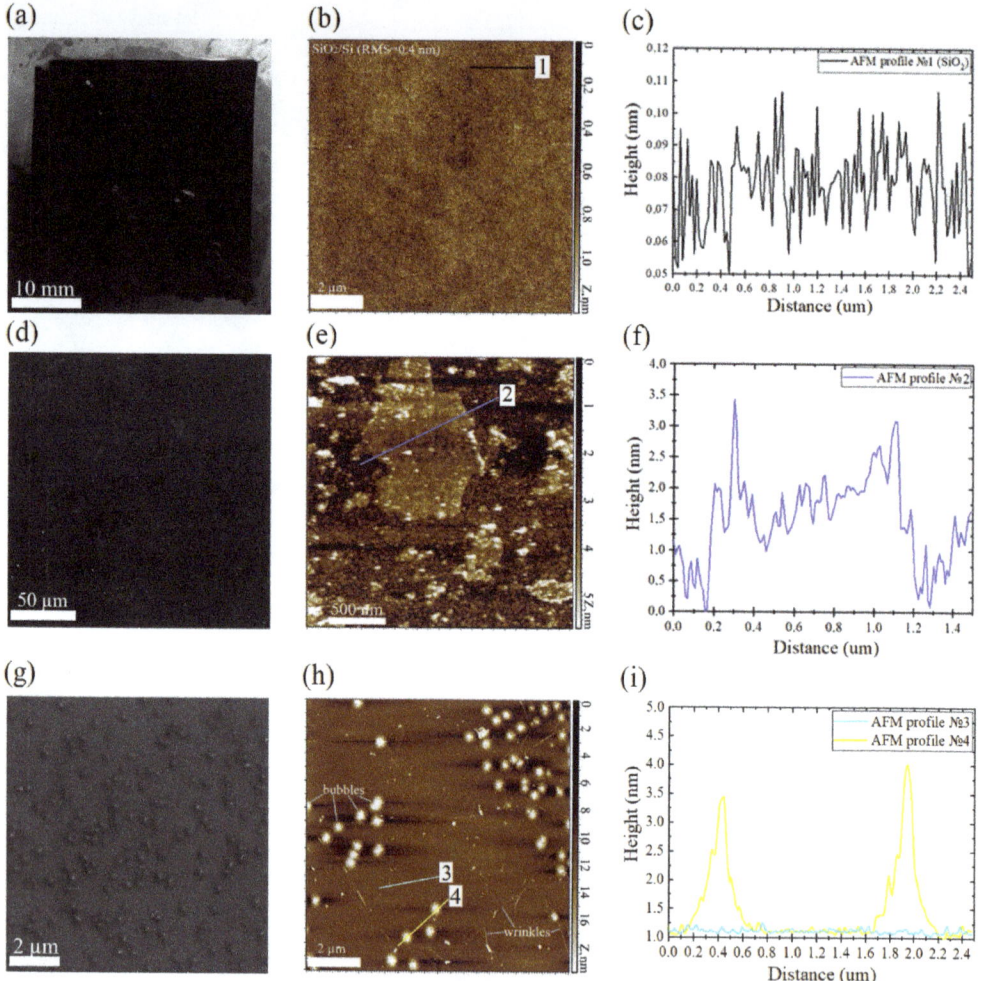

Figure 2. SEM images of graphene layers obtained on a 3-inch silicon wafer (with amorphous surface due to thermal oxide film) via PA-MBE epitaxial growth (**a**). Randomly selected regions of FLG (**d**) and MLG (**g**). AFM height images and line scan profiles: the oxide surface (**b,c**), the random regions of FLG (**e,f**) and MLG (**h,i**).

Contrast changes in the SEM images (Figure 2d) indicate the presence of SLG and few-layer graphene (FLG) regions, which were purposefully obtained in the process of graphene synthesis on Cu (100) by the CVD method. However, the AFM is the main method for directly determining and confirming the thickness and roughness of single- and few-layer graphene [36–38]. Figure 2e,h shows typical AFM results for transferred graphene on the SiO$_2$/Si substrate surface. As a result of the use of liquid transfer and subsequent drying of the sample, graphene can be raised several additional angstroms above the amorphous substrate [37]. According to the measurements of the AFM profiles (Figure 2f), the height of the areas of multi-layer graphene corresponds to 0.7–0.9 nm and can be identified as two-layer graphene [38]. The observed differences in surface morphology (Figure 2e,h) on the graphene samples are caused by the influence of the number of layers of crumpled graphene on the surface of the SiO$_2$/Si substrate [38]. When using a polymer framework and liquid transfer of graphene onto an amorphous substrate, a certain amount of polymer residues, wrinkles and tears are present on the surface, the number of which is significantly reduced after the process of thermal annealing in the PA-MBE chamber. The liquid process of transferring graphene to the substrate contributes to the deformation of graphene layers, such as bubbles and wrinkles, often formed by the capture of water, gas, or solid nanoparticles at the interface (Figure 2h,i) [39–41].

Analysis of the graphene samples' surface transfer to the SiO$_2$/Si substrate by the EDX method (Figure S2) demonstrates the presence of only three elements: carbon (C), oxygen (O) and silicon (Si). This indicates that we have carried out high-quality operations to cleanse graphene from contamination.

According to the light transmission measurements (Figure 3a), the transmission at 550 nm demonstrates a value of 96.3% for FLG (blue curve) and 60% for MLG (black curve). The first value corresponds to a number of graphene layers between 1 (97.5%) and 2 (95.5%) [42,43]. The second measurement indicates that we have obtained multilayer graphene with a large number of layers (more than 15) [43].

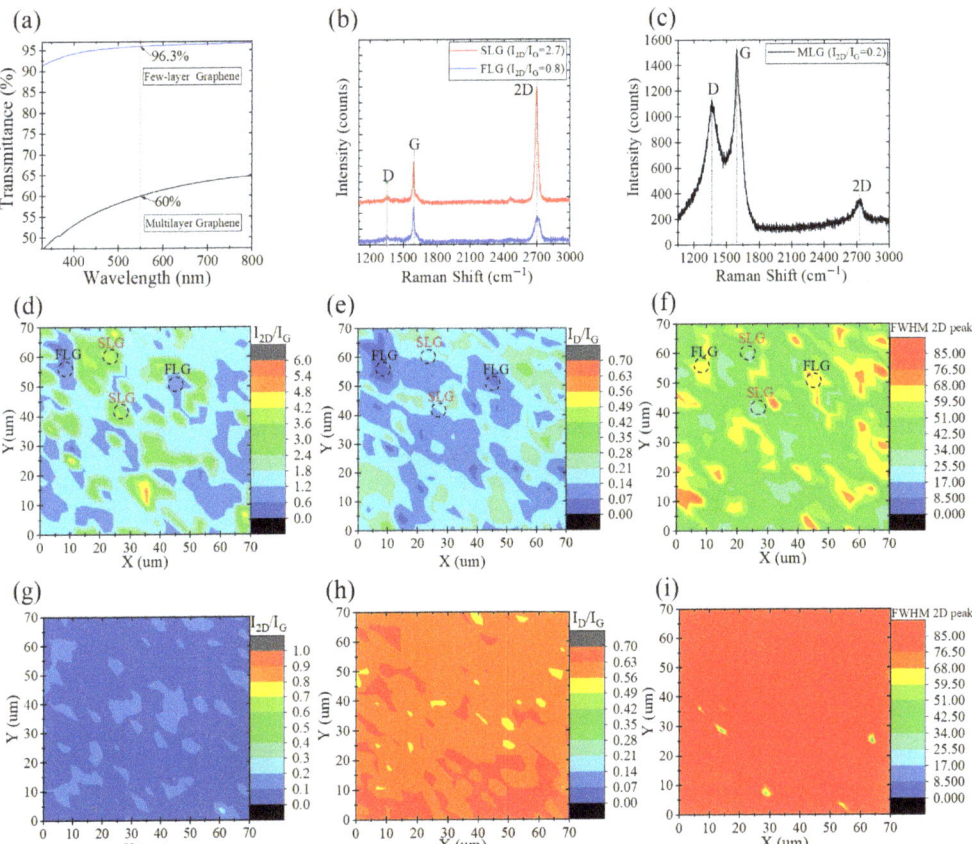

Figure 3. (a) The transmittance coefficient of graphene samples transferred on a SiO2/Si substrate as a function of light wavelength. Raman spectra (b,c), Raman maps of I2D/IG (d,g), ID/IG (e,h) ratios and 2D peak FWHM (f,i) from random areas on the continuous FLG (b,d–f) and MLG (c,g–i) transferred to a SiO2/Si substrate acquired at a laser excitation wavelength of 532 nm.

To confirm the morphology and structural quality of the FLG and MLG samples, we applied Raman spectroscopy associated with the emission (Stokes process) or absorption (anti-Stokes process) of phonons [44]. Micro-Raman spectroscopy is particularly useful in obtaining information on the quality, thickness, doping levels, and strain of graphene [45–51]. Figure 3b,c shows the typical Raman spectra from SLG (red line), FLG (blue line) and MLG (black line), measured at random sections of the substrates. Several main peaks appear on the Raman spectrum of the graphene: the primary mode of vibrations in the plane, G (~1580 cm^{-1}); another vibration in the plane, D (~1350 cm^{-1}); the second-order overtone, 2D (~2690 cm^{-1}) [45]. As known, the G peak corresponds to sp^2-hybridized graphene, whereas the D peak arises from defects induced in a perfect sp^2-hybridized structure [46]. Peak D is correlated with the disorder in graphene lattice. The positions of the peaks D and 2D are dispersive, so they depend on the energy of laser excitation [46]. In this paper, measurements were obtained using a laser with an excitation of 532 nm. Due to the additional forces arising from the interaction between graphene layers, as the number of layers increases, the spectrum will differ from the spectrum of single-layer graphene, namely the splitting of the 2D peak into an increasing number of modes [47]. Peak G also experiences displacement due to an increase in the number of layers [51]. Thus, to study the number of graphene layers, it is necessary to analyze the ratio of peak intensities 2D and G (I_{2D}/I_G), as well as the positions and shapes of these peaks (Figure 3d,g) [50]. In turn, the ratio of I_D to I_G (I_D/I_G) reflects the quality of the graphene (Figure 3e,h) [46].

Figure 3d–i shows the maps for the intensity ratio of 2D and G (I_{2D}/I_G), the intensity ratio of D and G (I_D/I_G) Raman bands and maps of 2D peak FWHM. From these maps we can identify areas with different values of the ratios I_{2D}/I_G and I_D/I_G. The value of the I_{2D}/I_G ratio greater than 1.3 apparently belongs to the areas of a single-layer graphene with characteristic values of FWHM 2D bands ~30 cm^{-1} [46,50–55]. In addition, these sections of graphene have a slight value of I_D/I_G (~0.175) (Figure S3c), which also corresponds to single-layer graphene with a small number of defects [46]. Besides the regions typical of SLG, there are areas of FLG graphene with a lower I_{2D}/I_G ratio. For such regions the value of the I_{2D}/I_G ratio is mainly from 0.5 to 1.25 (Figure 3d and Figure S3a). The value of the FWHM 2D-range is between 44 and 68 cm^{-1}. This indicates the presence of regions of few-layer graphene. It is noteworthy that for regions of FLG, the values of the I_D/I_G (~0.11) (Figure S3d) ratio are less than that for SLG (Figure S3c). In particular, an increase in the I_D/I_G ratio suggests that much higher defects prevail and are easily induced in thinner layers [55]. A comparative analysis of histograms of the I_{2D}/I_G intensity ratios and I_D/I_G for an area of 70 × 70 μm (Figure S3) indicates the predominance of single-layer over two-layer fractions in the resulting graphene, which is also confirmed by spectrophotometry data (see Figure 3a). The average values of the FWHM 2D bands for SLG and FLG are about 41 and 56 cm^{-1}, respectively (Figure S3e,f).

In turn, a sample of a multilayer graphene is characterized by a significant decrease in the ratio of I_{2D}/I_G (Figure 3g) and an increase in the ratio of I_D/I_G (Figure 3h). The value of the FWHM 2D bands are significantly increased (Figure 3i). A comparative analysis of the histograms (Figure S3) demonstrates the average value of the ratio I_{2D}/I_G is ~0.08 and ~0.61 for I_D/I_G, while the average value of the FWHM 2D bands is ~111 cm^{-1} (Figure S3b,g,h). These results indicate the similarity of the synthesized multilayer graphene with graphitized carbon [56].

Various types of defects play an important role in the process of the nucleation and growth of III-nitrides on the graphene surface [21,57]. The studied samples differed in the values of the RMS (Figure 2b,e,h) and the density of defects (Figure 3e,h). A Raman analysis confirmed this conclusion (we adopted the I_D/I_G ratio as a measure of the density of point defects in graphene [58–64]). The defect density (n_D) in the graphene samples used was quantified from the I_D/I_G ratio as $n_{D(SLG)}$ ~3.93 × 10^{10} cm^{-2}, $n_{D(FLG)}$ = 2.47 × 10^{10} cm^{-2} and $n_{D(MLG)}$ = 1.37 × 10^{11} cm^{-2} for SLG, FLG and MLG, respectively. The topic of distinguishing between the Raman spectra of one and another type of defective sp^2 carbon remains an area for future study.

Thus, for further studies of the epitaxy of nitrides, we obtained two separate graphene samples transferred to the amorphous surface of SiO$_2$/Si substrates. Based on the results of Raman spectroscopy and light transmission (Figure 3) of the obtained graphene samples, it is possible to identify SLG, FLG and MLG regions.

It was previously shown [28,65,66], the RHEED technique can be applied for visualization of graphene reciprocal structure and to analyze the ordering of its crystalline domains. Previously [38], it was shown that in the case of a single-crystal 2D material, its reciprocal space structure consists of vertical rods. Because of the relatively large wave vector of electrons in RHEED, the Ewald sphere is large and crosses the rods in reciprocal space. In general, bands form in RHEED patterns in 2D materials (Figure S4a). We obtained the reciprocal lattice of the 2D material by the measuring of streaks as a function of the momentum transfer parallel to the surface ($k_\parallel$) at different azimuthal angles φ [28,65,66]. We constructed a reverse space map for graphene with two types of crystallites with a mutual rotation of 30° (Figure S4c), which is synthesized on a Cu (100) catalyst using the CVD method [28].

The RHEED method was used to analyze the graphene in situ and control the processes of nucleation and subsequent stages of III-nitride growth (Figures 4 and 5). Figure 4 shows the RHEED patterns from the FLG (Figure 4a) and MLG (Figure 4d) transferred to a SiO$_2$/Si substrate. The in-plane structures of graphene and GaN/AlN films can be derived from their observed structures in 2D reciprocal space. In the reciprocal space, the radius and polar angle correspond to the reciprocal distance from the (00) spot and azimuthal angle, respectively. Figure 4b,e shows the intensity profiles of line scans as a function of $k_\parallel$ (the distances from (hk) to (00)) at a fixed value $k_\perp$ [28,65]. Since a sample with FLG is more characterized by SLG regions, the widened peaks characteristic of graphene are noticeable in the RHEED images (Figure 4a–c). In the case of MLG, the main peaks are characterized by a smaller FWHM (Figure 4e), which leads to the more contrasting RHEED reciprocal space structure (Figure 4f). A smaller half-width value is associated with a larger value of the thickness of the MLG film compared to FLG.

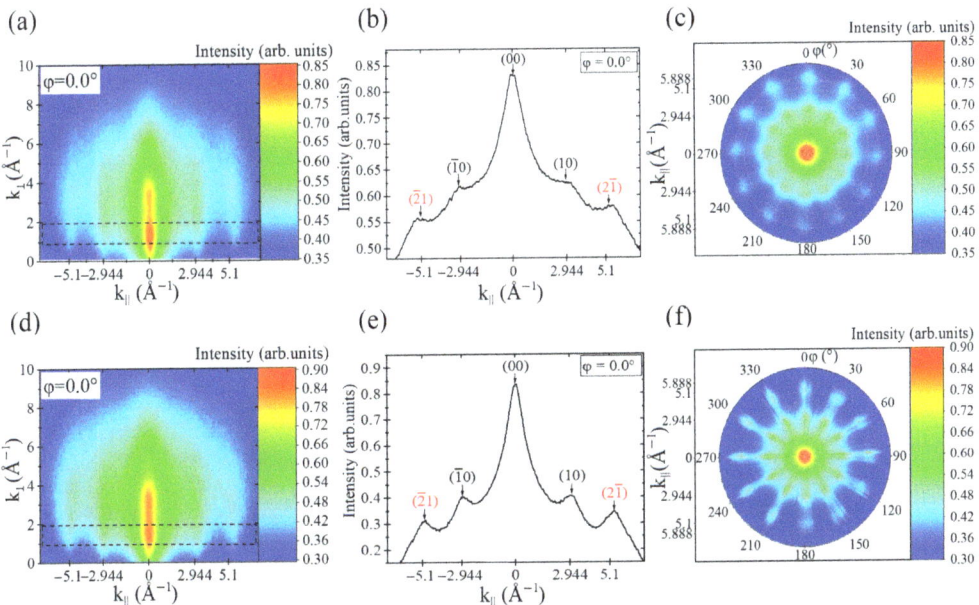

Figure 4. Images of RHEED patterns obtained of FLG (**a**) and MLG (**d**) at azimuthal angles $\varphi = 0°$ (**a**,**d**). Intensity profiles for azimuthal angles $\varphi = 0°$ (**b**,**e**) for FLG (**b**) and MLG (**e**), respectively. (**c**,**f**) The RHEED reciprocal space structures for the FLG (**c**) and MLG (**f**) samples.

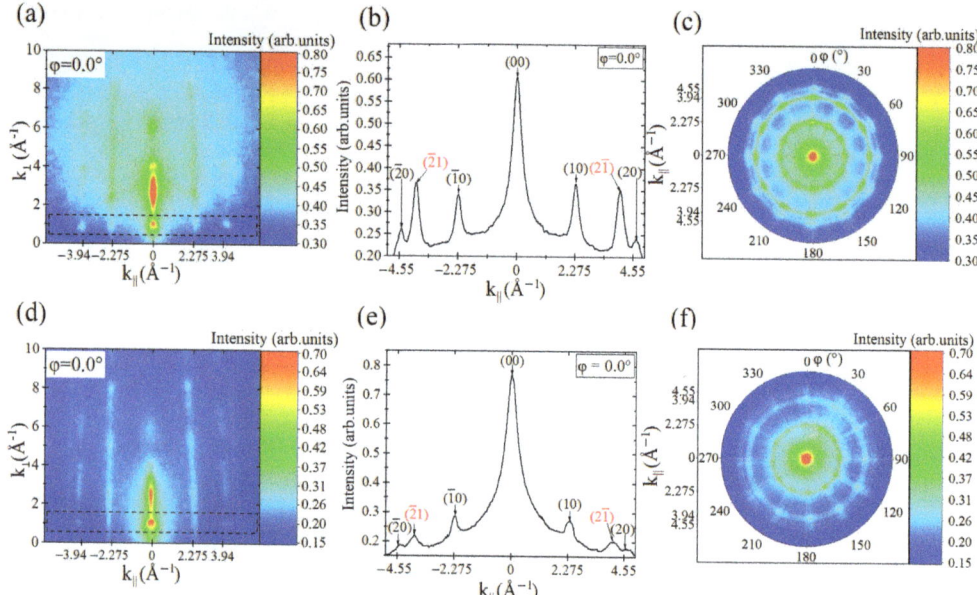

Figure 5. (**a,d**) Images of RHEED patterns from the GaN surface during PA-MBE growth on FLG (**a**) and MLG (**d**) at azimuthal angles φ = 0°. (**b,e**) Intensity profiles for azimuthal angles φ = 0° for the GaN layer growing on FLG (**b**) and MLG (**e**), respectively. (**c,f**) Structures of RHEED reciprocal spaces of the GaN layer grown on different graphene buffer layers: FLG (**c**) and MLG (**f**).

The growth kinetics of the AlN wetting layer on graphene to improve the quality of the GaN epitaxial layers is not well-studied. Nevertheless, in this work, we used the standard nucleation process, which showed itself well during the growth of heterostructures on SiC (30–40 nm-thickness of the AlN layer under a N_2-rich regime at T_S = 710 °C). PA-MBE growth at this stage was also controlled by the RHEED method (Figures S5.1 and S5.2).

The next RHEED patterns (Figure 5a,e) measured along the axes of the [01$\overline{1}$0] band follow the PA-MBE growth of GaN/AlN epilayers on different regions of the graphene buffer: FLG (Figure 5a) and MLG (Figure 5e). Note that the axes of the [01$\overline{1}$0] band correspond to φ = 0°.

Figure 4c,f shows the reciprocal space structures of the transferred FLG and MLG. Figure 4c,f, shows twelve mutually symmetrical spots at the reciprocal distances of 2.9 and 5.1 Å^{-1} from the center. These correspond to graphene domains randomly oriented with respect to each other. According to our previous study [28], the direction of graphene growth is rotated by 30° because of the 60° crossing of in-plane grain boundaries. Therefore, twelve observed peaks point to two graphene domains disoriented by 30°. Figure 5c,g shows constructed structures of the 2D reciprocal space derived from our experiments Figure 5c,g). Characteristic peaks from GaN grown on FLG (Figure 5a–c) demonstrate a greater intensity value with a smaller FWHM compared to nitride on MLG (Figure 5e–g). These results indicate the production of a higher-quality gallium nitride with a smooth surface morphology on FLG.

Figure 6 shows SEM images of the surface of GaN grown on an AlN/graphene/SiO$_2$/Si substrate. One can trace the difference between GaN growth on the SLG (Figure 6a,b) and FLG (Figure 6c) regions. The SEM images clearly show areas of different contrast (Figure 7a), differing in the quality of the epitaxial layers and surface morphology (Figure 6b,c). These areas correlate well with areas of single-layer and few-layer graphene (Figure 2d,e) before PA-MBE growth. Figure 6d shows a general AFM scan of the GaN/AlN film on SLG and FLG. Figure 6e,f shows AFM scans of GaN/AlN films on the SLG

(Figure 6e) and FLG (Figure 6f) areas separately. Figure 6g–i shows SEM and AFM images of GaN/AlN film grown on MLG.

Figure 6. SEM and AFM images of the surface of GaN/AlN (0001) film grown on SLG (**a,b,d,e**) with regions of FLG (**c,f**) and on MLG (**g–i**) in random regions of the samples.

In the PA-MBE process, epitaxial GaN/AlN layers grow directly on the graphene-coated regions; in other places, the nitride layers are polycrystalline. The formation of polycrystalline GaN/AlN occurs due to the low surface mobility of the adatom and rare nucleation sites. These regions possess increased reactivity and therefore facilitate nucleation of AlN. The improvement in the quality of the GaN layers in the sections of FLG indicates that the density of AlN cores correlates with the density of defects in graphene and its root-mean-square roughness. This correlates well when comparing the growth of GaN/AlN on graphene with different numbers of layers (Figure 6).

The thickness of the graphene significantly affects the selectivity of the growth of layers of III nitrides. Figure 6 confirms this conclusion for all the samples studied. A particularly clear contrast is observed in Figure 2e, where FLG regions are present. The contrast is due to the different number of graphene layers. The light and dark areas in the figure correspond to small and large numbers of graphene layers, respectively. Epitaxial layers of nitrides of the best quality are formed only in the darker regions (few-layer graphene), while in the light regions (low-layer graphene) the nucleation of nitrides is worse, as can be seen from Figure 6a–f. We assume that the role of graphene n-layers can be explained by differences in surface potential or chemical reactivity.

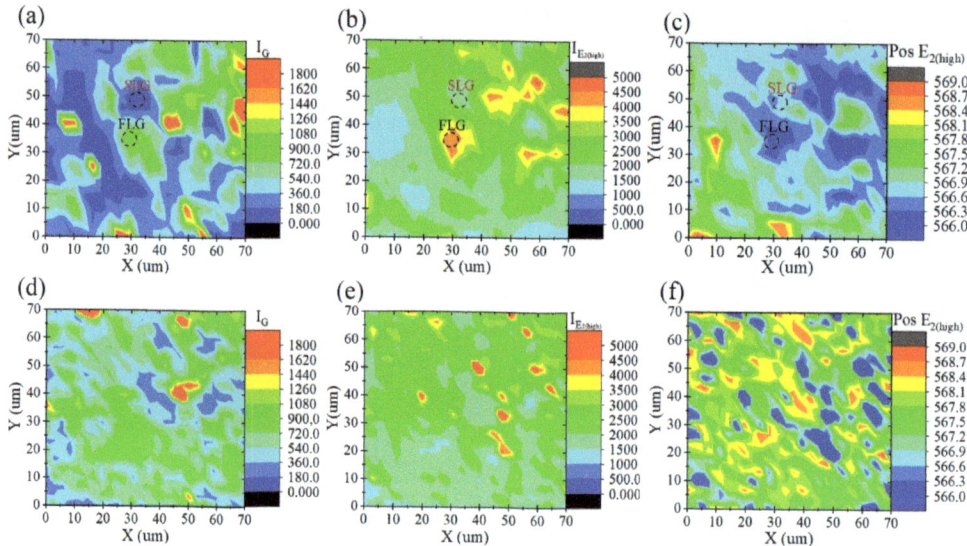

Figure 7. Raman analysis: (**a**,**d**) Raman mapping of G peak intensity and of the $E_{2(high)}$ peak intensity (**b**,**e**) and position (**c**,**f**) from random areas on the continuous FLG (**a**–**c**) and MLG (**d**–**f**) after nitrides PA-MBE growth acquired at a laser excitation wavelength of 532 nm.

In the case of the growth of nitrides on multilayer graphene, epitaxial layers are characterized by a developed surface morphology with the presence of a large number of structural defects (Figure 6g–i). There are also distinctly traced areas of a circular shape (Figure 6i), where bubbles with liquid were presumably present after the transfer of multilayer graphene to the surface of the amorphous substrate (Figure 2g–i). The surface of the bubbles was characterized by a convex shape (Figure 2i), which led to the formation of a polycrystalline layers of nitrides on its surface (Figure 6f). EDX analysis (Figure S6f) of nitride film on MLG indicates the presence of a small amount of copper, which confirms the nature of the observed bubbles due to fluid residues during the liquid transfer of graphene to the amorphous substrate.

By mapping Raman spectra and determining the intensity of the graphene and nitride peaks, we can identify regions of SLG and FLG graphene after PA-MBE of the III-nitride layers. Figure 7 shows the Raman maps for the obtained experimental samples. In the spectra of epitaxial layers, it is possible to identify the combination peaks of graphene (D-band, G-band, and 2D-band), as well as the characteristic $E_{2(high)}$-peak of GaN allowed for a c-oriented wurtzite structure. These data indicate that graphene remained under the nitride film. To characterize areas with a different number of graphene layers (SLG and FLG), maps of the G-peak intensity (Figure 7b,e) and maps of the $E_{2(high)}$ peak intensity (Figure 7b,e) and position (Figure 7c,f) were constructed. Raman spectral mapping of high peak GaN E_2 (Figure 7a,c) confirmed the selectivity of the GaN cores with respect to graphene-coated wafer areas. Analyzing the Raman spectral maps of the G-peak intensity (Figures 7a,d and S7a,b), it can be seen that the areas with a high intensity value (>200) of the G-peak can be attributed to FLG, and areas with a lower intensity value (<200) to SLG (Figures 7a and S7a). Analyzing the maps of the position and intensity of the $E_{2(high)}$-peak of GaN (Figures 7b,c and S7c–f), it can be concluded that the characteristic areas of SLG and FLG experience different stress values in the epitaxial layers of nitrides (Figure 7c). The difference in stress values may be associated with stress relaxation due to the occurrence of additional defects during growth on more defective sections of graphene that correspond to SLG. Thus, nitride nucleation occurs more qualitatively on the FLG regions, which are less defective than the SLG regions.

The observed differences in the surface morphology of the obtained epitaxial layers of III-nitrides can be explained by the initial difference between graphene areas. For SLG sections, the average value of the $E_{2(high)}$ band position is $\omega \sim 567.6$ (Figures 7b and S7e). At the same time, we observed a slight red shift of the $E_{2(high)}$ band ($\omega = 567.4$ cm^{-1}), as indicated in Figures 7b and S7f. Figure 7e and Figure S7h show a significant shift for GaN grown on the MLG buffer ($\omega = 568.4$ cm^{-1}). Red shift of $E_{2(high)}$ mode frequency usually indicates a tensile strain, typical for epitaxial GaN grown on Si (111) substrate. The coefficient 4.2 cm^{-1} GPa^{-1} describes the linear dependence between biaxial stress and spectral shift of the E_2 phonon mode. Therefore, we calculated a tensile stress of GaN equal to 0.05 ± 0.01 GPa and $\sim 0.2 \pm 0.01$ GPa for GaN grown on FLG and MLG, respectively. This value is much lower than the typical stress of GaN on silicon substrate (0.4–0.68 GPa). The results of the Raman spectroscopy show that the formation of nitrides on graphene with higher roughness leads to a significant increase in graphene defects and deterioration in the quality of the nitride layers.

We investigated the interaction of graphene with AlN crystal and the effect of defects on the interaction energy between the layers. As an atomistic model, a rectangular $C_{100}Al_{64}N_{64}H_{32}$ cell was considered, containing one graphene layer and two layers of aluminum nitride (see Figure 8). The outer surface of aluminum nitride was passivated by hydrogen atoms. The ratio of the AlN and graphene lattices constants is close to 5:4, therefore the aluminum nitride and graphene layers contained 4×4 and 5×5 elementary cells, respectively. The positions of atoms and the lattice vectors were optimized simultaneously. As a result, we obtained the optimal cell size of 12.37×21.33 Å^2. The energy of the interaction between graphene and aluminum nitride was calculated by the formula:

$$E_b = (E(C_{100}) + E(Al_{64}N_{64}H_{32}) - E(C_{100}Al_{64}N_{64}H_{32}))/S \qquad (1)$$

where S is the cell area. The resulting interaction energy was 0.194 J/m^2, which is very close to the energy of interlayer interaction in graphite. Thus, the graphene/graphene and graphene/AlN interfaces are almost equally energetically feasible. The optimal distance between graphene and AlN layers was equal to 3.66 Å.

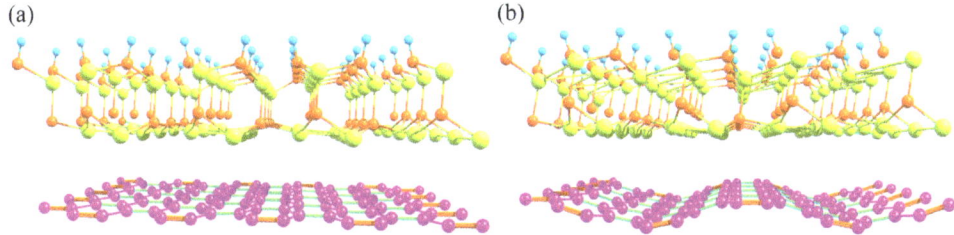

Figure 8. Periodic cell of the AlN–graphene heterostructure (**a**). Rippling of graphene under compression applied to the system (**b**). Purple, yellow, orange and blue balls represent C, Al, Ga and H atoms, respectively.

Next, we investigated the effect of uniaxial stretching and compression, as well as characteristic defects in graphene (vacancies, Stone–Wales defect and impurity atom of nitrogen) by E_b. Stretching and defects may occur as a result of the interaction of graphene with a silicon substrate if the number of graphene layers is not large enough. It is well known that graphene compression leads to its wave-like curvature without substantial shortening of C-C bonds. Therefore, the simulation of compressed graphene provides an estimation of the effect of sheet roughness on its interaction with AlN. In our study, the compression of the system by 5% led to roughness, at which the distance between the graphene and AlN layer was changed in the range of 3.2 to 4.3 Å.

The influence of various defects on the E_B value is represented in Table 1. Only one defect per cell shown in Figure 8 was introduced. Table 1 confirms that increased roughness due to graphene compression significantly enhances the interaction of graphene with AlN. In contrast, graphene stretching weakens this interaction. As for defects, only vacancies in graphene lead to a significant increase in the magnitude of E_B.

Table 1. Interaction energy of AlN surface with pristine, strained and defected graphene calculated with the density functional theory.

Type of Defect	E_b, J/m^2
No defects	0.194
Compression 5%	0.215
Stretching 5%	0.161
Stone-Wales defect	0.196
Vacancy	0.463
Nitrogen atom	0.187

On Figure 9a,c fragments of XRD spectra are shown, including GaN (0002) (2θ = 34.63°) and AlN (0002) (2θ = 36.06°) maxima for FLG (Figure 9a) and MLG (Figure 9c). The FWHM value of the (0002) rocking curve for GaN grown on an FLG buffer is 6.95 arc. min. For GaN on MLG the FWHM of the (0002) rocking curve was 7.848.

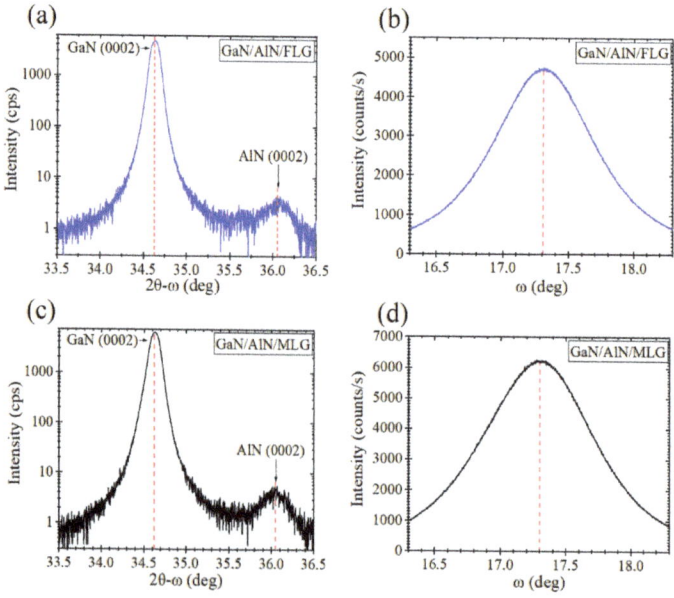

Figure 9. HR-XRD spectra of the experimental samples: (**a,c**) 2θ-ω scanning curve and (**b,d**) ω scanning curve of the GaN/AlN films grown on the FLG (**a,b**) and MLG (**c,d**).

In turn, Figure 9b,d shows the ω (rocking) scans for GaN/AlN samples, grown on FLG and MLG buffers, respectively. The peaks are centered at 34.56°, and a higher number of graphene layers results in the peaks broadening. The FWHM value increases from 55.55 to 60.00 arc. min. for samples grown on FLG and MLG buffers, respectively. The wider peaks observed in the case of using the MLG buffer confirm a poorer orientation and a large number of defects in the GaN grown on MLG.

4. Conclusions

In summary, we grew a bulk GaN/AlN on CVD graphene, transferred on silicon substrate coated by an amorphous oxide layer. In comparison with our previous works [18,28] on the study of nitrides epitaxy using a graphene buffer, in this research we clearly demonstrate that the number of graphene layers, the type of strain and defects are crucial for the high-quality epitaxy of GaN/AlN by the PA-MBE method. We found a strong correlation between the number of graphene layers and the quality of GaN/AlN grown on it. Bilayer and few-layer graphene were recognized as the most suitable buffers, which provide plasma-assisted molecular beam epitaxy of perfect GaN/AlN films. This fact can be explained by the difference in the FLG and MLG stiffness and defectiveness. We believe that our investigation contributes to the development of GaN-on-Si technology, which promises some benefits, such as low cost and high manufacturability. Due to its flat structure, moderate activity and tendency to non-covalent interaction, graphene probably can also be used for epitaxial growth of III-nitrides on other substrates which are amorphous of have a crystal lattice unsuitable for binding to III nitrides.

Supplementary Materials: The following supporting information can be downloaded at: https://www.mdpi.com/article/10.3390/app122211516/s1.

Author Contributions: Conceptualization, N.I.K., A.S.G. and D.P.B.; methodology, D.P.B., A.S.G., I.V.K. and K.P.K.; software, D.P.B. and K.P.K.; validation, D.P.B., A.S.G., I.V.K. and K.P.K.; formal analysis, D.P.B., A.S.G., I.V.K. and K.P.K.; investigation, D.P.B., A.S.G., P.L.D., A.A.T., M.M.M. (Mikhail Mikhailovich Mikhalik), K.P.K., M.M.M. (Mikhail Mikhailovich Maslov) and P.S.D.; resources, N.I.K., A.S.G., I.V.K. and K.P.K.; data curation, D.P.B., A.S.G. and I.V.K.; writing—original draft preparation, D.P.B.; writing—review editing, K.P.K., A.S.G. and D.P.B.; visualization, D.P.B., A.S.G. and K.P.K.; supervision, N.I.K. and V.A.L.; project administration, N.I.K.; funding acquisition, N.I.K. All authors have read and agreed to the published version of the manuscript.

Funding: This research received no external funding.

Institutional Review Board Statement: Not applicable.

Informed Consent Statement: Not applicable.

Data Availability Statement: The data presented in this study are available in Supplementary Materials.

Acknowledgments: This work was carried out using the equipment of the MEPhI Shared-Use Equipment Center (http://ckp-nano.mephi.ru, accessed on 31 October 2022).

Conflicts of Interest: The authors declare not conflict of interest.

References

1. Nakamura, S.; Krames, M.R. History of Gallium–Nitride-Based Light-Emitting Diodes for Illumination. *Proc. IEEE* **2013**, *101*, 2211–2220. [CrossRef]
2. Nakamura, S.; Mukai, T.; Senoh, M. Candela-class high-brightness InGaN/AlGaN double-heterostructure blue-light-emitting diodes. *Appl. Phys. Lett.* **1994**, *64*, 1687–1689. [CrossRef]
3. Nakamura, S.; Senoh, M.; Nagahama, S.; Iwasa, N.; Yamada, T.; Matsushita, T. and Sugimoto, Y. InGaN-Based Multi-Quantum-Well-Structure Laser Diodes. *Jpn. J. Appl. Phys.* **1996**, *35 Pt 2*, L74–L76. [CrossRef]
4. Watson, S.; Tan, M.; Najda, S.P.; Perlin, P.; Leszczynski, M.; Targowski, G.; Grzanka, S.; Kelly, A.E. Visible light communications using a directly modulated 422 nm GaN laser diode. *Opt. Lett.* **2013**, *38*, 3792–3794. [CrossRef] [PubMed]
5. Kneissl, M.; Seong, T.-Y.; Han, J.; Amano, H. The emergence and prospects of deep-ultraviolet light-emitting diode technologies. *Nat. Photonics* **2019**, *13*, 233–244. [CrossRef]
6. Asif Khan, M.; Bhattarai, A.; Kuznia, J.N.; Olson, D.T. High electron mobility transistor based on a GaN-$Al_xGa_{1-x}N$ heterojunction. *Appl. Phys. Lett.* **1993**, *63*, 1214–1215. [CrossRef]

7. Mishra, U.K.; Shen, L.; Kazior, T.E.; Wu, Y.-F. GaN-Based RF Power Devices and Amplifiers. *Proc. IEEE* **2008**, *96*, 287–305. [CrossRef]
8. Dadgar, A.; Poschenrieder, M.; Bläsing, J.; Contreras, O.; Bertram, F.; Riemann, T.; Krost, A. MOVPE growth of GaN on Si (111) substrates. *J. Cryst. Growth* **2003**, *248*, 556–562. [CrossRef]
9. Chen, K.J.; Häberlen, O.; Lidow, A.; Tsai, C.l.; Ueda, T.; Uemoto, Y.; Wu, Y. GaN-on-Si Power Technology: Devices and Applications. *IEEE Trans. Electron Devices* **2017**, *64*, 779–795. [CrossRef]
10. Morkoç, H. General properties of nitrides. In *Handbook of Nitride Semiconductors and Devices*; Wiley-VCH: Hoboken, NJ, USA, 2009.
11. Nikishin, S.A.; Faleev, N.N.; Antipov, V.G.; Francoeur, S.; Grave de Peralta, L.; Seryogin, G.A.; Temkin, H. High quality GaN grown on Si (111) by gas source molecular beam epitaxy with ammonia. *Appl. Phys. Lett.* **1999**, *75*, 2073–2075. [CrossRef]
12. Marchand, H.; Zhao, L.; Zhang, N.; Moran, B.; Coffie, R.; Mishra, U.K.; Speck, J.S.; DenBaars, S.P. Metalorganic chemical vapor deposition of GaN on Si (111): Stress control and application to field-effect transistors. *J. Appl. Phys.* **2001**, *89*, 7846–7851. [CrossRef]
13. Chung, K.; Lee, C.-H.; Yi, G.-C. Transferable GaN Layers Grown on ZnO-Coated Graphene Layers for Optoelectronic Devices. *Science* **2010**, *330*, 655–657. [CrossRef] [PubMed]
14. Chung, K.; Park, S.; Baek, H.; Chung, J.-S.; Yi, G.-C. High-quality GaN films grown on chemical vapor-deposited graphene films. *NPG Asia Mater.* **2012**, *4*, e24. [CrossRef]
15. Gupta, P.; Rahman, A.A.; Hatui, N.; Gokhale, M.R.; Deshmukh, M.M.; Bhattacharya, A. MOVPE growth of semipolar III-nitride semiconductors on CVD graphene. *J. Cryst. Growth* **2013**, *372*, 105–108. [CrossRef]
16. Li, T.; Liu, C.; Zhang, Z.; Yu, B.; Dong, H.; Jia, W.; Xu, B. GaN epitaxial layers grown on multilayer graphene by MOCVD. *AIP Adv.* **2018**, *8*, 045105. [CrossRef]
17. Araki, T.; Uchimura, S.; Sakaguchi, J. Radio-frequency plasma-excited molecular beam epitaxy growth of GaN on graphene/Si (100) substrates. *Appl. Phys. Express* **2014**, *7*, 071001. [CrossRef]
18. Borisenko, D.P.; Gusev, A.S.; Kargin, N.I.; Komissarov, I.V.; Kovalchuk, N.G.; Labunov, V.A. Plasma assisted-MBE of GaN and AlN on graphene buffer layers. *Jpn. J. Appl. Phys.* **2019**, *58*, SC1046. [CrossRef]
19. Yu, J.; Hao, Z.; Deng, J.; Li, X.; Wang, L.; Luo, Y.; Wang, J.; Sun, C.; Han, Y.; Xiong, B.; et al. Low-temperature van der Waals epitaxy of GaN films on graphene through AlN buffer by plasma-assisted molecular beam epitaxy. *J. Alloys Comp.* **2021**, *855*, 157508. [CrossRef]
20. Yan, H.; Ku, P.-C.; Gan, Z.-Y.; Liu, S.; Li, P. Strain Effects in Gallium Nitride Adsorption on Defective and Doped Graphene: First-Principles Calculations. *Crystals* **2018**, *8*, 58. [CrossRef]
21. Al Balushi, Z.Y.; Miyagi, T.; Lin, Y.-C.; Wang, K.; Calderin, L.; Bhimanapati, G.; Redwing, J.M.; Robinson, J.A. The impact of graphene properties on GaN and AlN nucleation. *Surf. Sci.* **2015**, *634*, 81–88. [CrossRef]
22. Li, T.; Liu, C.; Zhang, Z.; Yu, B.; Dong, H.; Jia, W.; Jia, Z.; Yu, C.; Gan, L.; Xu, B.; et al. Understanding the Growth Mechanism of GaN Epitaxial Layers on Mechanically Exfoliated Graphite. *Nanoscale Res. Lett.* **2018**, *13*, 130. [CrossRef] [PubMed]
23. Yu, Y.; Wang, T.; Chen, X.; Zhang, L.; Wang, Y.; Niu, Y.; Yu, J.; Ma, H.; Li, X.; Liu, F.; et al. Demonstration of epitaxial growth of strain-relaxed GaN films on graphene/SiC substrates for long wavelength light-emitting diodes. *Light. Sci. Appl.* **2021**, *10*, 117. [CrossRef] [PubMed]
24. Chen, Z.L.; Liu, Z.Q.; Wei, T.B.; Yang, S.; Dou, Z.; Wang, Y.; Ci, H.; Chang, H.; Qi, Y.; Yan, J.; et al. Improved Epitaxy of AlN Film for Deep-Ultraviolet Light-Emitting Diodes Enabled by Graphene. *Adv. Mater.* **2019**, *31*, 1807345. [CrossRef] [PubMed]
25. Fuke, S.; Sasaki, T.; Takahasi, M.; Hibino, H. In-situ X-ray diffraction analysis of GaN growth on graphene-covered amorphous substrates. *Jpn. J. Appl. Phys.* **2020**, *59*, 070902. [CrossRef]
26. Xu, Y.; Cao, B.; Li, Z.; Cai, D.; Zhang, Y.; Ren, G.; Xu, K. Growth Model of van der Waals Epitaxy of Films: A Case of AlN Films on Multilayer Graphene/SiC. *ACS Appl. Mater. Interf.* **2017**, *9*, 44001–44009. [CrossRef]
27. Lin, Y.-C.; Lu, N.; Perea-Lopez, N.; Li, J.; Lin, Z.; Peng, X.; Robinson, J.A. Direct Synthesis of van der Waals Solids. *ACS Nano* **2014**, *8*, 3715–3723. [CrossRef]
28. Borisenko, D.P.; Gusev, A.S.; Kargin, N.I.; Dobrokhotov, P.L.; Timofeev, A.A.; Labunov, V.A.; Kovalchuk, N.G.; Mikhalik, M.M.; Komissarov, I.V. Effect of graphene domains orientation on quasi van der Waals epitaxy of GaN. *J. Appl. Phys.* **2021**, *130*, 185304. [CrossRef]
29. Grimme, S. Semiempirical GGA-type density functional constructed with a long-range dispersion correction. *J. Comput. Chem.* **2006**, *27*, 1787–1799. [CrossRef]
30. Giannozzi, P.; Baroni, S.; Bonini, N.; Calandra, M.; Car, R.; Cavazzoni, C.; Ceresoli, D.; Chiarotti, G.L.; Cococcioni, M.; Dabo, I.; et al. QUANTUM ESPRESSO: A modular and open-source software project for quantum simulations of materials. *J. Phys. Condens. Matter* **2009**, *21*, 395502. [CrossRef]
31. Giannozzi, P.; Andreussi, O.; Brumme, T.; Bunau, O.; Buongiorno Nardelli, M.; Calandra, M.; Car, R.; Cavazzoni, C.; Ceresoli, D.; Cococcioni, M.; et al. Advanced capabilities for materials modelling with Quantum ESPRESSO. *J. Phys. Condens. Matter* **2017**, *29*, 465901. [CrossRef]
32. Perdew, J.P.; Burke, K.; Ernzerhof, M. Generalized Gradient Approximation Made Simple. *Phys. Rev. Lett.* **1996**, *77*, 3865–3868. [CrossRef] [PubMed]
33. Kresse, G.; Joubert, D. From ultrasoft pseudopotentials to the projector augmented-wave method. *Phys. Rev. B* **1999**, *59*, 1758–1775. [CrossRef]

34. Monkhorst, H.J.; Pack, J.D. Special points for Brillouin-zone integrations. *Phys. Rev. B* **1976**, *13*, 5188–5192. [CrossRef]
35. Methfessel, M.; Paxton, A.T. High-precision sampling for Brillouin-zone integration in metals. *Phys. Rev. B* **1989**, *40*, 3616–3621. [CrossRef]
36. Yao, Y.; Ren, L.; Gao, S.; Li, S. Histogram method for reliable thickness measurements of graphene films using atomic force microscopy (AFM). *J. Mater. Sci. Technol.* **2017**, *33*, 815–820. [CrossRef]
37. Shen, Z.; Li, J.; Yi, M.; Zhang, X.; Ma, S. Preparation of graphene by jet cavitation. *Nanotechnology* **2011**, *22*, 365306. [CrossRef]
38. Meyer, J.C.; Geim, A.K.; Katsnelson, M.I.; Novoselov, K.S.; Obergfell, D.; Roth, S.; Girit, C.; Zettl, A. On the roughness of single- and bi-layer graphene membranes. *Solid State Commum.* **2007**, *143*, 101–109. [CrossRef]
39. Pham, P.H.Q.; Quach, N.V.; Li, J.; Burke, P.J. Scalable and reusable micro-bubble removal method to flatten large-area 2D materials. *Appl. Phys. Lett.* **2018**, *112*, 163106. [CrossRef]
40. Ma, L.; Ren, W.; Cheng, H. Transfer Methods of Graphene from Metal Substrates: A Review. *Small Methods* **2019**, *3*, 1900049. [CrossRef]
41. Khestanova, E.; Guinea, F.; Fumagalli, L.; Geim, A.K.; Grigorieva, I.V. Universal shape and pressure inside bubbles appearing in van der Waals heterostructures. *Nat. Commum.* **2016**, *7*, 12587. [CrossRef]
42. Nair, R.R.; Blake, P.; Grigorenko, A.N.; Novoselov, K.S.; Booth, T.J.; Stauber, T.; Geim, A.K. Fine Structure Constant Defines Visual Transparency of Graphene. *Science* **2008**, *320*, 1308. [CrossRef] [PubMed]
43. Zhu, S.-E.; Yuan, S.; Janssen, G.C.A.M. Optical transmittance of multilayer graphene. *EPL (Europhys. Lett.)* **2014**, *108*, 17007. [CrossRef]
44. Cardona, M.; Güntherodt, G. (Eds.) *Light Scattering in Solids II*; Springer: Berlin/Heidelberg, Germany, 1982; Volume 50, pp. 19–176.
45. Saito, R.; Hofmann, M.; Dresselhaus, G.; Jorio, A.; Dresselhaus, M.S. Raman spectroscopy of graphene and carbon nanotubes. *Adv. Phys.* **2011**, *60*, 413–550. [CrossRef]
46. Shazni Mohammad Haniff, M.A.; Zainal Ariffin, N.H.; Ooi, P.C.; Mohd Razip Wee, M.F.; Mohamed, M.A.; Hamzah, A.A.; Syono, M.I.; Hashim, A.M. Practical Route for the Low-Temperature Growth of Large-Area Bilayer Graphene on Polycrystalline Nickel by Cold-Wall Chemical Vapor Deposition. *ACS Omega* **2021**, *6*, 12143–12154. [CrossRef] [PubMed]
47. Ferrari, A.C. Raman spectroscopy of graphene and graphite: Disorder, electron–phonon coupling, doping and nonadiabatic effects. *Solid State Commun.* **2007**, *143*, 47–57. [CrossRef]
48. Ferrari, A.C.; Meyer, J.C.; Scardaci, V.; Casiraghi, C.; Lazzeri, M.; Mauri, F.; Piscanec, S.; Jiang, D.; Novoselov, K.S.; Roth, S.; et al. Raman Spectrum of Graphene and Graphene Layers. *Phys. Rev. Lett.* **2006**, *97*, 187401. [CrossRef] [PubMed]
49. Gupta, A.; Chen, G.; Joshi, P.; Tadigadapa, S.; Eklund, P.C. Raman Scattering from High-Frequency Phonons in Supported n-Graphene Layer Films. *Nano Lett.* **2006**, *6*, 2667–2673. [CrossRef]
50. Li, J.; Ji, H.; Zhang, X.; Wang, X.; Jin, Z.; Wang, D.; Wan, L.-J. Controllable atmospheric pressure growth of monolayer, bilayer and trilayer graphene. *Chem. Commun.* **2014**, *50*, 11012–11015. [CrossRef]
51. Karamat, S.; Sonuşen, S.; Dede, M.; Uysallı, Y.; Özgönül, E.; Oral, A. Coalescence of few layer graphene grains grown by chemical vapor deposition and their stacking sequence. *J. Mater. Res.* **2015**, *31*, 46–54. [CrossRef]
52. Malard, L.M.; Pimenta, M.A.; Dresselhaus, G.; Dresselhaus, M.S. Raman spectroscopy in graphene. *Phys. Rep.* **2009**, *473*, 51–87. [CrossRef]
53. Zhao, H.; Lin, Y.-C.; Yeh, C.-H.; Tian, H.; Chen, Y.-C.; Xie, D.; Yang, Y.; Suenaga, K.; Ren, T.-L.; Chiu, P.-W. Growth and Raman Spectra of Single-Crystal Trilayer Graphene with Different Stacking Orientations. *ACS Nano* **2014**, *8*, 10766–10773. [CrossRef] [PubMed]
54. Kim, K.; Coh, S.; Tan, L.Z.; Regan, W.; Yuk, J.M.; Chatterjee, E.; Crommie, M.F.; Cohen, M.L.; Louie, S.G.; Zettl, A. Raman Spectroscopy Study of Rotated Double-Layer Graphene: Misorientation-Angle Dependence of Electronic Structure. *Phys. Rev. Lett.* **2012**, *108*, 246103. [CrossRef]
55. Ni, Z.; Wang, Y.; Yu, T.; Shen, Z. Raman spectroscopy and imaging of graphene. *Nano Res.* **2008**, *1*, 273–291. [CrossRef]
56. Goldie, S.J.; Bush, S.; Cumming, J.A.; Coleman, K.S. A Statistical Approach to Raman Analysis of Graphene-Related Materials: Implications for Quality Control. *ACS Appl. Nano Mater.* **2020**, *3*, 11229–11239. [CrossRef]
57. Chen, Q.; Yin, Y.; Ren, F.; Liang, M.; Yi, X.; Liu, Z. Van der Waals Epitaxy of III-Nitrides and Its Applications. *Materials* **2020**, *13*, 3835. [CrossRef] [PubMed]
58. Cançado, L.G.; Jorio, A.; Ferreira, E.H.M.; Stavale, F.; Achete, C.A.; Capaz, R.B.; Moutinho, M.V.O.; Lombardo, A.; Kulmala, T.S.; Ferrari, A.C. Quantifying Defects in Graphene via Raman Spectroscopy at Different Excitation Energies. *Nano Lett.* **2011**, *11*, 3190–3196. [CrossRef]
59. Banhart, F.; Kotakoski, J.; Krasheninnikov, A.V. Structural Defects in Graphene. *ACS Nano* **2010**, *5*, 26–41. [CrossRef]
60. Jia, Y.; Zhang, L.; Du, A.; Gao, G.; Chen, J.; Yan, X.; Brown, C.L.; Yao, X. Defect Graphene as a Trifunctional Catalyst for Electrochemical Reactions. *Adv. Mater.* **2016**, *28*, 9532–9538. [CrossRef]
61. Zandiatashbar, A.; Lee, G.-H.; An, S.J.; Lee, S.; Mathew, N.; Terrones, M.; Hayashi, T.; Picu, C.R.; Hone, J.; Koratkar, N. Effect of defects on the intrinsic strength and stiffness of graphene. *Nat. Commun.* **2014**, *5*, 3186. [CrossRef]
62. Eckmann, A.; Felten, A.; Mishchenko, A.; Britnell, L.; Krupke, R.; Novoselov, K.S.; Casiraghi, C. Probing the Nature of Defects in Graphene by Raman Spectroscopy. *Nano Lett.* **2012**, *12*, 3925–3930. [CrossRef]

63. Sato, K.; Saito, R.; Oyama, Y.; Jiang, J.; Cançado, L.G.; Pimenta, M.A.; Jorio, A.; Samsonidze, G.; Dresselhaus, G.; Dresselhaus, M. D-band Raman intensity of graphitic materials as a function of laser energy and crystallite size. *Chem. Phys. Lett.* **2006**, *427*, 117–121. [CrossRef]
64. Ferrari, A.C.; Basko, D.M. Raman spectroscopy as a versatile tool for studying the properties of graphene. *Nat. Nanotechnol.* **2013**, *8*, 235–246. [CrossRef] [PubMed]
65. Lu, Z.; Sun, X.; Xiang, Y.; Washington, M.A.; Wang, G.-C.; Lu, T.-M. Revealing the Crystalline Integrity of Wafer-Scale Graphene on SiO_2/Si: An Azimuthal RHEED Approach. *ACS Appl. Mater. Interf.* **2017**, *9*, 23081–23091. [CrossRef] [PubMed]
66. Ichimiya, A.; Cohen, P.I. *Reflection High-Energy Electron Diffraction*; Cambridge University Press: Cambridge, UK, 2004.

Article

Synthesis of C/SiC Mixtures for Composite Anodes of Lithium-Ion Power Sources

Anastasia M. Leonova [1,*], Oleg A. Bashirov [1], Natalia M. Leonova [1], Alexey S. Lebedev [2], Alexey A. Trofimov [1,3] and Andrey V. Suzdaltsev [1,3,*]

1. Scientific Laboratory of the Electrochemical Devices and Materials, Ural Federal University, Mira St. 28, 620002 Yekaterinburg, Russia
2. South Urals Research Center of Mineralogy and Geoecology UB RAS, Ilmen Reservation, 456317 Miass, Russia
3. Scientific-Research Department of Electrolysis, Institute of High-Temperature Electrochemistry UB RAS, Academicheskaya St. 20, 620137 Yekaterinburg, Russia
* Correspondence: a.m.leonova@urfu.ru (A.M.L.); a.v.suzdaltsev@urfu.ru (A.V.S.)

Abstract: Nowadays, research aimed at the development of materials with increased energy density for lithium-ion batteries are carried out all over the world. Composite anode materials based on Si and C ultrafine particles are considered promising due to their high capacity. In this work, a new approach for carbothermal synthesis of C/SiC composite mixtures with SiC particles of fibrous morphology with a fiber diameter of 0.1–2.0 μm is proposed. The synthesis was carried out on natural raw materials (quartz and graphite) without the use of complex equipment and an argon atmosphere. Using the proposed method, C/SiC mixture as well as pure SiC were synthesized and used to manufacture anode half-cells of lithium-ion batteries. The potential use of the resulting mixtures as anode material for lithium-ion battery was shown. Energy characteristics of the mixtures were determined. After 100 cycles, pure SiC reached a discharge capacity of 180 and 138 mAh g^{-1} at a current of C/20 and C, respectively, and for the mixtures of (wt%) 29.5C–70.5 SiC and 50Si–14.5C–35.5SiC discharge capacity of 328 and 400 mAh g^{-1} at a current of C/2 were achieved. The Coulombic efficiency of the samples during cycling was over 99%.

Keywords: lithium-ion battery; silicon carbide; electrodeposited silicon; composite material; anode material; Coulombic efficiency

Citation: Leonova, A.M.; Bashirov, O.A.; Leonova, N.M.; Lebedev, A.S.; Trofimov, A.A.; Suzdaltsev, A.V. Synthesis of C/SiC Mixtures for Composite Anodes of Lithium-Ion Power Sources. *Appl. Sci.* **2023**, *13*, 901. https://doi.org/10.3390/app13020901

Academic Editor: Gaind P. Pandey

Received: 15 December 2022
Revised: 2 January 2023
Accepted: 6 January 2023
Published: 9 January 2023

Copyright: © 2023 by the authors. Licensee MDPI, Basel, Switzerland. This article is an open access article distributed under the terms and conditions of the Creative Commons Attribution (CC BY) license (https:// creativecommons.org/licenses/by/ 4.0/).

1. Introduction

Graphite is traditionally used as an anode material of lithium-ion batteries (LIBS) due to its relatively low cost, low volume expansion (up to 10%), high electrical conductivity, charge rate. On the other hand, this material capacity (up to 372 mAh g^{-1}) [1–3] no longer meets the requirements of modern devices and machines for an increased energy density.

Promising anode materials with a higher specific capacity for LIBS are transition metal oxides [4–9], silicon [10–14], germanium [15–17], SiC [18,19], as well as various composite mixtures of the above materials with carbon [20–27]. Transition metal oxides seem to be cheap, easy to synthesize. They provide a relatively high capacity (theoretical up to 718 mAh g^{-1} [4]; experimental up to 1150 mAh g^{-1} [4]) and charge rate, but suffer from high volume expansion (about 96% [8]). Silicon also seems to be a suitable material that provides maximum capacity (theoretical up to 4200 mAh g^{-1} [9–12], experimentally achieved—3900 mAh g^{-1} [13]) and a sufficiently high charge rate, but drastic volume expansion (up to 300% [10–12]) still presents the biggest challenge to the realization of Si anodes. Germanium is a less accessible material with a lower capacity (theoretical, up to 1624 mAh g^{-1}; experimental, up to 1248 mAh g^{-1} [16]) compared to silicon, but it provides advantages such as higher electronic conductivity. Moreover, the diffusion coefficient of lithium ions in germanium is 400 times higher than in silicon [17].

Silicon carbide is a perspective easily available anode material with high mechanical, chemical and thermal stability [18,19]. However, for lithiation, pure SiC must have a certain structure and size. Therefore, it is considered more as a matrix or substrate that allows to compensate the volume expansion of silicon. Nevertheless, several works have reported on the LIBs with SiC anode. In [28,29], a decrease in the thickness of SiC films or particles leads to an increase in their capacity from 300 to 1370 mAh g^{-1} with a Coulombic efficiency of about 90%. This also explains the significant scatter of the SiC anode capacity in the available papers. Finally, recently a significant number of articles have been devoted to the development of LIBs with composite anode materials represented by silicon, graphite, SiC, SiO$_x$, etc., including various sizes and structures: core–shell wires, tubes, needles, fibers, multilayer films of graphene and silicene [30–34]. Calculations showed improved electrochemical parameters of such structures during lithiation/delithiation [35]. Despite the advantages of such composite anodes, their relatively complex production should be noted. Often, single samples are presented in the existing works and nothing is reported about the large-scale production of the materials. At present, the carbothermal method is mainly used for the large-scale production of silicon carbide [36,37]. At the same time, many works are devoted to the metallothermic preparation of SiC [38,39], although these methods require further purification of carbide from intermediate synthesis products.

Previously, we showed the possibility of synthesizing ultrafine SiC fibers from cheap materials using carbothermal synthesis [40,41]. The feature of our method is the use of natural samples of pure quartz and graphite with a certain morphology and particle size. In this work, in addition to the obtaining of SiC, this approach was also used for obtaining of C/SiC mixtures suitable for use as composite anodes of lithium-ion power sources.

2. Materials and Methods

Scheme of the synthesis: The synthesis of SiC powder was carried out in a graphite crucible, which was placed in a protective alumina crucible. Mixture of SiO$_2$ and graphite powder with molar ratio of 1–3 was prepared by hand mixing in agate mortar and placed in a crucible. Graphite crucible was covered with graphite cap and buried under extra graphite layer. Synthesis was conducted at a temperature of 1600 °C for 5 h under CO atmosphere. In such reactor and conditions, the atmosphere maintained stable due to the oxidation of graphite powder [40,41]. The scheme for the synthesis of SiC-based mixtures with different compositions is shown in Figure 1. It was shown experimentally that mixtures with 40–95 wt% SiC and unreacted graphite are formed during synthesis depending on the given SiO$_2$:C ratio, temperature and synthesis duration; the products may also contain traces of silicon and its nonstoichiometric oxides, which are removed by additional treatment in the HF solution. Therefore, the primary synthesis products are the C/SiC mixtures. The ratio of components in such mixtures can be controlled in order to obtain required mixture.

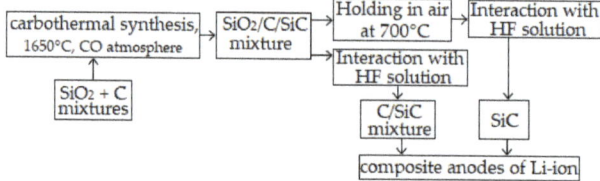

Figure 1. Process flow diagram for obtaining C/SiC mixtures for the manufacture of LIBs anodes.

In this work we synthesized a C/SiC mixture and pure SiC that were studied in the anode half-cell of LIBs. Along with these materials, a composite anode based on the resulting C/SiC mixture with the addition of electrtodeposited silicon fibers from the KCl–K$_2$SiF$_6$ melt was also tested [42].

Analysis of the morphology and composition: The chemical and phase composition of the reagents and products was determined by inductively coupled plasma atomic emission spectroscopy (AES-ICP) using iCAP 6300 Duo Spectrometer (Thermo Scientific, Waltham, MA, USA), X-ray phase analysis (XRD) using Rigaku D/MAX-2200VL/PC diffractometer (Rigaku, Tokyo, Japan) and Raman spectroscopy using U1000 Raman spectrometer (Renishaw, New Mills, UK). The morphology and elemental composition of the obtained samples were studied using a Tescan Vega 4 (Tescan, Brno–Kohoutovice, Czech Republic) scanning electron microscope with Xplore 30 EDS detector (Oxford, UK).

Electrochemical performance: Electrochemical performance of SiC, C/SiC and Si/C/SiC anodes were investigated in a 3-electrode half-cell. The obtained compositions were mixed with 10 wt% polyvinylidene fluoride dissolved in N-methyl-2-pyrollidone without any other additives. LIBs anode half-cell fabrication was performed in an argon-filled glove box (O_2, H_2O < 0.1 ppm). Stainless steel mesh with the applied composite anode material was used as the working electrode and two separate lithium strips as the counter and reference electrodes. All electrodes were divided by 2 layers of polypropylene separator and tightly placed in the cell. The cell was filled with 1 mL of electrolyte—1 M $LiPF_6$ in a mixture of ethylene carbonate/dimethyl carbonate/diethyl carbonate (1:1:1 by volume). Electrochemical measurements and cycling experiments were performed using a Zive-SP2 potentiostat (WonATech, Seoul, Republic of Korea).

3. Results

3.1. Samples Characterization

C/SiC mixtures: Figure 2 shows a SEM-image of a C/SiC mixture and maps of elemental distribution after carbothermal synthesis and treatment in HF solution. The resulting mixture is represented by particles with a size of about 20–40 microns (graphite) and smaller fibers (SiC). According to X-ray microanalysis, the content of the elements was (wt%) silicon 47.5–51.3; carbon 47.7–51.7; oxygen—up to 1.6. The presence of oxygen could be due to both the insufficient treatment time of the mixture in the HF solution and the subsequent oxidation of the silicon or SiC presented in the mixture. According to ICP analysis, the resulting mixture contained 48.8–49.4 wt% silicon (the rest was carbon) and no more than 0.4 ppm of such impurities as Fe, Al, Ti and Ca. If we do not take into account the presence of oxygen, the ratio of components corresponds to a mixture of (wt%): 70.5SiC-29.5C.

 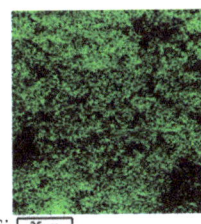 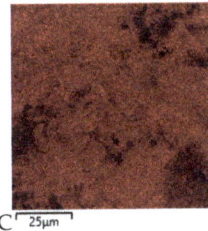 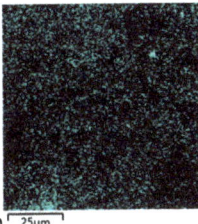

Figure 2. SEM-images and elements distribution in a C/SiC mixture after carbothermal synthesis and treatment in HF solution.

SiC: Figure 3 shows a SEM-image of a typical SiC agglomerate obtained after annealing of residual carbon from a C/SiC mixture. Its total size is about 50–60 µm. It consists of fibers with a diameter of 0.1 to 2 µm. The obtained morphology is similar to the previously obtained samples of ultrafine SiC [40] since the reagents and the synthesis procedure were reproduced almost completely. The distribution of elements in the resulting SiC are also shown. Uniform distribution of silicon, carbon and oxygen is observed for this sample. According to X-ray microanalysis, the average content of elements at different points of the sample was (wt%): 68–69 silicon, 29–30 carbon, up to 1.6 oxygen. This ratio is close to the SiC stoichiometric composition.

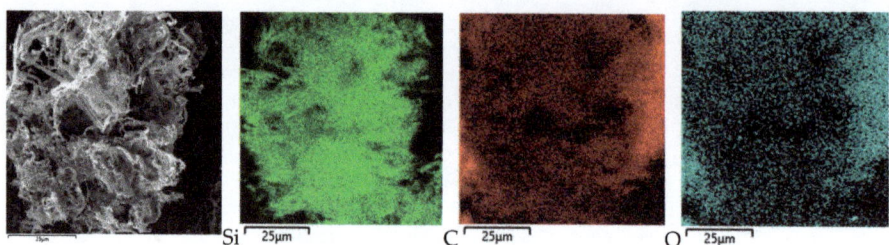

Figure 3. SEM-images and elements distribution in a SiC sample after carbothermal synthesis, oxidation of unreacted graphite, and treatment in HF solution.

Figure 4 shows the X-ray diffraction patterns and Raman spectra for C/SiC and SiC samples. For the C/SiC sample there are peaks of residual SiO_2 (21–22°), carbon (26–27°), as well as peaks indicating presence of two carbide modifications (α-SiC, β-SiC) in sample (35–36°, 41.5°, 44.5°, 55° and 60°). For the SiC sample, there are no peaks of C and SiO_2; only additional signals of the β-SiC phase appear at 64.5°. A similar picture is also observed in Raman spectra, in which responses of α-SiC and β-SiC carbide modifications at 770–793 cm^{-1} were fixed. In this case, for the SiC sample, there is also a response at 502 cm^{-1} that can correspond to both residual silicon and the β-SiC modification [43]. Moreover, there are lines near 1370, 1510 and 1585 cm^{-1} on both spectra, which indicate the C–C bonds. In this work, the structure of the used graphite powder was not studied in detail. However, based on the absence of pronounced peaks and intensity ratio of carbon lines, one can only note the presence of different carbon structures [44–47]. For the SiC sample, the intensities of these lines are less pronounced.

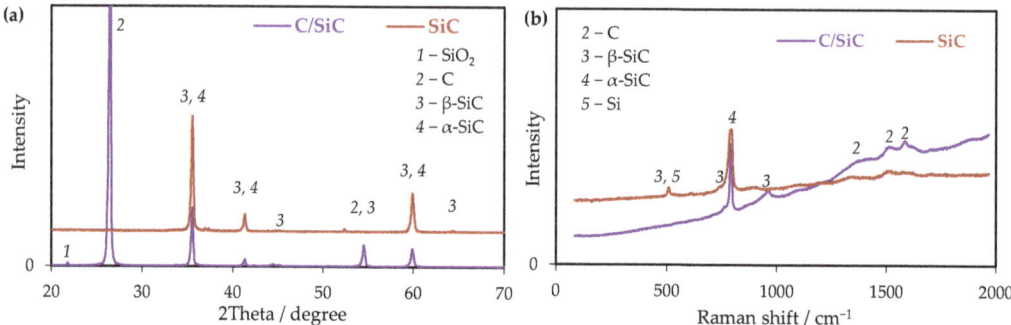

Figure 4. X-ray diffraction patterns (**a**) and Raman spectra (**b**) for the obtained C/SiC and SiC samples.

Electrodeposited Si: Figure 5 shows a SEM-image of silicon deposits obtained by electrolysis of the (wt%) 98KCl–2K_2SiF_6 melt at a temperature of 780 °C and a cathode current density of 25 mA cm^{-2}. A detailed procedure and parameters for silicon synthesis were provided earlier [42]. The resulting silicon deposits are arbitrarily shaped fibers with a diameter in the range of 0.45–0.55 μm and a length of up to 20–25 μm. According to X-ray microanalysis data, the oxygen content in the obtained silicon was from 1.2 to 1.5 wt% and other impurities did not exceed 0.18 ppm (mainly iron and nickel). Figure 6 shows the X-ray diffraction pattern and the Raman spectra of silicon deposit. As can be seen, the sample is represented by polycrystalline silicon with SiO_2 impurities. This is also indicated by the Raman spectra of the sample in which only the Si–Si bond at 510 cm^{-1} was found [43].

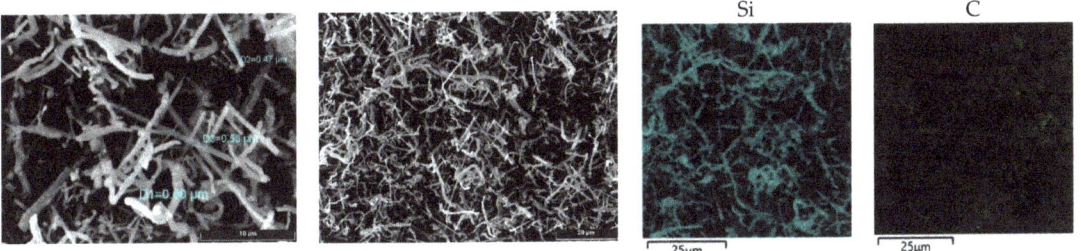

Figure 5. SEM-image and the elements distribution in a Si sample obtained by electrolysis of the (wt%) 98KCl-2K$_2$SiF$_6$ melt at a temperature of 780 °C and a cathode current density of 25 mA cm^{-2}.

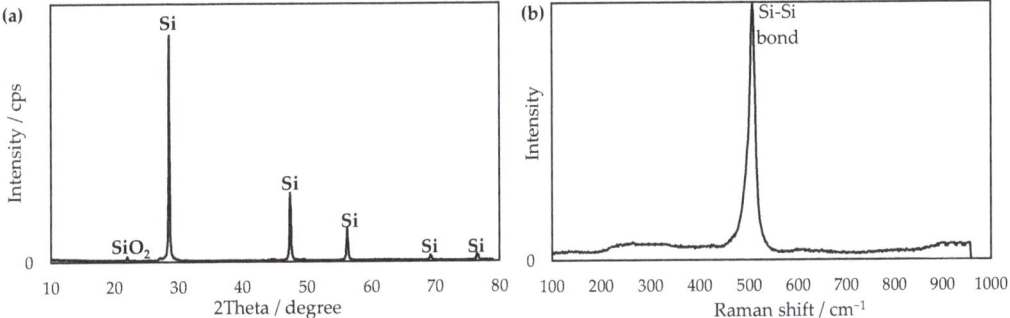

Figure 6. X-ray diffraction pattern (**a**) and the Raman spectra (**b**) of silicon deposit.

3.2. Electrochemical Performance of the Comosite Anodes

C/SiC composite anode: Figure 7 shows the change in the C/SiC composite anode potential during lithiation/delithiation, as well as changes in the discharge capacity and Coulombic efficiency of the sample during cycling. When the C/SiC composite anode sample was initially charged with a current of 0.1 C (first cycle) its charging capacity was 658 and discharge capacity was 322 mAh g^{-1} (Coulombic efficiency is 49%). Such capacity loss is usually attributed to a solid–electrolyte interface (SEI) layer formation. All subsequent cycling was carried out at a current of 0.5 C. In the second cycle, the discharge capacity was 308 mAh g^{-1} (Coulombic efficiency is 92%), and after the 100th cycle the discharge capacity was 328 mAh g^{-1} (see Figure 7c). During cycling, the Coulombic efficiency of the C/SiC anode reached 99% and remained at this level even after capacity began to fade.

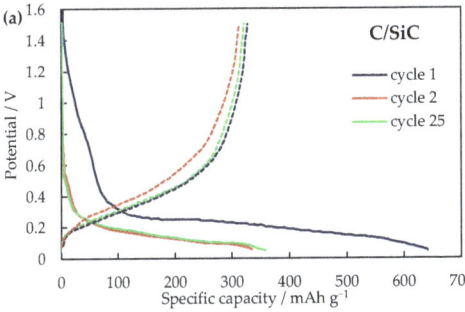
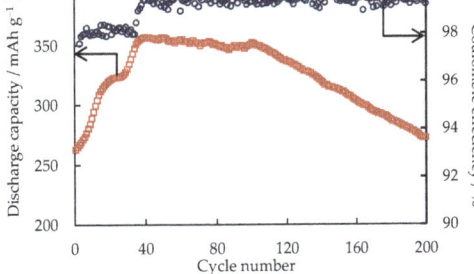

Figure 7. *Cont.*

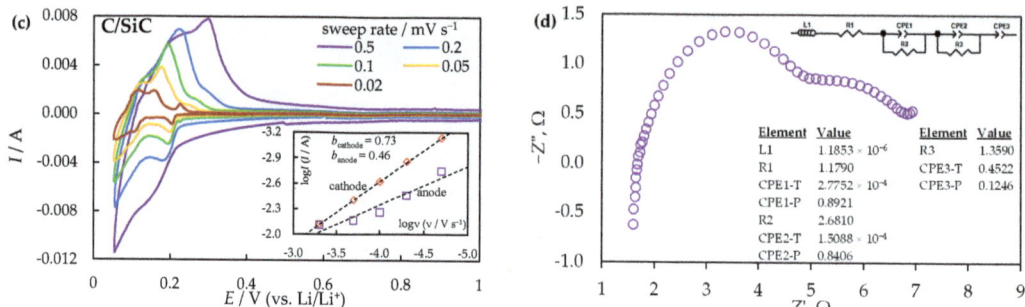

Figure 7. Cycling of C/SiC electrode: (**a**) potential change during lithiation (dashed line) and delithiation (solid line); (**b**) discharge capacity and Coulombic efficiency; (**c**) CVs at different scan rates and log(I)–log(v) dependencies; (**d**) EIS data with equivalent circuit diagram.

Figure 7c shows current–voltage dependences (CVs) characterizing the kinetics of charge and discharge of a C/SiC composite anode material. It can be noted that the beginning of the charge occurs at an electrode potential negatively than 0.22 V relative to the potential of the lithium electrode. This behavior may be due to the interaction of lithium ions with the electrode material, which is accompanied by the chemical formation of compounds. In this case the presence of several peaks in the cathode region of the CVs indicates the occurrence of several lithiation reactions. At a potential of about 0.05 V a lithium reduction wave is formed. During discharge in the anodic region of the CVs in the potential range from 0.05 to 0.5 V there are corresponding discharge (oxidation) peaks of lithium. Moreover, the shift in the peak potentials of the lithium reduction and oxidation indicates that the processes is not electrochemically reversible. Obviously, the irreversibility may be due to the interaction of lithium with the anode material. Since the LIBs capacity includes the contribution of both diffusion and capacitance reactions the expression [48] is valid for the usual lithiation/delithiation process:

$$\log(I) = \log(a) + b\log(v) \tag{1}$$

where I—peak current, A; v—potential sweep rate, V s^{-1}; a and b—constants. In our case, the calculated b value was of 0.73 (see Figure 7c), which indicates the pseudocapacitive behavior of the C/SiC anode material [48]. This may be associated with a slow chemical reaction of silicon carbide with lithium. Figure 7d shows the electrochemical impedance spectra (frequency range from 100,000 to 1 Hz) of the sample after the forming cycle. The EIS contains two arcs corresponding to the processes of charge transfer through the electrolyte layer and between the electrode and electrolyte [49–51]. The obtained data can be described by a typical for LIBs equivalent circuit. It has two RC circuits connected in series and a resistance (Figure 7e). Parameters changes in this scheme during cycling have not yet been studied.

SiC anode: The cycling results of the SiC electrode are shown in Figure 8. When the SiC electrode sample was initially charged at a current of C/20 (first cycle) its charge capacity was only 78 and the discharge capacity was of 42 mAh g^{-1} (Coulombic efficiency is 54%). During further cycling, the capacity gradually increased. After the 60 cycles at C/20, capacity increased up to 180 mAh g^{-1}. This situation can be caused by gradual activation of the anode material accompanied by partial destruction of SiC and the formation of Li–C and Li–Si compounds [28,29] via the reactions:

$$SiC + xLi^+ + x e^- \rightarrow Li_xSi_yC + (1-y)Si \ (y < 1) \tag{2}$$

$$Si + zLi^+ + ze^- \leftrightarrow Li_zSi \tag{3}$$

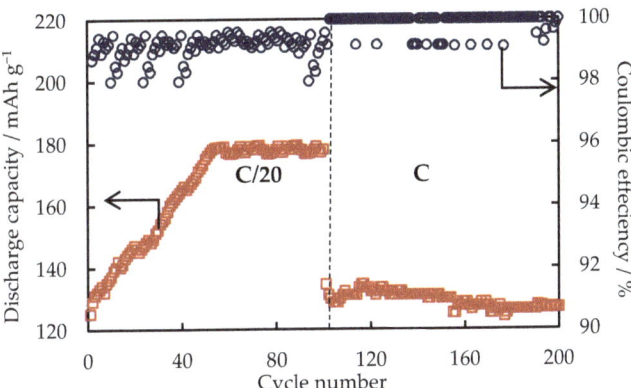

Figure 8. Changes in the discharge capacity and Coulombic efficiency of the SiC sample during cycling at a current of C/20 and C.

As a result, a gradual conversion of SiC to Si and C occurs [28,29]. In turn, reactions (1) and (2) also explain the reason for the onset of lithium discharge on anodes with SiC at a potential 0.22 V more positive than the lithium potential (see Figure 7b). With an increase in the charge current to C, the discharge capacity decreased, while its value was stable during 100 cycle and the Coulombic efficiency was more than 99.2%. Relatively low capacity can be explained by the presence of voids (see Figure 3) and a lack of the electrical contact between stiff SiC wires.

Si/C/SiC anode: Figure 9 shows the changes of the Si/SiC/C electrode potential during lithiation and delithiation as well as changes in its discharge capacity and Coulombic efficiency during cycling. A voltage plateau below 0.1 V (Figure 9a) indicates lithiation of graphite and silicon; delithiation occurs at 0.15–0.4 V. Similar results were obtained for lithiation of the silicon anode [52]. In the first cycle at a current of C/20, the discharge capacity was 225 mAh g^{-1} and the initial Coulombic efficiency was 54%. Coulombic efficiency gradually increased during further cycling at a current of 0.5C and remained above 98% after 20 cycles (Figure 9b). The discharge capacity gradually increased up to 525 mAh g^{-1} after 40 cycles. The higher capacity value can be explained by the silicon lithiation in the anode material and the additional increase during cycling can be explained by the gradual activation of the electrode, as in the case of the SiC anode (see Figure 8a). The subsequent decrease in the discharge capacity to 400 mAh g^{-1} by the 100th cycle may be due to the contact loss of part of the anode material with the substrate due to the local expansion and cracking of silicon.

Figure 9c shows the CVs obtained at different scan rates for the Si/SiC/C electrode. There are clear redox peaks indicating the charge and discharge of the sample. In this case the calculated value of b was 0.63 (see Figure 9c). This indicates that the operation of the Si/C/SiC electrode mainly proceeds under lithium diffusion conditions. A distinctive feature of the obtained CVs is also the fact that the charge and discharge currents of the anode sample are observed in a wider potential range (0.8–0.1 and 0.1–1.4 V respectively), which is due to the higher bonding energy of lithium with silicon and a larger number of compounds in the Li–Si system [52–54].

Figure 9d shows the electrochemical impedance spectrum of the Si/C/SiC sample after the forming cycle. One can note the relatively high resistance R1 (7.7 Ω), which can result in a significant change in the parameters of two series-connected RC chains (R2, C2, R3 and W1) [49,50]. The decrease in R1 will be the subject of our further research.

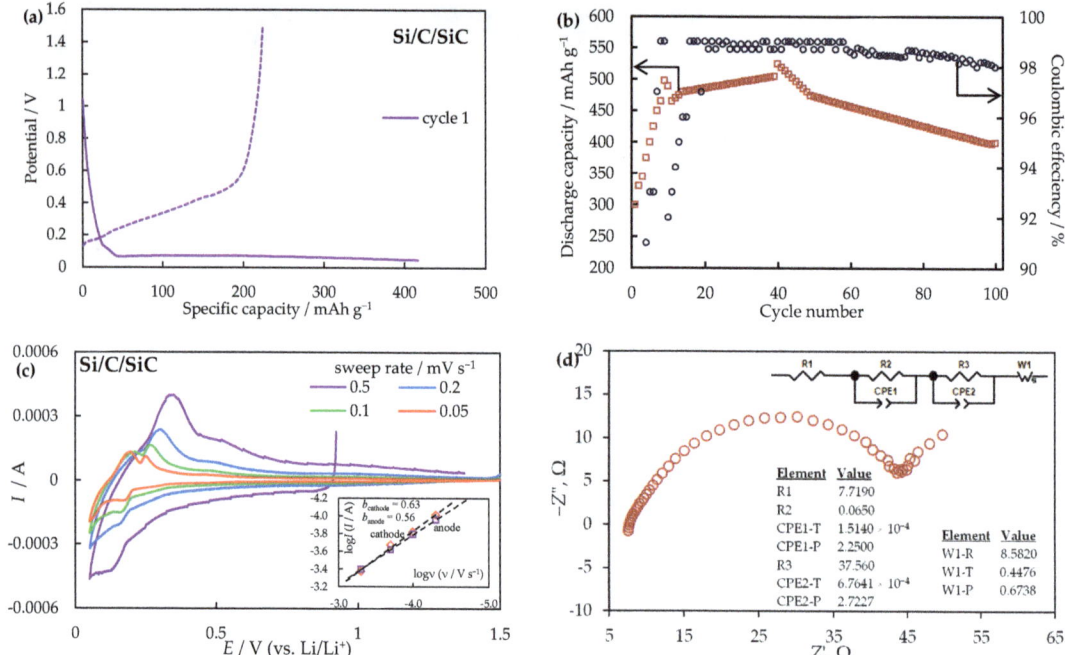

Figure 9. Cycling of Si/C/SiC electrode: (**a**) potential change during first cycle lithiation (dashed line) and delithiation (solid line); (**b**) discharge capacity and Coulombic efficiency; (**c**) CVs at different scan rates and log(I)–log(ν) dependencies; (**d**) EIS data with equivalent circuit diagram.

The lithiation mechanism of the Si/C/SiC anode as a whole can be represented by the parallel flow of reactions (2) and (3) as well as reactions (4)–(6):

$$Si + xLi^+ + xe^- \leftrightarrow Li_xSi \qquad (4)$$

$$Li_xSi + yLi^+ + ye^- \leftrightarrow Li_{(x+y)}Si \qquad (5)$$

$$C + zLi^+ + ze^- \leftrightarrow Li_zC \qquad (6)$$

Accurate estimation of the lithiated products and intermediate products (or its absence) on the basis of peak potentials is difficult. In the literature, there is a very wide spread of the potentials of the occurring reactions and the available analysis methods (EDX, XRD) do not allow one to make a local assessment with the required accuracy when a thick SEI layer is formed.

The partial replacement of SiC with electrodeposited Si leads to an increase in the discharge capacity of the anode, while the cycling stability of the Si/C/SiC composite electrode is lower. In this regard, further work will be aimed at studying the morphology and composition of samples after cycling in order to identify ways to optimize anode composition.

4. Conclusions

Nowadays, one of the most popular issues is connected with the search of anode materials for high energy density lithium-ion batteries. One of the promising materials that meet these requirements are composite materials based on Si/C/SiC mixtures. In such materials, silicon provides increased capacitance, graphite provides high electrical conductivity and SiC provides strength and thermal stability.

In this work, we proposed a new approach for the fabrication of composite anodes based on SiC as well as mixtures of C/SiC and Si/C/SiC. The proposed approach includes carbothermal synthesis and makes it possible to exclude complex equipment and expensive reagents for the anode materials synthesis. Using the proposed method, samples of C/SiC and SiC were synthesized and investigated. A sample of Si/C/SiC was fabricated with the addition of electrodeposited silicon fibers. It was shown that the synthesized SiC is represented by agglomerates of carbide fibers with diameter ranging from 0.1 to 2 μm; the C/SiC mixture is represented by evenly distributed fibers over the matrix of unreacted graphite; silicon is represented by arbitrary shape fibers with a diameter ranging from 0.45 to 0.55 μm.

Electrochemical behavior of the synthesized samples was studied as part of the anode half-cell of a lithium-ion battery. The possibility of using the obtained samples as part of the composite anode is shown. After 100 cycles pure SiC reached a discharge capacity of 180 and 138 mAh g^{-1} at a current of C/20 and C, respectively. The mixtures of (wt%) 29.5C-70.5 SiC and 50Si-14.5C-35.5SiC reached a discharge capacity of 328 and 400 mAh g^{-1} respectively at a C/2 current. The Coulombic efficiency of sample cycling was over 99%. The parameters of the equivalent circuits of the half-cells were estimated and ways to optimize their manufacturing process are noted.

Author Contributions: Conceptualization, A.V.S.; methodology; A.M.L., A.S.L. and N.M.L.; validation, A.S.L. and A.V.S.; formal analysis, A.M.L., O.A.B. and N.M.L.; investigation, A.M.L., O.A.B., A.S.L. and N.M.L.; writing—original draft preparation, A.M.L., A.A.T. and A.V.S.; writing—review and editing, A.V.S.; supervision, A.V.S.; project administration, A.V.S. All authors have read and agreed to the published version of the manuscript.

Funding: This research received no external funding.

Institutional Review Board Statement: Not applicable.

Informed Consent Statement: Not applicable.

Data Availability Statement: The data presented in this study are available on request from the corresponding author.

Acknowledgments: This work is performed in the frame of the State Assignment number 075-03-2022-011 dated 01/14/2022 (the theme number FEUZ-2020-0037). The elemental composition of the samples was analyzed using the equipment of the Central Research Laboratory "Composition of compounds" (IHTE UB RAS).

Conflicts of Interest: The authors declare no conflict of interest.

References

1. Kulova, T.L. New Electrode Materials for Lithium-Ion Batteries (Review). *Russ. J. Electrochem.* **2013**, *49*, 1–25. [CrossRef]
2. Luo, H.; Wang, Y.; Feng, Y.-H.; Fan, X.-Y.; Han, X.; Wang, P.-F. Lithium-Ion Batteries under Low-Temperature Environment: Challenges and Prospects. *Materials* **2022**, *15*, 8166. [CrossRef] [PubMed]
3. Mitsch, T.; Kramer, Y.; Feinauer, J.; Gaiselmann, G.; Markotter, H.; Manke, I.; Hintennach, A.; Schmidt, V. Preparation and Charactherization of Li-Ion Graphite Anodes Using Synchrotron Tomography. *Materials* **2014**, *7*, 4455–4472. [CrossRef]
4. Liu, R.-J.; Yang, L.-X.; Lin, G.-Q.; Bu, H.-P.; Wang, W.-J.; Liu, H.-J.; Zeng, C.-L. Superior electrochemcial performances of core-shell structured vanadium oxide@vanadium carbide composites for Li-ion storage. *Appl. Surf. Sci.* **2022**, *588*, 152904. [CrossRef]
5. Bini, M.; Ambrosetti, M.; Spada, D. ZnFe$_2$O$_4$, a Green and High-Capacity Anode Material for Lithium-Ion Batteries: A Review. *Appl. Sci.* **2021**, *11*, 11713. [CrossRef]
6. Purwanto, A.; Muzayanha, S.U.; Yudha, C.S.; Widiyandari, H.; Jumari, A.; Dyartanti, E.R.; Nizam, M.; Putra, M.I. High Performance of Salt-Modified–LTO Anode in LiFePO$_4$ Battery. *Appl. Sci.* **2020**, *10*, 7135. [CrossRef]
7. Zhang, Y.; Wang, N.; Bai, Z. The Progress of Cobalt-Based Anode Materials for Lithium Ion Batteries and Sodium Ion Batteries. *Appl. Sci.* **2020**, *10*, 3098. [CrossRef]
8. Feng, Y.; Zhang, H.; Li, W.; Fang, L.; Wang, Y. Targeted Synthesis of Novel Hierarchical Sandwiched NiO/C Arays as High-Efficiency Lithium Ion Batteries Anode. *J. Power Sources* **2016**, *301*, 78–86. [CrossRef]
9. Spinner, N.; Mustain, W.E. Nanostructural Effects on the Cycle Life and Li$^+$ Diffusion Coefficient of Nickel Oxide Anodes. *J. Electroanal. Chem.* **2013**, *711*, 8–16. [CrossRef]

10. Gevel, T.; Zhuk, S.; Leonova, N.; Leonova, A.; Trofimov, A.; Suzdaltsev, A.; Zaikov, Y. Electrochemical Synthesis of Nano-Sized Silicon from KCl–K$_2$SiF$_6$ Melts for Powerful Lithium-Ion Batteries. *Appl. Sci.* **2021**, *11*, 10927. [CrossRef]
11. Feng, K.; Li, M.; Liu, W.; Kashkooli, A.G.; Xiao, X.; Cai, M.; Chen, Z. Silicon-Based Anodes for Lithium-Ion Batteries: From Fundamentals to Practical Applications. *Small* **2018**, *14*, 1702737. [CrossRef] [PubMed]
12. Ha, Y.; Schulze, M.C.; Frisco, S.; Trask, S.E.; Teeter, G.; Neale, N.R.; Veith, G.M.; Johnson, C.S. Li$_2$O-Based Cathode Additives Enabling Prelithiation of Si Anodes. *Appl. Sci.* **2021**, *11*, 12027. [CrossRef]
13. Suzdaltsev, A. Silicon Electrodeposition for Microelectronics and Distributed Energy: A Mini-Review. *Electrochem* **2022**, *3*, 760–768. [CrossRef]
14. Yang, X.; Tachikawa, N.; Katayama, Y.; Li, L.; Yan, J. Effect of the Pillar Size on the Electrochemical Performance of Laser-Induced Silicon Micropillars as Anodes for Lithium-Ion Batteries. *Appl. Sci.* **2019**, *9*, 3623. [CrossRef]
15. Stokes, K.; Flynn, G.; Geaney, H.; Bree, G.; Ryan, K.M. Axial Si–Ge Heterostructure Nanowires as Lithium-Ion Battery Anodes. *Nano Lett.* **2018**, *18*, 5569–5575. [CrossRef] [PubMed]
16. Chockla, A.M.; Klavetter, K.C.; Mullins, C.B.; Korgel, B.A. Solution-Grown Germanium Nanowire Anodes for Lithium-Ion batteries. *ACS Appl. Mater. Interfaces* **2012**, *4*, 4658–4664. [CrossRef]
17. Kulova, T.L.; Skundin, A.M. Germanium in Lithium-Ion and Sodium-Ion Batteries (A Review). *Russ. J. Electrochem.* **2022**, *57*, 1105–1137. [CrossRef]
18. Liu, X.; Zhu, X.; Pan, D. Solutions for the Problems of Silicon–Carbon Anode Materials for Lithium-Ion Batteries. *R. Soc. Open Sci.* **2018**, *5*, 172370. [CrossRef]
19. Kolosov, D.A.; Glukhova, O.E. Theoretical Study of a New Porous 2D Silicon-Filled Composite Based on Graphene and Single-Walled Carbon Nanotubes for Lithium-Ion Batteries. *Appl. Sci.* **2020**, *10*, 5786. [CrossRef]
20. Abe, Y.; Saito, I.; Tomioka, M.; Kabir, M.; Kumagai, S. Effects of Excessive Prelithiation on Full-Cell Performance of Li-Ion Batteries with a Hard-Carbon/Nanosized-Si Composite Anode. *Batteries* **2022**, *8*, 210. [CrossRef]
21. Hou, Y.; Yang, Y.; Meng, W.; Lei, B.; Ren, M.; Yang, X.; Wang, Y.; Zhao, D. Core-Shell Structured Si@Cu Nanoparticles Encapsulated in Carbon Cages as High-Performance Lithium-Ion Battery Anodes. *J. Alloys Compd.* **2021**, *874*, 159988. [CrossRef]
22. Fan, Z.; Wang, Y.; Zheng, S.; Xu, K.; Wu, J.; Chen, S.; Liang, J.; Shi, A.; Wang, Z. A Submicron Si@C Core-Shell Intertwined with Carbon Nanowires and Graphene Nanosheet as a High-Performance Anode Material for Lithium Ion Battery. *Energy Storage Mater.* **2021**, *39*, 1–10. [CrossRef]
23. Zhang, X.; Min, B.-I.; Wang, Y.; Hayashida, R.; Tanaka, M.; Watanabe, T. Preparation of Carbon-Coated Silicon Nanoparticles with Different Hydrocarbon Gases in Induction Thermal Plasma. *J. Phys. Chem. C* **2021**, *125*, 15551–15559. [CrossRef]
24. Liu, G.; Yang, Y.; Lu, X.; Qi, F.; Liang, Y.; Trukhanov, A.; Wu, Y.; Sun, Z.; Lu, X. Fully Active Bimetallic Phosphide Zn$_{0.5}$Ge$_{0.5}$P: A Novel High-Performance Anode for Na-Ion Batteries Coupled with Diglyme-Based Electrolyte. *ACS Appl. Mater. Interfaces* **2022**, *14*, 31803–31813. [CrossRef] [PubMed]
25. Liu, G.; Wang, N.; Qi, F.; Lu, X.; Liang, Y.; Sun, Z. Novel Ni–Ge–P Anodes for Lithium-Ion Batteries with Enhanced Reversibility and Reduced Redox Potential. *Inorg. Chem. Front.* **2023**, *in press*. [CrossRef]
26. Sierra, L.; Gibaja, C.; Torres, I.; Salagre, E.; Avilés Moreno, J.R.; Michel, E.G.; Ocón, P.; Zamora, F. Alpha-Germanium Nanolayers for High-Performance Li-ion Batteries. *Nanomaterials* **2022**, *12*, 3760. [CrossRef] [PubMed]
27. Liang, Y.; Chen, Y.; Ke, X.; Zhang, Z.; Wu, W.; Lin, G.; Zhou, Z.; Shi, Z. Coupling of Triporosity and Strong Au–Li Interaction to Enable Dendrite-Free Lithium Plating/Stripping for Long-Life Lithium Metal Anodes. *J. Mater. Chem. A* **2020**, *8*, 18094. [CrossRef]
28. Sun, X.; Shao, C.; Zhang, F.; Li, Y.; Wu, Q.-H.; Yang, Y. SiC Nanofibers as Long-Life Lithium-Ion Battery Anode Materials. *Front. Chem.* **2018**, *6*, 166. [CrossRef]
29. Huang, X.D.; Zhang, F.; Gan, X.F.; Huang, Q.A.; Yang, J.Z.; Lai, T.; Tang, W.M. Electrochemical Characteristics of Amorphous Silicon Carbide Film as a Lithium-Ion Battery Anode. *RSC Adv.* **2018**, *8*, 5189–5196. [CrossRef]
30. Schmidt, H.; Jerliu, B.; Hüger, E.; Stahn, J. Volume Expansion of Amorphous Silicon Electrodes During Potentiostatic Lithiation of Li-ion Batteries. *Electrochem. Comm.* **2020**, *115*, 106738. [CrossRef]
31. Liu, G.; Wei, Y.; Li, T.; Gu, Y.; Guo, D.; Wu, N.; Qin, A.; Liu, X. Green and Scalable Fabrication of Sandwich-like NG/SiO$_x$/NG Homogenous Hybrids for Superior Lithium-Ion Batteries. *Nanomaterials* **2021**, *11*, 2366. [CrossRef] [PubMed]
32. Lou, D.; Chen, S.; Langrud, S.; Razzaq, A.A.; Mao, M.; Younes, H.; Xing, W.; Lin, T.; Hong, H. Scalable Fabrication of Si-Graphene Composite as Anode for Li-ion Batteries. *Appl. Sci.* **2022**, *12*, 10926. [CrossRef]
33. Jo, M.; Sim, S.; Kim, J.; Oh, P. Micron-Sized SiO$_x$–Graphite Compound as Anode Materials for Commercializable Lithium-Ion Baterries. *Nanomaterials* **2022**, *12*, 1956. [CrossRef] [PubMed]
34. Jumari, A.; Yudha, C.S.; Widiyandari, H.; Lestari, A.P.; Rosada, R.A.; Santosa, S.P.; Purwanto, A. SiO$_2$/C Composite as a High Capacity Anode Material of LiNi$_{0.8}$Co$_{0.15}$Al$_{0.05}$O$_2$ Battery Derived from Coal Combustion Fly Ash. *Appl. Sci.* **2020**, *10*, 8428. [CrossRef]
35. Galashev, A.; Vorob'ev, A. An Ab Initio Study of Lithization of Two-Dimensional Silicon–Carbon Anode Material for Lithium-Ion Batteries. *Materials* **2021**, *14*, 6649. [CrossRef] [PubMed]
36. Shao, Y.; Wang, S.; Li, X. The Effect of Silicon-Containing Minerals on Coal Evolution at High-Temperature Pre-Graphitization Stage. *Minerals* **2023**, *13*, 20. [CrossRef]
37. Lee, K.-J.; Kang, Y.; Kim, Y.H.; Baek, S.W.; Hwang, H. Synthesis of Silicon Carbide Powders from Methyl-Modified Silica Aerogels. *Appl. Sci.* **2020**, *10*, 6161. [CrossRef]

38. Xing, Z.; Lu, J.; Ji, X. A Brief Review of Metallothermic Reduction Reactions for Materials Preparation. *Small Methods* **2018**, *2*, 1800062. [CrossRef]
39. Zhang, Y.; Hu, K.; Zhou, Y.; Xia, Y.; Yu, N.; Wu, G.; Zhu, Y.; Wu, Y.; Huang, H. A Facile, One-Step Synthesis of Silicon/Silicon Carbide/Carbon Nanotube Nanocomposite as a Cycling-Stable Anode for Lithium Ion Batteries. *Nanomaterials* **2019**, *9*, 1624. [CrossRef]
40. Lebedev, A.S.; Suzdaltsev, A.V.; Anfilogov, V.N.; Farlenkov, A.S.; Porotnikova, N.M.; Vovkotrub, E.G.; Akashev, L.A. Carbothermal Synthesis, Properties, and Structure of Ultrafine SiC Fibers. *Inorg. Mat.* **2020**, *56*, 20–27. [CrossRef]
41. Anfilogov, V.N.; Lebedev, A.V.; Ryzhkov, V.M.; Blinov, I.A. Carbothermal Synthesis of Nanoparticulate Silicon Carbide in a Self-Contained Protective Atmosphere. *Inorg. Mat.* **2016**, *52*, 655–660. [CrossRef]
42. Gevel, T.; Zhuk, S.; Suzdaltsev, A.V.; Zaikov, Y.P. Study into the Possibility of Silicon Electrodeposition from a Low-Fluoride KCl–K_2SiF_6 Melt. *Ionics* **2022**, *28*, 3537–3545. [CrossRef]
43. Nakashima, S.; Harima, H. Raman Investigation of SiC Polytypes. *Phys. Status Solidi A* **1997**, *162*, 39–64. [CrossRef]
44. Xing, Z.; Wang, B.; Halsted, J.K.; Subashchandrabose, R.; Stickleb, W.F.; Ji, X. Direct Fabrication of Nanoporous Graphene from Graphene Oxide by Adding a Gasification Agent toa Magnesiothermic Reaction. *Chem. Commun.* **2015**, *51*, 1969. [CrossRef] [PubMed]
45. Brodova, I.; Yolshina, L.; Razorenov, S.; Rasposienko, D.; Petrova, A.; Shirinkina, I.; Shorokhov, E.; Muradymov, R.; Garkushin, G.; Savinykh, A. Effect of Grain Size on the Properties of Aluminum Matrix Composites with Graphene. *Metals* **2022**, *12*, 1054. [CrossRef]
46. Hossain, S.T.; Johra, F.T.; Jung, W.-G. Fabrication of Silicon Carbide from Recycled Silicon Wafer Cutting Sludge and Its Purification. *Appl. Sci.* **2018**, *8*, 1841. [CrossRef]
47. Martínez-Periñán, E.; Foster, C.W.; Down, M.P.; Zhang, Y.; Ji, X.; Lorenzo, E.; Kononovs, D.; Saprykin, A.I.; Yakovlev, V.N.; Pozdnyakov, G.A.; et al. Graphene Encapsulated Silicon Carbide Nanocomposites for High and Low Power Energy Storage Applications. *J. Carbon Res.* **2017**, *3*, 20. [CrossRef]
48. Masias, A.; Marcicki, J.; Paxton, W.A. Opporttunities and Challenges of Lithium Ion Batteries in Automotive Applications. *ACS Energy Lett.* **2021**, *6*, 621–630. [CrossRef]
49. Hosen, M.S.; Gopalakrishnan, R.; Kalogiannis, T.; Jaguemont, J.; Van Mierlo, J.; Berecibar, M. Impact of Relaxation Time on Electrochemical Impedance Spectroscopy Characterization of the Most Common Lithium Battery Technologies—Experimental Study and Chemistry-Neutral Modeling. *World Electr. Veh. J.* **2021**, *12*, 77. [CrossRef]
50. Maddipatla, R.; Loka, C.; Choi, W.J.; Lee, K.-S. Nanocomposite of Si/C Anode Material Prepared by Hybrid Process of High-Energy Mechanical Milling and Carbonization for Li-Ion Secondary Batteries. *Appl. Sci.* **2018**, *8*, 2140. [CrossRef]
51. Choi, J.-H.; Choi, S.; Cho, J.S.; Kim, H.-K.; Jeong, S.M. Efficient Synthesis of High Areal Capacity Si@Graphite@SiC Composite Anode Material via One-Step Electro-deoxidation. *J. Alloys Compd.* **2022**, *896*, 163010. [CrossRef]
52. Galashev, A.Y.; Vorob'ev, A.S. First Principle Modeling of a Silicene Anode for Lithium Ion Batteries. *Electrochim. Acta* **2021**, *378*, 138143. [CrossRef]
53. Uxa, D.; Huger, E.; Dorrer, L.; Schmidt, H. Lihium-Silicon Compounds as Electrode Material for Lithium-Ion batteries. *J. Electrochem. Soc.* **2020**, *167*, 130522. [CrossRef]
54. Jiang, Y.; Offer, G.; Jiang, J.; Marinescu, M.; Wang, H. Voltage Hysteresis Model for Silicon Electrodes for Lithium Ion Batteries, Including Multi-Step Phase Transformations, Cristallization and Amorphization. *J. Electrochem. Soc.* **2020**, *167*, 130533. [CrossRef]

Disclaimer/Publisher's Note: The statements, opinions and data contained in all publications are solely those of the individual author(s) and contributor(s) and not of MDPI and/or the editor(s). MDPI and/or the editor(s) disclaim responsibility for any injury to people or property resulting from any ideas, methods, instructions or products referred to in the content.

Article

Disks of Oxygen Vacancies on the Surface of TiO$_2$ Nanoparticles

Vladimir B. Vykhodets [1,*], Tatiana E. Kurennykh [1] and Evgenia V. Vykhodets [2]

[1] Institute of Metal Physics UB RAS, 18 S.Kovalevskaya Street, 620108 Ekaterinburg, Russia
[2] Ural Federal University, 19 Mira Street, 620002 Ekaterinburg, Russia
* Correspondence: vykhod@imp.uran.ru

Abstract: Oxide nanopowders are widely used in engineering, and their properties are largely controlled by the defect structure of nanoparticles. Experimental data on the spatial distribution of defects in oxide nanoparticles are unavailable in the literature, and in the work presented, to gain such information, methods of nuclear reactions and deuterium probes were employed. The object of study was oxygen-deficient defects in TiO$_2$ nanoparticles. Nanopowders were synthesized by the sol–gel method and laser evaporation of ceramic targets. To modify the defect structure in nanoparticles, nanopowders were subjected to vacuum annealings. It was established that in TiO$_2$ nanoparticles there form two-dimensional defects consisting of six titanium atoms that occupy the nanoparticle surface and result in a remarkable deviation of the chemical composition from the stoichiometry. The presence of such defects was observed in two cases: in TiO$_2$ nanoparticles alloyed with cobalt, which were synthesized by the sol–gel method, and in nonalloyed TiO$_2$ nanoparticles synthesized by laser evaporation of ceramic target. The concentration of the defects under study can be varied in wide limits via vacuum annealings of nanopowders which can provide formation on the surface of oxide nanoparticles of a solid film of titanium atoms 1–2 monolayers in thickness.

Keywords: nanoparticles; TiO$_2$; oxygen vacancies; nuclear reactions; deuterium probes

Citation: Vykhodets, V.B.; Kurennykh, T.E.; Vykhodets, E.V. Disks of Oxygen Vacancies on the Surface of TiO$_2$ Nanoparticles. *Appl. Sci.* **2022**, *12*, 11963. https://doi.org/10.3390/app122311963

Academic Editor: Suzdaltsev Andrey

Received: 27 October 2022
Accepted: 18 November 2022
Published: 23 November 2022

Publisher's Note: MDPI stays neutral with regard to jurisdictional claims in published maps and institutional affiliations.

Copyright: © 2022 by the authors. Licensee MDPI, Basel, Switzerland. This article is an open access article distributed under the terms and conditions of the Creative Commons Attribution (CC BY) license (https://creativecommons.org/licenses/by/4.0/).

1. Introduction

The physico-chemical and functional properties of oxide nanoparticles, which are widely employed in engineering, are largely dependent on the defect structure of nanoparticles, namely, the concentration of point defects, their type, and spatial distribution. In this field, works have mainly been devoted to defects that condition oxygen deficiency in nanoparticles compared to the stoichiometry [1,2]. In addition, oxygen vacancies and clusters based on them strongly affect the electrical, magnetic, diffusion, catalytic, and other properties of oxide nanopowders [3–5]. For example, a concept of d^0-magnetism was proposed [6], which helped to explain ferromagnetic properties observed on oxide nanopowders that did not contain any magnetic dopants by an oxygen deficiency in nanoparticles. Such results were obtained on the oxide powders of TiO$_2$, Al$_2$O$_3$, ZnO, In$_2$O$_3$, HfO$_2$, CuO, and others [7–11].

The majority of results concerning the defect structure of oxide nanoparticles are obtained for the TiO$_2$ nanoparticles with the use of a large number of techniques [3,5,8,12–15]. They highlight a wide variety of types of point defects and the evolution of the defect structure under different treatments of nanopowders, say, oxygen annealing [3]. When investigating the defect structure, the most informative were the methods of positron annihilation [3,5,12–15], X-ray photoelectron spectroscopy [3,5,13], and photoluminescence [14]; as defects, there were specified oxygen and titanium vacancies [3,5], interstitial titanium atoms [12], 3D vacancy clusters [8], complexes consisting of oxygen and titanium vacancies [3], Ti^{3+} ions and oxygen vacancies [15], and others [12]. The diversity of defect types in the oxide nanoparticles is conditioned by a large number of atoms on the sample surface, the presence of defects - charge carriers, mixed cation valence, changing oxygen concentration upon different powder treatments, etc.

With the bulk of important results accumulated on the types of defects in the oxide nanoparticles and their impact on the magnetic properties of nanoparticles [3,16–20], the state-of-art in this field can hardly be considered sufficient. First, no measurements of the oxygen concentration were performed and, hence, it was not accounted for in the concentration balance. For this reason, it cannot be excluded that information on several defect types is lacking. The more probable it seems to take into account the selective sensitivity of many techniques to defects of different types. Second, there are no available experimental data on the spatial distribution of defects in nanoparticles gained by direct observations. Such experiments are necessary since the energy states of atoms near the surface and in the bulk are drastically different.

In view of the above, this work is devoted to studying the defect structure of oxide nanoparticles via measurements of the oxygen concentration and using a technique that allows the determination of the defect concentration on the surface and in the bulk of nanoparticles. The choice of TiO_2 nanoparticles as the object for investigation is conditioned by a wide scope of practical applications of the dioxide titanium nanopowders [21–25]. The oxygen concentration was measured by the method of nuclear reactions (NRA); the error of measurements being about 0.5–1.0%, which makes it applicable only for nanopowders with rather a large oxygen deficiency compared to stoichiometry, at a level of several percent. Nanopowders of TiO_2 meet this condition [26]. The defect concentration was determined by the method of deuterium probes (DP) [2]. It is based on the dependence of the deuterium solubility in the oxide particles on the concentration of defects of a specified type. The application of the DP method was reasonable since, first, it is sensitive to defects bringing about oxygen deficiency [2] and, second, owing to a drastic difference in its basic principles from those of the earlier applied techniques, it enables one to detect defects of unknown types. The NRA and DP techniques have already been used to measure the oxygen and defect concentrations in oxide nanoparticles [2,26–28], but in these works, the type of defects and their spatial distribution in nanoparticles were not the subject of research.

2. Experimental
2.1. Synthesis of Oxide Nanopowders

Nanopowders were produced by the technology of laser evaporation of ceramic target (LE) [26] and sol–gel method (SG) [16,29]. As an LE, pills of titanium dioxide 60 mm in diameter were produced by pressing at room temperature, with the specific surface area of micropowder being 2 m^2/g. For vaporizing, a fiber-ytterbium laser was used with a wavelength of 1.07 µm and a maximal power output of 1 kW; the frequency of laser pulses was 5 kHz, duration was 50 µs. To produce nanopowders with a different specific area, the type of inert gas (argon or helium), gaseous pressure, and laser power were varied. For the synthesis of titanium oxide nanoparticles doped with cobalt ions by the SG method [29], acid hydrolysis of organometallic precursors was employed; in our case, titanium isopropyl oxide (Titanium (IV) isopropoxide) and cobalt (II) acetylacetonate as a source of metal ions. All reagents used were obtained from Sigma Aldrich. The preliminary required amounts of titanium isopropoxide (2 mL) and cobalt acetylacetonate complex were placed in a 5 mL test tube; then, 2 mL of acetone was poured into it. After vigorous shaking, a bluish-brown solution was formed. The solution was left overnight. Then, 1 mL of hydrochloric acid (0.1M) was poured into the solution. The hydrolysis reaction occurs almost instantly; however, for the process to run uniformly throughout the entire volume of the colloid, it was subjected to intensive ultrasonic action using an ultrasonic activator with a submerged titanium probe. Ultrasound exposure was 30 s. Then, the particles were separated by centrifugation (10,000 RPM) for 10 min and washed with acetone. The washing process was repeated three times. The resulting precipitate was dried at 70 °C. The precursor was further calcined in air at different temperatures (from 300 to 500 °C). Figure 1 shows micrographs of nanoparticles calcined at 400 and 500 °C, the image was obtained using transmission electron microscopy (TEM).

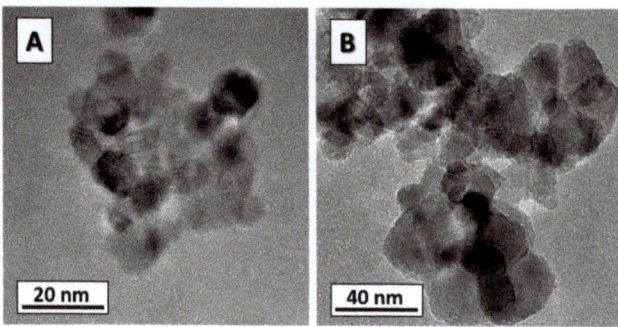

Figure 1. TEM micrographs of the TiO$_2$ nanoparticles calcined at 400 °C (**A**) and 500 °C (**B**).

2.2. Methods of Nuclear Reactions and Deuterium Probes

All nanopowders predominantly contained such phases of titanium dioxide as anatase, rutile, brookite, and their mixtures. Data on the phase composition are given in more detail in Section 3.1. In the current approach, the defect structure is characterized by the oxygen concentration C_O and defect concentration C_d and real defects contain a certain number of elementary point defects. Hence, a real defect can comprise vacancies and interstitial atoms of oxygen and titanium. Within the approach accepted, an oxygen vacancy cannot be distinguished from an interstitial titanium atom, the same as a titanium vacancy from an interstitial oxygen atom. When interpreting the results of studying the thus prepared samples, we considered solely oxygen vacancies as the most realistic elementary point defects. In the DP technique, after annealing in gaseous deuterium, nanopowders become three-component and their chemical composition can be presented by formula TiO$_{2-x}$D$_{2x/N}$, which accounts for the presence of 2 deuterium atoms per one defect [2]. Moreover, oxygen vacancies are supposed to be prone to joining in clusters, and N designates the amount of vacancies in a cluster. This formula enables easily obtaining that concentrations C_O and C_d are connected as follows

$$\frac{C_O}{C_O^0} = 1 - \frac{N+4}{3C_O^0}C_d, \qquad (1)$$

where C_O^0 is the oxygen concentration in a defectless oxide; herefrom, concentrations are expressed in fractions of the number of atoms in a nanoparticle. It is seen from Expression (1) that when treating the dependences $C_O(C_d)$, the number of vacancies in a defect N can be determined if oxygen vacancies are combined in complexes. Expression (1) is obtained for nanoparticles of arbitrary shape and does not depend on the spatial distribution of defects and mutual arrangement of vacancies in a cluster. For example, it is valid for cases when defects are distributed in a nanoparticle uniformly or only in its thin surface layer with vacancies forming in a cluster of both 2D and 3D complexes.

When using the method of LE, the average concentrations C_O and C_d were varied via the synthesis of nanopowders with different specific surface areas S. This approach makes it possible to produce nanopowders with different C_O and C_d, if the concentrations of defects in the bulk and near the surface are different. In the SG technology, nanopowders with different defect concentrations were produced using vacuum annealings. In the preliminary experiments, it was established that vacuum annealings are by no means a universal way of decreasing the oxygen concentration in the titanium dioxide nanopowders. In particular, in the case of undoped TiO$_2$, vacuum annealings exert a poor effect on the concentrations of oxygen and defects, which did not allow the analysis of the $C_O(C_d)$ dependence with reliable accuracy. At the same time, Co-doped nanopowders of TiO$_2$ with widely ranging oxygen concentrations were prepared using vacuum annealings. The same result was obtained in work [5], where it was shown that alloying of TiO$_2$ with cobalt leads to increasing vacancy concentration. In view of this, in the case of the SG method, investigation of the dependence $C_O(C_d)$ was performed on powders alloyed with cobalt, its concentration in the cation sublattice made up of 3 at. %. Vacuum annealings were carried out after completion of the synthesis routine at temperatures varying from 200 to 500 °C and holding time, 15 to 30 min, 500 °C.

To measure the defect concentration using the DP method, first, nanopowders are annealed in gaseous deuterium, and then, the concentration of dissolved deuterium C_D is measured using the NRA technique. The DP method is shown in work [2] to be sensitive to defects bringing about oxygen deficiency in nanoparticles and is based on the existence of unambiguous relation of the C_d and C_D concentrations, which is provided by the formation of nanoparticles annealed in deuterium of clusters of a strictly specified composition. In particular, in the TiO_2 nanoparticles, one cluster consists of a defect and 2 deuterium atoms. The cluster composition was determined using the dependence of C_D on the dose of deuterium irradiation. In the current work, such a procedure was applied to all powders after synthesis and vacuum annealings, and the cluster composition was the same. In general, nanoparticles can contain defects of different types simultaneously. For example, in study [2], in the TiO_2 particles, along with defects detectable for the DP method, one more type of defect was observed, which consisted of Ti^{3+} ions and oxygen vacancies. The duration of deuterium annealing we used was, just as in [2], 1 h, with deuterium pressure being 0.6 atm. In the DP method, the determination of the defect concentration C_d is reduced to measuring the deuterium concentration C_D. For this reason, the metrological characteristics of the NRA technique for the reaction $^2H(d,p)^3H$ controlled the accuracy and sensitivity in the determination of the defect concentration C_d: statistical error made up several percent of the measured value and measurement sensitivity was about 0.01 at. %.

As was noted in Section 1, no experimental data on the spatial arrangement of defects in oxide nanoparticles are available in the literature. Such a task is objectively complicated because of the small size of nanoparticles. In the current work, a particular solution was obtained. It is based on an approach that involves the characterization of the defect structure using only two defect concentrations, namely, at the surface of nanoparticle C_{ds} and average concentration in the nanoparticle bulk C_{dv}. For the powders synthesized using LE, we investigated the dependence of defect concentration on the specific surface area $C_d(S)$, which allowed us to obtain, in the model with two concentration values C_{ds} and C_{dv}, information on the spatial distribution of defects in nanoparticles. For example, the growth of C_d with increasing S means higher defect concentration near the surface then in the bulk and vice versa. If $C_{ds} = C_{dv}$, the defect concentration C_d evidently is independent of S. In this work, of special interest was the case when defects to which the DP method is sensitive are localized near the nanoparticle surface. In this case, the chemical composition of nanopowders can be expressed by the formula $(1 - \Delta\rho S)TiO_2(\Delta\rho S)\left[(TiO_{2-y})D_{2y/N}\right]$, and the following relationship works:

$$C_d = \frac{\Delta y \rho S}{N(3 - \Delta y \rho S)},\qquad(2)$$

where ρ is the oxide density, Δ is the thickness of the oxygen-deficient layer whose chemical composition is denoted as TiO_{2-y}. Expression (2) is valid for particles of arbitrary shape. Hence, with equal parameters N, Δ, and y, in the powders with different specific area S, the dependence $C_d(S)$ will be close to linear, with the extrapolated value $C_d(0) = 0$. Since the DP method provides a high accuracy in measuring the defect concentration, in some particular cases the study of $C_d(S)$ can give reliable information on the ratio of defect concentrations in the bulk and near the nanoparticle surface. Note that application of this approach is bound to solely the case when powders are synthesized by LE technology. It favors independence of the parameters, Δ, and y of S, which will be considered in Section 3.2.

The average concentrations of oxygen C_O and deuterium C_D in nanopowders were determined by the NRA technique at a 2 MB van de Graaf accelerator using reactions $^{16}O(d,p_1)^{17}O^*$ and $^2H(d,p)^3H$ with the deuteron energy of 900 and 650 keV, respectively. In the majority of experiments in the accelerator, the samples were at room temperature and nanoparticles of powders were pressed into an indium plate. When the concentration C_O was measured at the temperature of 400 °C, nanoparticles were pressed into copper powder. Using Rutherford backscattering spectroscopy, it was established that indium or copper atoms were absent in the zone under analysis. The sample surface was set perpendicular to the incident beam, nuclear reaction products were registered with the use of a silicon surface barrier detector. The angle of proton registration was 160°, and the diameter of the incident beam was 1 or 2 mm. The irradiation dose was measured on the samples using a secondary monitor, with the statistical error being 0.2 to 1.0%. When mathematically processing the data, spectra from the samples under study were compared with those from reference samples having constant-in-depth isotope concentrations. For ^{16}O, it was CuO, whereas for deuterium, $ZrCr_2D_{0.12}$. In more detail, the experimental setting and processing procedure are described in work [2].

When doing this, it was necessary to provide highly accurate measurements of concentrations C_O and C_D, which was not trivial as the data on C_O depended on the presence on the nanoparticle surface of water molecules. With increasing the thickness of the water layer, the oxygen concentration in factions of the number of atoms in a nanoparticle decreases since the H_2O molecule contains fewer oxygen atoms than TiO_2. In view of this, the powders were dried under vacuum pumping-off. To monitor the oxygen concentration C_O, sampling measurements of the powders were performed in the accelerator chamber at 400 °C, which allows the elimination of the adsorbed water molecules. The results of the concentration C_D and C_d, unlike C_O, were not changed when varying experimental conditions, temperature, pressure, and annealing time. In addition, these results did not depend on the conditions of storing the powders for a year and more, since the deuterium annealing provided the elimination of water molecules from the nanoparticle surface. Figure 2 shows the spectra of protons p for the nuclear reactions $^{16}O(d,p_1)^{17}O^*$ and $^2H(d,p)^3H$, which were used to determine the oxygen and defect concentrations. The C_O and C_d concentrations are directly proportional to the number of registered protons, the shape of the spectra does not depend on C_O and C_d.

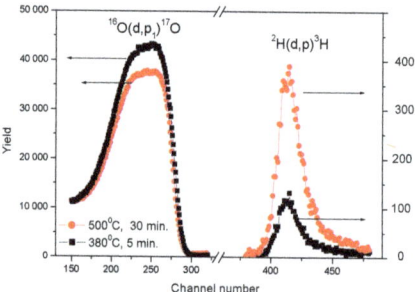

Figure 2. The spectra of products of the nuclear reactions $^{16}O(d,p_1)^{17}O^*$ and $^2H(d,p)^3H$ for the nanopowders synthesized by sol–gel method and annealed in vacuum. The temperatures and annealing times are shown in the figure.

3. Results and Discussion

3.1. Number of Oxygen Vacancies in a Defect

The dependence $C_O(C_d)$ shown in Figure 3 agrees with Expression (1), which testifies to the formation in the TiO_2 nanoparticles of defects consisting of oxygen vacancies. Processing by the least-square method gave the number of vacancies in a defect $N = 12.5 \pm 0.9$.

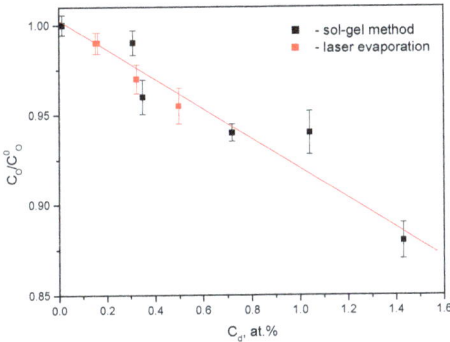

Figure 3. Dependence of the oxygen concentration C_O in TiO$_2$ nanopowders on the defect concentration C_d. Black points show results for powders synthesized by sol–gel method; red points, laser evaporation of ceramic target.

As follows from Figure 3, the formation in the dioxide titanium nanoparticles of defects containing $N = 12$ oxygen vacancies is invariant with respect to experimental conditions. First, it takes place in the nanopowders synthesized by different technologies and characterized by different mechanisms of defect formation. In the technology of SG method, in the as-synthesized powders of composition close to stoichiometric, the formation of defects occurs in the course of vacuum annealing through the oxygen depletion of nanoparticles. Upon LE, oxygen-deficient defects were formed directly in the course of synthesizing. Second, the powders strongly differ in the number of defects per unit volume of nanoparticles. This follows from Figure 2 since this parameter decreases with decreasing concentration C_O. Finally, the powders under study differed in phase composition. In the case of the SG method, it was anatase, whereas upon LE, the mixture of TiO$_2$ phases. In particular, in the powders with $S > 200$ m^2/g, brookite with small additions of anatase and rutile predominates, whereas at $S < 200$ m^2/g, quite the opposite, brookite was not observed at all, only anatase and rutile. Since the number of vacancies in a defect $N = 12$ does not depend on the synthesis technology, mechanism of defect formation, phase composition, and other characteristics of nanoparticles, one can suggest that the formation of such defects, denoted as Ti$_6$, is conditioned by thermodynamic reasons. This question needs further investigation.

3.2. Spatial Distribution of Defects Ti$_6$ in Nanoparticles

In the investigation of spatial distribution and size of defects consisting of 6 titanium atoms, different versions are considered, namely, defects located in the bulk and on the surface of nanoparticles and two-dimensional or three-dimensional ones. To obtain information on these issues, the dependence $C_d(S)$ was studied for the powders synthesized by the LE. As is seen in Figure 4, the experimental dependence $C_d(S)$ matches Expression (2) at constant parameters N, Δ, and y, which indicates that the defects consisting of 12 oxygen vacancies are located on the surface and absent in the bulk of TiO$_2$ nanoparticles.

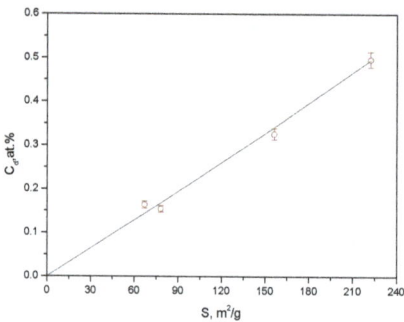

Figure 4. Dependence of the defect concentration C_d on the specific surface area S of TiO$_2$ nanoparticles. Points are experimental data for the powders synthesized by laser evaporation of ceramic targets. Line is calculated by Expression (2) at $N = 12$ and $\Delta \rho$ y = 7.56×10^{-8} g/cm^2.

The independence of the parameter N on S was experimentally shown in Section 3.1. As for the parameters Δ and y, it is conditioned by the application of technology of LE. In this case, the formation of the defect structure of the surface layer proceeds in two steps. In the first step, at high temperatures in a vacuum, there form nanoparticles of stoichiometric composition in the bulk with the surface atomic layer virtually depleted of oxygen. It was established that the thickness of the oxygen-depleted layer does not depend on S and is equal to 0.36 nm [26], which approximately matches up an oxide monolayer. In works fulfilled within the theory of density functional, the following theoretical explanation of this result was given, namely, the existence of an oxygen-free layer on the oxide surface is energetically more favorable than of stoichiometric composition TiO$_2$ [26,30,31]. In the second step, the powder is cooled to room temperature in air and oxygen enters into the surface layer of nanoparticles. Evidently, in this case, the thickness of the oxygen-deficient layer does not depend on S. The amount of oxygen entering into nanoparticles depends on the regime of cooling the nanopowders. In this work, these regimes were the same for powders with different S.

We already mentioned above that the thickness of defects Ti$_6$ in nanoparticles synthesized by LE technology is within the atomic scale, whereas other linear dimensions are several times as high, i.e., the defects are two-dimensional. For nanoparticles synthesized by the SG method, no data on the spatial distributions of Ti$_6$ defects and inference on their two-dimensional or three-dimensional shape are presented in this work because of the lack of appropriate techniques. Yet, there are grounds to conclude that the Ti$_6$ defects in nanoparticles synthesized by the SG method also are two-dimensional and localized on the nanoparticle surface. In our opinion, these defects cannot occupy internal bulk regions of nanoparticles since the crystal structure of titanium dioxide with inner defects consisting of 12 oxygen vacancies will hardly be stable. At the same time, the presence of such defects on the surface of nanoparticles does not cause destruction of the crystal lattice of the oxide, which is evidenced by the results of works [26,30,31].

In the surface layer of TiO$_2$ particles, there are present two types of regions: (i) consisting of 6 titanium atoms and (ii) having composition close to stoichiometric TiO$_2$. It is worth estimating fractions of these regions. If to take up a constancy of distance between the titanium atoms upon the formation of the Ti$_6$ defect, the fraction of nanoparticle surface occupied by regions consisting of 6 titanium atoms is expressed as

$$\propto = \frac{1 - C_O/C_O^0}{\Delta \rho S (1 - C_O)} \leq 1. \qquad (3)$$

In Figure 5, the dependences $\propto (C_O/C_O^0)$ calculated by Expression (3) for several values of S are shown. Points in Figure 5 show sampling experimental results, data on y

(C_O/C_O^0) are also given. It is easy to show that the values of α and y obey the expression $y = 2\alpha$. Bends on the dependences α (C_O/C_O^0) correspond to the formation on the nanoparticle surface of a solid monolayer in which oxygen atoms are absent. The horizontal portions of the dependences α (C_O/C_O^0) correspond to the formation in nanoparticles of one more monolayer deficient in oxygen. It follows from Figure 5 that the fraction α that is taken by regions consisting of 6 titanium atoms can be an easily controlled parameter. In particular, this refers to nanoparticles of TiO$_2$ produced by the SG method. In the as-synthesized state, they are virtually stoichiometric ($\alpha \approx 0$), and by vacuum annealings and alloying with cobalt, this value can be increased to 1.

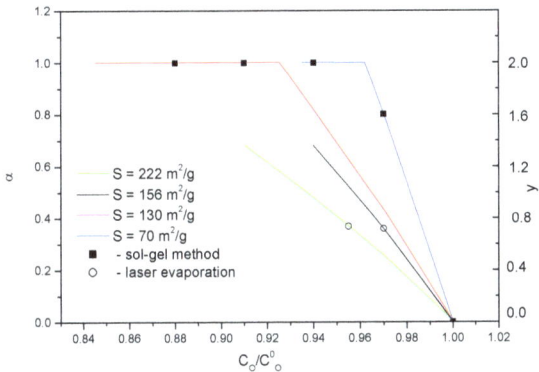

Figure 5. Dependence of the parameters α and y for the surface monolayer of the titanium dioxide nanoparticles on the oxygen concentration C_O/C_O^0 in nanopowders. Lines show the calculation results by Expression (3): green—for nanopowders with the specific area 222 m^2/g; black, 156 m^2/g, red, 130 m^2/g, and blue, 70 m^2/g. Points are sampling results for the nanoparticles under study: filled ones refer to the powders synthesized by the sol-gel method; empty, laser evaporation of ceramic target.

It is shown that under certain conditions, in the TiO$_2$ nanoparticles there form two-dimensional defects consisting of 6 titanium atoms and localized on the nanoparticle surface, which results in a significant deviation of the chemical composition from the stoichiometry. Since, as noted in Section 1, in the TiO$_2$ nanoparticles there are present defects of other types, of interest is to compare the data on the concentration of surface two-dimensional defects with those obtained in previous studies. Unfortunately, in view of the absence of such information in works on oxide nanoparticles, we have to restrict ourselves to rough estimates. In a general case, defects of almost all types cause changes in the chemical composition of nanoparticles and their presence in the samples gives rise to a deviation from the linear dependence in Figure 3. Since in the work no deviations beyond the statistical experimental errors were registered, one can conclude that the concentration of other defects was far smaller than that of defects consisting of 6 titanium atoms.

4. Conclusions

Thus, new data on the defects that condition the oxygen deficit in the oxide nanoparticles of TiO$_2$ have been gained. The formed two-dimensional defects of atomic thickness are established to consist of six titanium atoms. They are free of oxygen atoms and located on the nanoparticle surface. The presence of such defects is detected in two cases: in the cobalt-doped nanoparticles of TiO$_2$ synthesized by the sol–gel method and subjected to vacuum annealings and undoped nanoparticles synthesized by laser evaporation of ceramic target. The data on the defect structure were obtained by the method of deuterium probes, DP, which was applied to measure the defect concentrations, and by nuclear reactions, NRA, to determine the oxygen concentration in nanopowders. The concentration of the defects under study can be varied in wide limits; in particular, using vacuum annealings can provide gradual growth from 0 to 1.5 at. %, including the formation of a solid film of titanium atoms on the nanoparticle surface. The concentration of defects of other types in the nanoparticles under study was lower than that of two-dimensional defects located on the nanoparticle surface.

Author Contributions: Conceptualization, V.B.V.; Investigation, T.E.K. and E.V.V.; Writing—original draft, V.B.V. All authors have read and agreed to the published version of the manuscript.

Funding: The research was carried out within the state assignment of the Ministry of Science and Higher Education of the Russian Federation (theme "Function" no. AAAA-A19-119012990095-0).

Institutional Review Board Statement: Not applicable.

Informed Consent Statement: Not applicable.

Acknowledgments: Authors are very grateful to A.S. Minin, A.E. Yermakov, and M.A. Uimin for useful discussion and valuable remarks and to A.S. Minin for the synthesis of nanopowders by the sol–gel method.

Conflicts of Interest: The authors declare no conflict of interest.

References

1. Diebold, U. The surface science of titanium dioxide. *Surf. Sci. Rep.* **2003**, *48*, 53–229. [CrossRef]
2. Vykhodets, V.B.; Kurennykh, T.E. Characterization of the defect structure of oxide nanoparticles with the use of deuterium probes. *RSC Adv.* **2020**, *10*, 3837–3843. [CrossRef] [PubMed]
3. Ghosh, S.; Nambissan, P. Evidence of oxygen and Ti vacancy induced ferromagnetism in post-annealed undoped anatase TiO_2 nanocrystals: A spectroscopic analysis. *J. Solid State Chem.* **2019**, *275*, 174–180. [CrossRef]
4. Mikhalev, K.N.; Germov, A.Y.; Ermakov, A.E.; Uimin, M.A.; Buzlukov, A.L.; Samatov, O.M. Crystal structure and magnetic properties of Al_2O_3 nanoparticles by ^{27}Al NMR data. *Phys. Solid State* **2017**, *59*, 514–519. [CrossRef]
5. Qin, X.B.; Zhang, P.; Liang, L.H.; Zhao, B.Z.; Yu, R.S.; Wang, B.Y.; Wu, W.M. Vacancy identification in Co+doped rutile TiO_2 crystal with positron annihilation spectroscopy. *J. Physics: Conf. Ser.* **2011**, *262*, 012051. [CrossRef]
6. Bouzerar, G.; Ziman, T. Model for Vacancy-Inducedd0d0Ferromagnetism in Oxide Compounds. *Phys. Rev. Lett.* **2006**, *96*, 207602–207605. [CrossRef]
7. Sundaresan, A.; Bhargavi, R.; Rangarajan, N.; Siddesh, U.; Rao, C.N.R. Ferromagnetism as a universal feature of nanoparticles of the otherwise nonmagnetic oxides. *Phys. Rev. B* **2006**, *74*, 161306. [CrossRef]
8. Grosh, S.; Khan, G.G.; Mandal, K.; Samanta, A.; Nambissan, M.G. Evolution of Vacancy-Type Defects, Phase Transition, and Intrinsic Ferromagnetism during Annealing of Nanocrystalline TiO_2 Studied by Positron Annihilation Spectroscopy. *J. Phys. Chem. C* **2013**, *117*, 8458–8467. [CrossRef]
9. Sudakar, C.; Kharela, P.; Suryanarayanan, R.; Thakurc, J.S.; Naikd, V.M.; Naika, R.; Lawesa, G. Room temperature ferromagnetism in vacuum-annealed TiO_2 thin films. *J. Magn. Magn. Mater.* **2008**, *320*, L31–L36. [CrossRef]
10. Hong, N.H.; Sakai, J.; Poirot, N.; Brizé, V. Room-temperature ferromagnetism observed in undoped semiconducting and insulating oxide thin films. *Phys. Rev. B* **2006**, *73*, 132404. [CrossRef]
11. Coey, J.M.D. High-temperature ferromagnetism in dilute magnetic oxides. *J. Appl. Phys.* **2005**, *97*, 10D313. [CrossRef]
12. Murakami, H.; Onizuka, N.; Sasaki, J.; Thonghai, N. Ultra-fine particles of amorphous TiO_2 studied by means of positron annihilation spectroscopy. *J. Mater. Sci.* **1998**, *33*, 5811–5814. [CrossRef]
13. Gaberle, J.; Shluger, A. The role of surface reduction in the formation of Ti interstitials. *RSC Adv.* **2019**, *9*, 12182–12188. [CrossRef] [PubMed]
14. Jiang, X.; Zhang, Y.; Jiang, J.; Rong, Y.; Wang, Y.; Wu, Y.; Pan, C. Characterization of Oxygen Vacancy Associates within Hydrogenated TiO_2: A Positron Annihilation Study. *J. Phys. Chem. C* **2012**, *116*, 22619–22624. [CrossRef]
15. Xu, N.N.; Li, G.P.; Pan, X.D.; Wang, Y.B.; Chen, J.S.; Bao, L.M. Oxygen vacancy-induced room-temperature ferromagnetism in D—D neutron irradiated single-crystal TiO_2 (001) rutile. *Chin. Phys. B* **2014**, *23*, 106101. [CrossRef]
16. Doeuff, S.; Henry, M.; Sanchez, C.; Livage, J. The gel route to Cr^{3+}-doped TiO_2, an ESR study. *J. Non-Cryst. Solids* **1987**, *89*, 84–97. [CrossRef]
17. Coey, J.M.D.; Venkatesan, M.; Stamenov, P. Surface magnetism of strontium titanate. *J. Phys.: Condens. Matter.* **2016**, *28*, 485001. [CrossRef]
18. Ermakov, A.E.; Uimin, M.A.; Korolev, A.V.; Volegov, A.S.; Byzov, I.V.; Shchegoleva, N.N.; Minin, A.S. Anomalous Magnetism of the Surface of TiO_2 Nanocrystalline Oxides. *Fizika Tverdogo Tela* **2017**, *59*, 458–473. [CrossRef]
19. Ahmed, S.A. Annealing effects on structure and magnetic properties of Mn-doped TiO_2. *J. Magn. Magn. Mater.* **2016**, *402*, 178–183. [CrossRef]

20. Sokovnin, S.Y.; Balezin, M.E. Production of nanopowders using nanosecond electron beam. *Ferroelectrics* **2012**, *436*, 108–111. [CrossRef]
21. Kamat, P.V. TiO$_2$ Nanostructures: Recent Physical Chemistry Advances. *J. Phys. Chem. C* **2012**, *116*, 11849. [CrossRef]
22. Akdogan, N.; Nefedov, A.; Zabel, H.; Westerholt, K.; Becker, H.-W.; Somsen, C.; Gök, S.; Bashir, A.; Khaibullin, R.; Tagirov, L. High-temperature ferromagnetism in Co-implanted TiO$_2$ rutile. *J. Phys. D Appl. Phys.* **2009**, *42*, 115005. [CrossRef]
23. Chen, Z.; Zhao, Y.S.; Ma, J.; Liu, C.; Ma, Y. Detailed XPS analysis and anomalous variation of chemical state for Mn- and V-doped TiO$_2$ coated on magnetic particles. *Ceram. Int.* **2017**, *43*, 16763–16772. [CrossRef]
24. Cuaila, J.L.S.; Alayo, W.; Avellaneda, C. Ferromagnetism in spin-coated cobalt-doped TiO$_2$ thin films and the role of crystalline phases. *J. Magn. Magn. Mater.* **2017**, *442*, 212. [CrossRef]
25. Li, D.; Li, D.K.; Wu, H.Z.; Liang, F.; Xie, W.; Zou, C.W.; Chao, L.X. Defects related room temperature ferromagnetism in Cu-implanted ZnO nanorod arrays. *J. Alloy. Comp.* **2014**, *591*, 80–84. [CrossRef]
26. Vykhodets, V.; Johnson, K.G.; Kurennykh, T.E.; Beketov, I.V.; Samatov, O.M.; Medvedev, A.I.; Jarvis, E.A.A. Direct Observation of Tunable Surface Structure and Reactivity in TiO$_2$ Nanopowders. *Surf. Sci.* **2017**, *665*, 10–19. [CrossRef]
27. Vykhodets, V.B.; Jarvis, E.A.A.; Kurennykh, T.E.; Beketov, I.V.; Obukhov, S.I.; Samatov, O.M.; Medvedev, A.I.; Davletshin, A.E.; Whyte, T. Inhomogeneous depletion of oxygen ions in oxide nanoparticles. *Surf. Sci.* **2016**, *644*, 141–147. [CrossRef]
28. Vykhodets, V.B.; Jarvis, E.A.A.; Kurennykh, T.E.; Davletshin, A.E.; Obukhov, S.I.; Beketov, I.V.; Samatov, O.M.; Medvedev, A.I. Extreme deviations from stoichiometry in alumina nanopowders. *Surf. Sci.* **2014**, *630*, 182–186. [CrossRef]
29. Uymin, M.A.; Minin, A.S.; Yermakov, A.Y.; Korolyov, A.V.; Balezin, M.Y.; Sokovnin, S.Y.; Konev, A.S.; Konev, S.F.; Molochnikov, L.S.; Gaviko, V.S.; et al. Magnetic properties and structure of TiO$_2$ -Mn (0.73%) nanopowders: The effects of electron irradiation and vacuum annealing. *Lett. Mater.* **2019**, *9*, 91–96. [CrossRef]
30. Jarvis, E.A.A.; Carter, E.A. Metallic Character of the Al$_2$O$_3$(0001)-($\sqrt{31} \times \sqrt{31}$)R ± 9° Surface Reconstruction. *J. Phys. Chem. B* **2001**, *105*, 4045–4052. [CrossRef]
31. Jarvis, E.A.A.; Carter, E.A. A nanoscale mechanism of fatigue in ionic solids. *Nano Lett.* **2006**, *6*, 505–509. [CrossRef] [PubMed]

Article

Neural Network Modeling of Microstructure Formation in an AlMg6/10% SiC Metal Matrix Composite and Identification of Its Softening Mechanisms under High-Temperature Deformation

Alexander Smirnov *, Vladislav Kanakin and Anatoly Konovalov

Institute of Engineering Science, UB RAS, 34 Komsomolskaya St., 620049 Ekaterinburg, Russia
* Correspondence: smirnov@imach.uran.ru

Citation: Smirnov, A.; Kanakin, V.; Konovalov, A. Neural Network Modeling of Microstructure Formation in an AlMg6/10% SiC Metal Matrix Composite and Identification of Its Softening Mechanisms under High-Temperature Deformation. *Appl. Sci.* **2023**, *13*, 939. https://doi.org/10.3390/app13020939

Academic Editor: Yuzo Nakamura

Received: 28 November 2022
Revised: 7 January 2023
Accepted: 8 January 2023
Published: 10 January 2023

Copyright: © 2023 by the authors. Licensee MDPI, Basel, Switzerland. This article is an open access article distributed under the terms and conditions of the Creative Commons Attribution (CC BY) license (https://creativecommons.org/licenses/by/4.0/).

Abstract: The paper investigates the rheological behavior and microstructuring of an AlMg6/10% SiC metal matrix composite (MMC). The rheological behavior and microstructuring of the AlMg6/10% SiC composite is studied for strain rates ranging between 0.1 and 4 s^{-1} and temperatures ranging from 300 to 500 °C. The microstructure formation is studied using EBSD analysis, as well as finite element simulation and neural network models. The paper proposes a new method of adding data to a training sample, which allows neural networks to correctly predict the behavior of microstructure parameters, such as the average grain diameter, and the fraction and density of low-angle boundaries with scanty initial experimental data. The use of neural networks has made it possible to relate the thermomechanical parameters of deformation to the microstructure parameters formed under these conditions. These dependences allow us to establish that, at strain rates ranging from 0.1 to 4 s^{-1} and temperatures between 300 to 500 °C, the main softening processes in the AlMg6/10% SiC MMC are dynamic recovery and continuous dynamic recrystallization accompanied, under certain strain and strain rate conditions at 300 and 350 °C, by geometric recrystallization.

Keywords: metal matrix composite; high temperature; aluminum; simulation; rheology; neural network; relaxation; rheological behavior; flow stress; Al-Mg alloy; Al-Mg-Sc-Zr alloy

1. Introduction

Aluminum and aluminum alloys have become widely used in industry due to their high specific mechanical properties, high thermal conductivity, and corrosion resistance, as well as other important technological parameters. Numerous aluminum-based alloys doped with various alloying elements offering the required physical and mechanical properties of an alloy have been created. These materials include Al-Mg alloys, which have good corrosion resistance, ductility, and weldability [1–6]. The latter alloys are used as structural materials for cryogenic structures, and they are used in the aerospace industry. Magnesium additives in aluminum strengthen it significantly. Each percent of Mg content increases the tensile strength by 25–30 MPa [7]. At the same time, Al-Mg alloys can be considered as deformable with a Mg content of up to 11–12 wt%. With a Mg content of up to 8 wt%, these alloys cannot be hardened by heat treatment [7]. However, with a magnesium content of more than 7 wt%, its anticorrosive properties deteriorate sharply; therefore, Al-Mg alloys with this Mg content are hardly ever used. This leads to the fact that Al-Mg alloys used in industry are not amenable to heat treatment, and they have medium strength. As a result, these alloys find limited use only in some areas of the industry.

One of the ways to increase the strength properties of Al-Mg alloys is their alloying with the Sc rare earth element, which, even in a small content, increases the strength of alloys [8,9]. The hardening of Al-Mg alloys by adding scandium is a combined effect of dispersion hardening and structural hardening [10,11]. These alloys are generally alloyed with Sc in an amount of up to 0.3% and modified with Mn and Zr [11–14]. However, despite the low content of scandium due to its high cost, the use of Al-Mg-Sc alloys is also limited.

Another way to obtain increased material properties, which has recently gained increasing attention from the industry and researchers, is the creation of metal matrix composites based on an aluminum alloy matrix reinforced with various fillers in various percentages [15–20]. The development of a new material consists not only in synthesis but also in the need to form the required properties after its manufacture. The required properties of alloys and alloy-based composites, as well as products made of them, are formed using heat treatment and machining in a wide temperature range. In particular, for structural materials, these types of processing make it possible to obtain materials and products with compromised plastic and strength properties. Thus, it becomes possible to control the properties using controlled thermomechanical action consisting of mechanical action on the workpiece under certain temperature conditions [21–23]. As a rule, the thermomechanical processing of alloys and alloy-based composites is carried out at elevated temperatures, at which the alloy structure is actively rearranged.

During plastic deformation, these materials undergo competing processes associated with hardening due to the increased dislocation density and phase transformations, as well as with softening due to dynamic recrystallization, recovery, and increased damage [23–32]. The interaction of hardening and softening affects the rheological behavior of the material. In particular, the flow stress curves at certain temperatures and strain rates may have a peak [26,33–35], a steady-state portion [25–27,34,35], and several hardening and softening sections [10,11,36]; in addition, there may be an inverse strain rate dependence [37–39], when an increase in strain rate causes a decrease in flow stress. To determine the conditions for the formation of the required shape of the workpiece from a material with the necessary properties, one is to relate the thermomechanical parameters of deformation to flow stress and the microstructure parameters. These relationships are generally established by means of mathematical models based on the functional relation of microstructure parameters to thermomechanical parameters [40–43], as well as using physically based models [35,44–49] and computational models, including models based on the cellular automata method [34,50–54] and molecular dynamics [55]. Neural network models for describing microstructure formation under high-temperature deformation have not found wide application due to the need to obtain a large amount of experimental data which is time-consuming and leads to the inexpediency of using neural networks. The problem of the scanty experimental data is solved by simulating microstructure formation using cellular automata or embedding functions into finite element programs [56–59]. Simulation enables computational experiments to be conducted, thus making it possible to reduce the time spent on forming a sample of the necessary size for training neural networks. Nevertheless, this approach is rather time-consuming as it requires first developing or applying a model that adequately describes the microstructure formation in the material and then using the results of calculations based on the model to train neural networks. This paper proposes a new and relatively simple method for processing scarce experimental data relating microstructure parameters to the parameters of thermomechanical action on a metal material for subsequent neural network training. The neural networks constructed in this study are used to identify the softening mechanisms in an AlMg6/10% SiC metal matrix composite at temperatures from 300 to 500 °C and strain rates ranging between 0.1 and $4\ s^{-1}$, as well as to relate the microstructure parameters of the metal matrix composite to the thermomechanical parameters of deformation.

2. Materials and Methods

2.1. Material and Research Technique

The AlMg6/10% SiC metal matrix composite (MMC) [60] produced using powder technology is used as the material for the study. The initial components of the AlMg6/10% SiC MMC were mixed in a vibratory mixer in an argon atmosphere. Sintering was carried out for 60 min at a temperature of 420 °C. The pressure at which sintering occurred was 30 MPa. No additives modifying the surface of SiC particles were used.

The size of the SiC particles corresponds to the F1500 standard, with an average diameter of 2.0 ± 0.4 µm. Figure 1 shows an EBSD image and SEM image of the composite before deformation. As can be seen from this figure, the matrix particles are a polycrystalline material. After sintering of the composite, the SiC reinforcer particles are located along the boundaries of the matrix particles. In this paper, the composite was studied without pre-extrusion. Its mechanical properties at room temperature are shown in Table 1. For comparison, this table shows the properties of AlMg6 and Al-Mg-Sc-Zr alloys, which are similar in use to the composite.

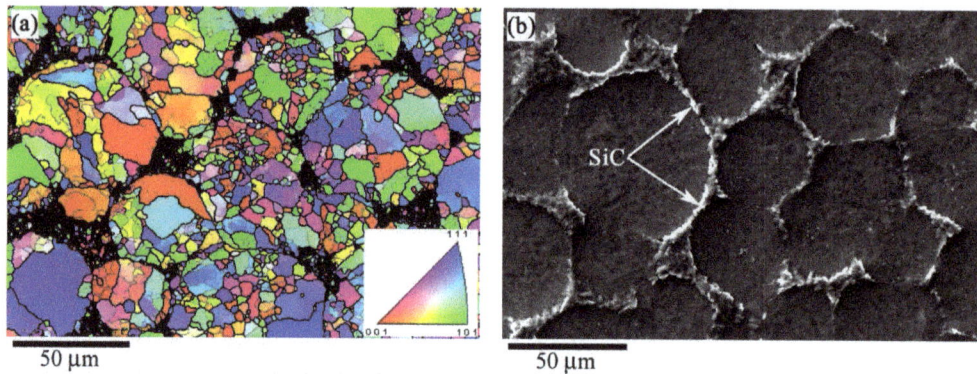

Figure 1. (a) The EBSD image after electropolishing and (b) the SEM image of the AlMg6/10% SiC MMC microstructure.

Table 1. Material properties.

Material	Yield Strength, MPa	Elastic Modulus, GPa	Density, g/cm^3
AlMg6/10%SiC	179	93	2.67
Al-Mg-Sc-Zr alloy	164	66	2.65
AlMg6 alloy	141	76	2.62

Before testing, all the materials were identically annealed for 15 h at 520 °C. It can be seen from this table that the yield strength and elastic modulus of the composite are higher than those of the AlMg6 alloy and its more expensive Al-Mg-Sc-Zr analog. Table 2 shows the chemical composition of the Al-Mg-Sc-Zr alloy, the AlMg6 alloy, and the matrix of the AlMg6/10% SiC metal matrix composite. The chemical compositions were determined by means of a Spectromaxx LMF04 optical emission spectrometer, and the granulometric composition of the powders was determined by laser diffraction using a LaSca-TD.

Table 2. Chemical composition of materials, wt%.

Material	Al	Mg	Mn	Sc	Fe	Zr	Si	Cu	Ti	Zn	Be
AlMg6 alloy and the matrix of AlMg6/10% SiC	balance	6.56	0.5	-	0.27	-	0.16	0.013	0.04	0.02	0.0012
Al-Mg-Sc-Zr alloy	balance	5.18	0.36	0.23	0.12	0.07	0.01	0.022	0.02	0.02	0.003

The densities of the materials were determined using the hydrostatic method according to ASTM B311-13, by weighing the specimens in air and distilled water on an Ohaus Pioneer PA 214 analytical balance.

Experiments on the determination of the mechanical properties and rheological behavior of the composite at room and high temperatures were performed for cylindrical specimens. Compression specimens had a diameter d_0 of 6 ± 0.05 mm and a height h_0 of 9 ± 0.05 mm. Cylindrical tensile specimens had the following dimensions: $d_0 = 5 \pm 0.05$ mm, and the length of the gauge part $l_0 = 25 \pm 0.1$ mm. The data obtained from high-temperature compression tests of specimens were used to construct flow stress curves depending on temperature, strain, and strain rate. These curves were used for finite element simulation of specimen compression. Compression experiments were carried out using an automated plastometric installation designed at the Institute of Engineering Science, UB RAS [11,39].

In compression experiments, a graphite-containing lubricant was used to reduce friction between the punch and specimen. The lubricant provided the coefficient of Coulomb friction between the flat die and aluminum alloy equal to 0.09 at 300 °C and equal to 0.1 and 0.13 at 400 and 500 °C, respectively. The values of the friction coefficients were obtained using the procedure described in [61], the essence of which is to compress specimens to different heights and select the friction coefficient in the Coulomb friction law according to the results of finite element simulation in such a way that the maximum and minimum diameters of the specimen coincide as closely as possible with the simulated one at different strains.

In the temperature range between 300 and 500 °C, the specimens became barrel-shaped despite the use of the lubricant (Figure 2). The specimens were cooled immediately after the end of deformation, and in 2 s, the specimen temperature did not exceed 70 °C. After deformation, the specimens were cut in a longitudinal section (parallel to the compression axis) using electric spark cutting. Then, thin sections for EBSD analysis were made in the plane of the longitudinal section of the specimens. Zones of the specimens for EBSD analysis are highlighted in orange in Figure 2. For the same zones, the strain ε and strain rate $\dot{\varepsilon}$ were calculated on the basis of finite element simulation of compression of specimens at the studied range of temperatures, strains and strain rates. The formulation of the finite element simulation problem is given in Section 2.2.

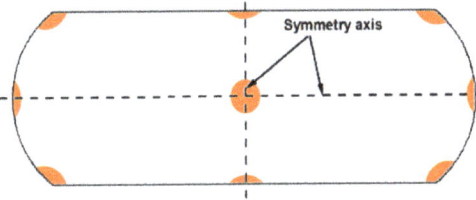

Figure 2. The longitudinal section of a cylindrical specimen after compression. The zones of the EBSD analysis of the specimen microstructure are highlighted in orange.

The microstructures of the specimens before and after deformation were studied by electron backscattered diffraction (EBSD) using a Tescan Vega II scanning electron microscope with an Oxford HKL Nordlys F+ EBSD analysis accessory. For the EBSD analysis, the specimens were first mechanically polished and then electropolished by 90% CH_3COOH +10% $HClO_4$ electrolyte cooled to 8 °C. The polishing time averaged 6 s at a voltage of 40 V and a current density of 0.3 A/mm^2. The scanning step during the EBSD analysis was equal to 0.3 µm. The size of the scanned area was 300×150 µm. It was assumed that the grain misorientation exceeded 15° and that the subgrain misorientation ranged between 2 and 15°.

2.2. Formulation of the Finite Element Problem for Specimen Compression

In order to determine the stress–strain state in the specimen, its finite element model was constructed in the Deform program. For the specimen material, an isothermal viscoplastic model of isotropic strain hardening was used. The flow stress was set in a tabular form based on cylindrical specimen compression tests (Figure 3a). The calculations were performed under the assumption of the axisymmetric stress–strain state and deformation symmetry about the horizontal geometric symmetry axis of the specimen (Figure 2). Thus, only a quarter of the specimen cross-section was simulated (Figure 3b).

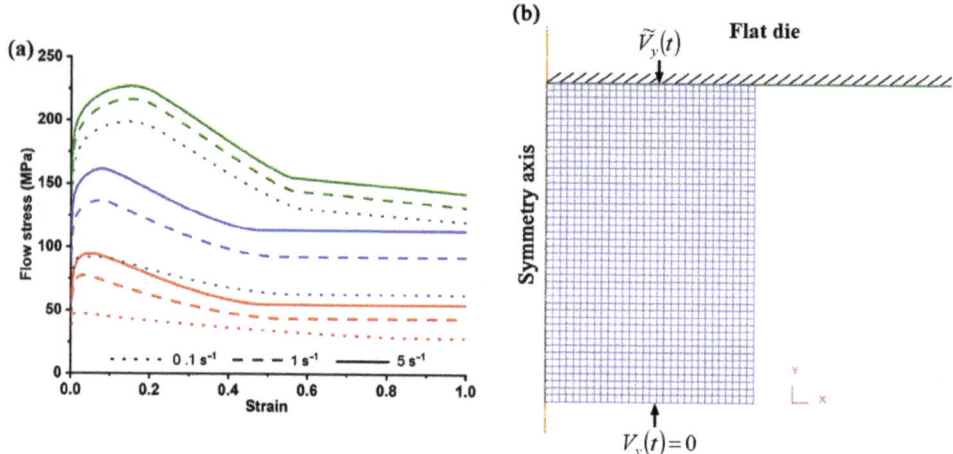

Figure 3. (**a**) Flow stress as dependent on strain for 300 °C (green curve), 400 °C (blue curve), 500 °C (red curve), and strain rates ranging between 0.1 and 10 s^{-1}; (**b**) the finite element grid used for the simulation.

The simulation was performed in accordance with the experimental conditions of specimen loading in which the time dependence of die speed $\widetilde{V}_y(t)$ (Figure 3b) was taken for each specific tested specimen. It was assumed that the friction between the specimen and the die followed Coulomb's friction law and depended on temperature. In the simulation of specimen deformation, the Coulomb friction coefficient was set to 0.09 at a temperature of 300 °C and to 0.1 and 0.13 at 400 and 500 °C, respectively.

2.3. Neural Network Models for Forming Microstructure Parameters

The average grain diameter D, the fraction of low-angle boundaries P_L, and their density S_L were the analyzed parameters of the composite matrix microstructure depending on temperature, strain, and strain rate. The fraction and density of low-angle boundaries were calculated using the formulas

$$P_L = \frac{L}{L+H} \text{ and } S_L = \frac{L}{F} \qquad (1)$$

Here, L is the sum of the lengths of all the low-angle boundaries; H is the sum of the lengths of all the high-angle boundaries; and F is the area of the microstructure image on which the sum of the lengths of all the low-angle boundaries L was determined. These parameters were determined from the data obtained by EBSD. The average grain diameter D was calculated based on a sample that included grains containing at least 3 indexed points. To describe the evolution of the microstructure parameters, we used one-, two-, and three-layer neural networks. Their general scheme is shown in Figure 4. The neural networks were built using the scikit-learn library.

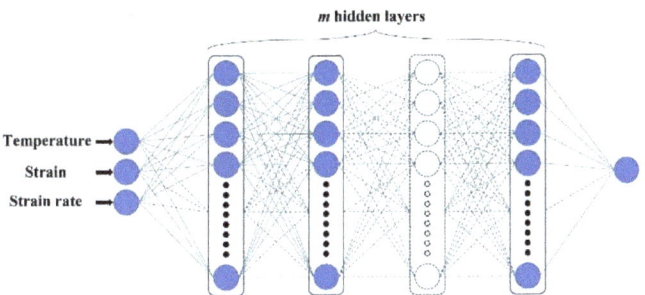

Figure 4. General scheme of the neural network model.

3. Results

3.1. The Rheological Behavior and Microstructuring of the Composite

Under high deformation temperatures, relaxation processes occur in metallic materials and composites based on them, which affect the form of flow stress curves. The flow stress curves for the AlMg6/10% SiC MMC (Figure 3a) can be divided into several portions (stages). At stage I, the composite undergoes hardening until the peak (maximum) flow stress value is reached. Moreover, the strain corresponding to the peak stress decreases with increasing temperature and increases with an increasing strain rate. The hardening stage is followed by a portion corresponding to material softening, where the flow stress value decreases with increasing strain. At stage III (the steady-state portion), the hardening and softening rates are close, and the flow stress value remains almost unchanged with increasing strain. This rheological behavior of the composite at high temperatures means that dynamic recrystallization occurred during deformation, which can follow the discontinuous, continuous, or geometric recrystallization mechanisms [23]. In order to identify the recrystallization mechanism, an EBSD analysis of the specimens was performed after various thermomechanical loading conditions.

Figures 5–7 show the EBSD images of the microstructures of the AlMg6/10% SiC MMC after deformation as dependent on temperature, strain, and strain rate. The black spots in the figures correspond to non-indexed zones, which generally are zones with a high content of silicon carbide falling out from the matrix during electrochemical polishing. The values of strain ε and the strain rate $\dot{\varepsilon}$ were determined from the results of the finite element simulation of specimen compression. The formulation of the finite element problem is described in Section 2.

Although the specimens have the same height after compression, the accumulated strain in the same zones varies at different temperatures (Figure 8). Thus, it is difficult to analyze the occurring softening processes depending on temperature and strain-rate loading conditions without using any approximating equations. Table 3 shows the experimentally obtained values of the average grain diameter D, and the fraction P_L and density S_L of low-angle boundaries depending on temperature, strain, and strain rate. According to these data, maps of the formation of the average grain diameter, and the fraction and density of low-angle boundaries are plotted against the thermomechanical parameters of deformation in Figures 9–11. As can be seen from these maps, increasing strain decreases the average grain diameter, while an increasing deformation temperature forms larger grains in the composite matrix. The effect of strain rate on grain evolution is nonmonotonic. In the grain map (Figure 9), there are portions where the diameter of the formed grains decreases with an increasing strain rate, and there are portions where, in contrast, the average grain diameter decreases with a decreasing strain rate.

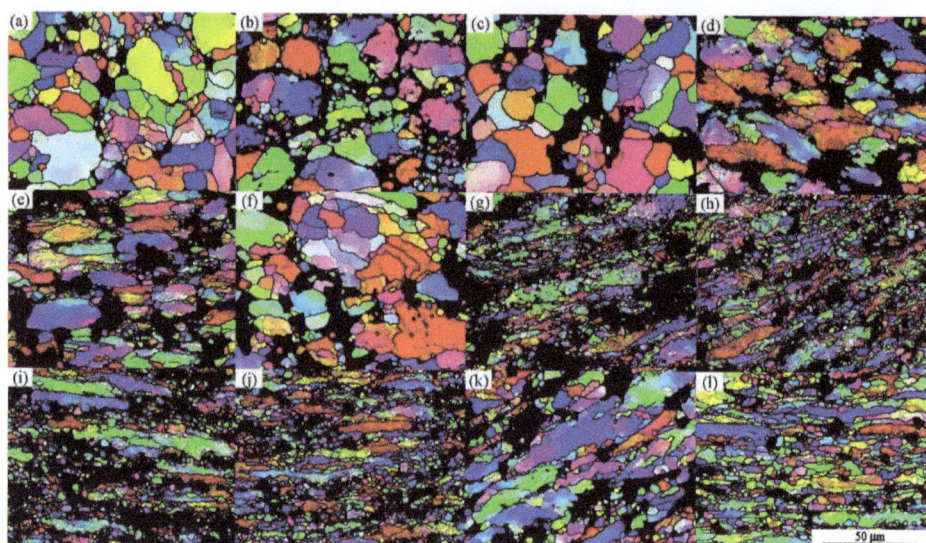

Figure 5. An EBSD image of the composite matrix microstructure at a deformation temperature of 300 °C for (**a**) $\varepsilon = 0.12$ and $\dot{\varepsilon} = 0.83\ \mathrm{s}^{-1}$; (**b**) $\varepsilon = 0.13$ and $\dot{\varepsilon} = 0.26\ \mathrm{s}^{-1}$; (**c**) $\varepsilon = 0.17$ and $\dot{\varepsilon} = 0.05\ \mathrm{s}^{-1}$; (**d**) $\varepsilon = 0.37$ and $\dot{\varepsilon} = 2.48\ \mathrm{s}^{-1}$; (**e**) $\varepsilon = 0.40$ and $\dot{\varepsilon} = 0.8\ \mathrm{s}^{-1}$; (**f**) $\varepsilon = 0.45$ and $\dot{\varepsilon} = 0.15\ \mathrm{s}^{-1}$; (**g**) $\varepsilon = 0.58$ and $\dot{\varepsilon} = 3.85\ \mathrm{s}^{-1}$; (**h**) $\varepsilon = 0.63$ and $\dot{\varepsilon} = 1.25\ \mathrm{s}^{-1}$; (**i**) $\varepsilon = 0.63$ and $\dot{\varepsilon} = 4.18\ \mathrm{s}^{-1}$; (**j**) $\varepsilon = 0.66$ and $\dot{\varepsilon} = 1.32\ \mathrm{s}^{-1}$; (**k**) $\varepsilon = 0.73$ and $\dot{\varepsilon} = 0.24\ \mathrm{s}^{-1}$; (**l**) $\varepsilon = 0.76$ and $\dot{\varepsilon} = 0.26\ \mathrm{s}^{-1}$.

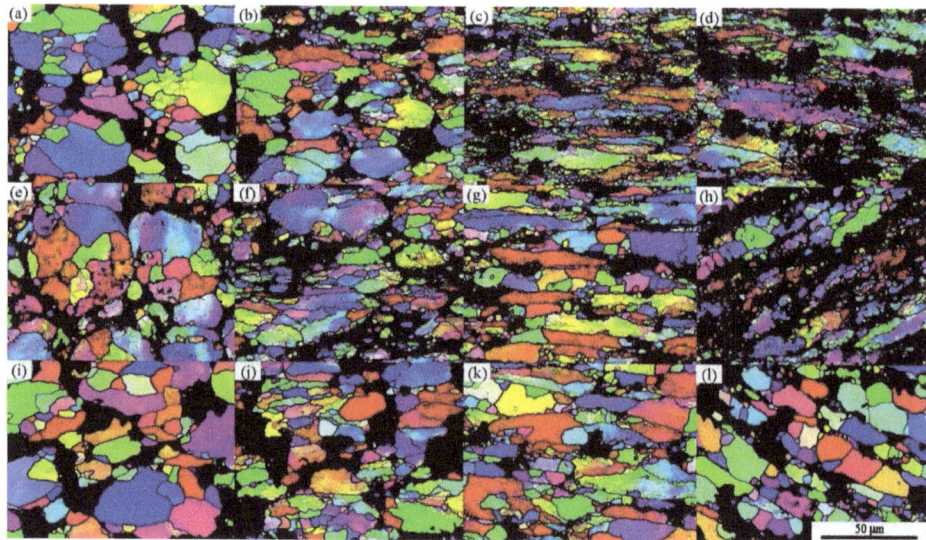

Figure 6. An EBSD image of the composite matrix microstructure at a deformation temperature of 400 °C for (**a**) $\varepsilon = 0.07$ and $\dot{\varepsilon} = 0.54\ \mathrm{s}^{-1}$; (**b**) $\varepsilon = 0.11$ and $\dot{\varepsilon} = 0.22\ \mathrm{s}^{-1}$; (**c**) $\varepsilon = 0.17$ and $\dot{\varepsilon} = 0.05\ \mathrm{s}^{-1}$; (**d**) $\varepsilon = 0.34$ and $\dot{\varepsilon} = 2.32\ \mathrm{s}^{-1}$; (**e**) $\varepsilon = 0.43$ and $\dot{\varepsilon} = 0.81\ \mathrm{s}^{-1}$; (**f**) $\varepsilon = 0.49$ and $\dot{\varepsilon} = 0.15\ \mathrm{s}^{-1}$; (**g**) $\varepsilon = 0.60$ and $\dot{\varepsilon} = 4.06\ \mathrm{s}^{-1}$; (**h**) $\varepsilon = 0.63$ and $\dot{\varepsilon} = 4.29\ \mathrm{s}^{-1}$; (**i**) $\varepsilon = 0.79$ and $\dot{\varepsilon} = 1.47\ \mathrm{s}^{-1}$; (**j**) $\varepsilon = 0.83$ and $\dot{\varepsilon} = 0.25\ \mathrm{s}^{-1}$; (**k**) $\varepsilon = 0.88$ and $\dot{\varepsilon} = 1.64\ \mathrm{s}^{-1}$; (**l**) $\varepsilon = 0.92$ and $\dot{\varepsilon} = 0.28\ \mathrm{s}^{-1}$.

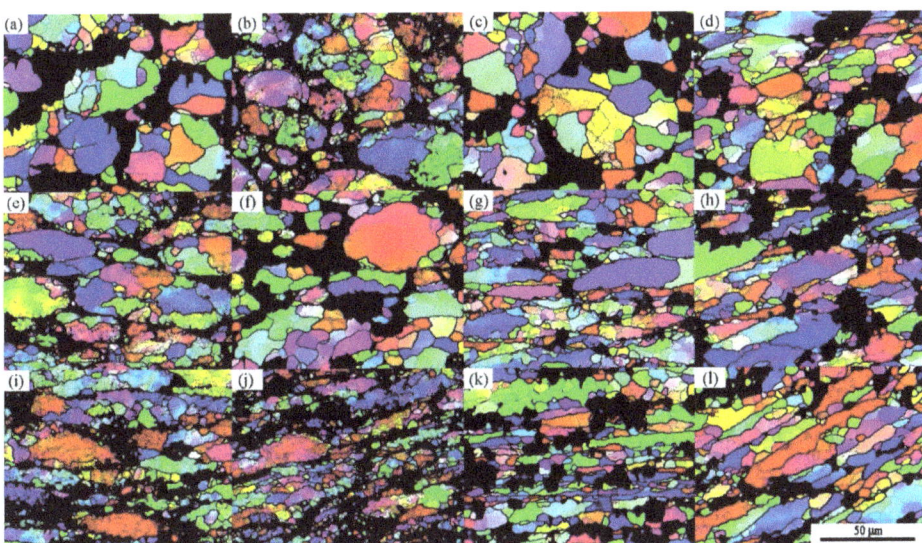

Figure 7. An EBSD image of the composite matrix microstructure at a deformation temperature of 500 °C for (**a**) $\varepsilon = 0.02$ and $\dot{\varepsilon} = 0.3$ s^{-1}; (**b**) $\varepsilon = 0.05$ and $\dot{\varepsilon} = 0.1$ s^{-1}; (**c**) $\varepsilon = 0.1$ and $\dot{\varepsilon} = 0.03$ s^{-1}; (**d**) $\varepsilon = 0.33$ and $\dot{\varepsilon} = 2.36$ s^{-1}; (**e**) $\varepsilon = 0.4$ and $\dot{\varepsilon} = 0.81$ s^{-1}; (**f**) $\varepsilon = 0.51$ and $\dot{\varepsilon} = 0.16$ s^{-1}; (**g**) $\varepsilon = 0.66$ and $\dot{\varepsilon} = 4.57$ s^{-1}; (**h**) $\varepsilon = 0.73$ and $\dot{\varepsilon} = 5.0$ s^{-1}; (**i**) $\varepsilon = 0.77$ and $\dot{\varepsilon} = 1.54$ s^{-1}; (**j**) $\varepsilon = 0.92$ and $\dot{\varepsilon} = 1.84$ s^{-1}; (**k**) $\varepsilon = 0.97$ and $\dot{\varepsilon} = 0.29$ s^{-1}; (**l**) $\varepsilon = 1.15$ and $\dot{\varepsilon} = 0.35$ s^{-1}.

Figure 8. Accumulated strain distribution in the AlMg6/10% SiC MMC specimens under 50% compression at 300, 400, and 500 °C.

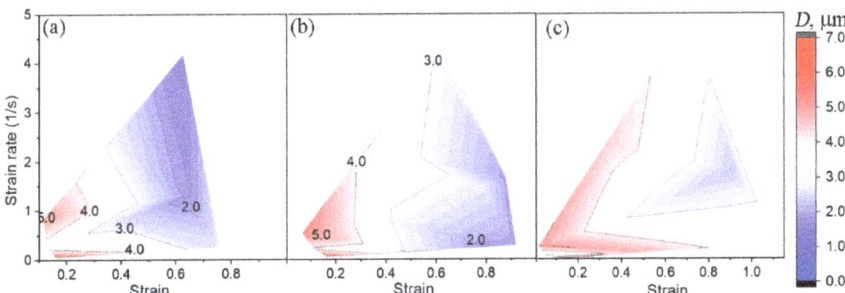

Figure 9. The effect of strain and strain rate on the average grain diameter depending on temperature, °C: (**a**) 300, (**b**) 400, (**c**) 500. The dependences are plotted from the results of the EBSD analysis of the microstructure.

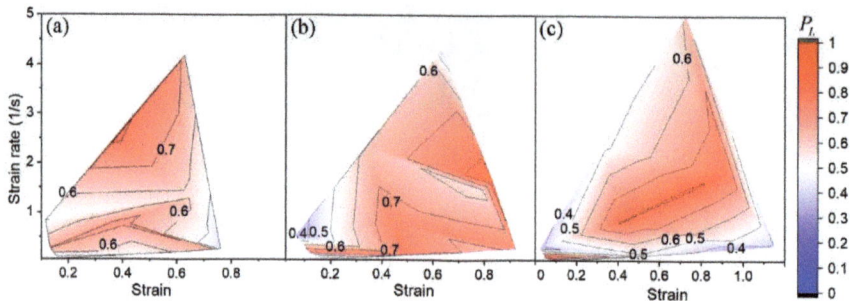

Figure 10. The effect of strain and strain rate on the fraction of low-angle boundaries depending on temperature, °C: (**a**) 300, (**b**) 400, (**c**) 500. The dependences are plotted from the results of the EBSD analysis of the microstructure.

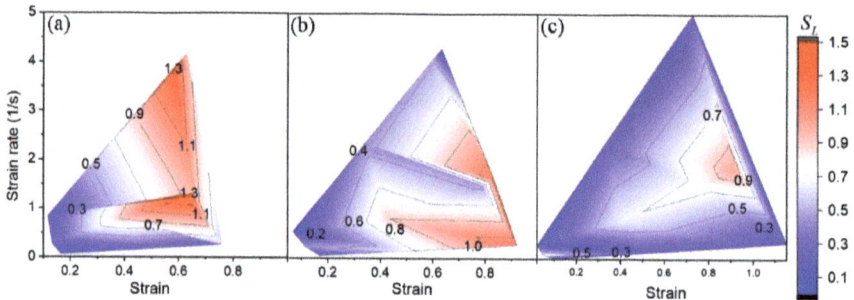

Figure 11. The effect of strain and strain rate on the density of low-angle boundaries depending on temperature, °C: (**a**) 300, (**b**) 400, (**c**) 500. The dependences are plotted from the results of the EBSD analysis of the microstructure.

Table 3. Microstructure parameters of the AlMg6/10 % SiC MMC at temperatures of 300, 400, and 500 °C.

Temp., °C	300											
Strain	0.12	0.13	0.17	0.37	0.40	0.45	0.58	0.63	0.63	0.66	0.73	0.76
Strain rate, s^{-1}	0.83	0.26	0.05	2.48	0.8	0.15	3.85	1.25	4.18	1.32	0.24	0.26
D, μm	5.07	3.52	5.45	2.73	2.56	4.49	1.67	1.82	1.61	1.83	3.33	2.68
P_L	0.5	0.7	0.43	0.82	0.73	0.55	0.76	0.64	0.67	0.57	0.58	0.43
S_L, μm^{-1}	0.13	0.17	0.1	0.73	0.97	0.23	1.30	1.50	0.97	0.97	0.20	0.53
Temp., °C	400											
Strain	0.07	0.11	0.17	0.34	0.43	0.49	0.60	0.63	0.79	0.88	0.83	0.92
Strain rate, s^{-1}	0.54	0.22	0.05	2.32	0.81	0.15	4.06	4.29	1.47	0.25	1.64	0.28
D, μm	5.46	3.89	5.6	3.63	2.9	2.89	2.98	3.66	2.73	1.81	1.92	1.86
P_L	0.35	0.77	0.38	0.63	0.76	0.79	0.60	0.44	0.57	0.65	0.85	0.76
S_L, μm^{-1}	0.07	0.43	0.13	0.43	0.9	0.63	0.47	0.13	0.47	1.10	1.20	1.20
Temp., °C	500											
Strain	0.02	0.05	0.1	0.33	0.4	0.51	0.66	0.73	0.77	0.92	0.97	1.15
Strain rate, s^{-1}	0.3	0.1	0.03	2.36	0.81	0.16	4.57	5.00	1.54	1.84	0.29	0.35
D, μm	5.13	2.85	5.27	4.56	3.04	4.76	3.37	3.53	2.43	2.27	3.55	3.91
P_L	0.38	0.83	0.39	0.49	0.8	0.49	0.46	0.62	0.81	0.79	0.34	0.39
S_L, μm^{-1}	0.10	0.70	0.10	0.27	0.67	0.17	0.20	0.27	0.77	1.03	0.20	0.13

The comparison of the initial microstructure (Figure 1a) with the microstructure formed during deformation (Figures 5–7) testifies that, after high-temperature deformation, small grains appear in the microstructure. It can also be seen that inside the initial large grains, there are subgrains with low-angle boundaries (highlighted in gray in Figures 5–7). At the same time, new elongated grains and nearly equiaxed ones are formed in the initial elongated grains along the composite matrix flow direction; the prevailing majority of them retain their geometry throughout the range of temperatures and strain rates under study. At a strain of at least 0.7, some subgrains have an almost equiaxed shape, and the other ones are elongated but with a greater equiaxiality coefficient than the grains containing these subgrains. Another experimentally determined fact is that the matrix grains are formed inside large, deformed grains. Thus, we can say that the grain formation process in the composite matrix is associated with the mechanism of continuous dynamic recrystallization [23,43]. At the same time, the subboundary migration rate in the matrix is high enough for a significant number of the subgrains to remain equiaxed. Note also that the high-angle boundaries of the new grains also have a migration rate sufficient for a significant part of the recrystallized grains to be equiaxed at strains of at least 0.7. The obtained experimental data confirm the need to construct physically based mathematical models of microstructure evolution that take into account the migration of low- and high-angle boundaries during continuous dynamic recrystallization [43,44,62,63].

It was reported in [11,62,64] that, during continuous dynamic recrystallization, the fraction of low-angle boundaries changes nonmonotonically with increasing strain. That is, the curve representing the strain dependence of the fraction of low-angle boundaries may have portions of decreasing and increasing fractions of low-angle boundaries. It is obvious from the plotted map of the formation of low-angle boundaries (Figure 10) that the composite under study has similar nonmonotonic dependencies.

It can be seen from the microstructure image (Figures 5–7) that increasing temperature induces the formation of larger subgrains and that, at a temperature of 500 °C, in some large grains, there are no subgrains at all (Figure 7). This temperature effect stems from the increasing rate of dislocation annihilation due to an increase in the velocity of their chaotic motion caused by increasing temperature [23].

It was noted above that, in real experiments at high temperatures and large plastic deformations, due to friction, the strain and strain rate loading conditions can hardly be maintained constant. As a result, at the same speed of the punch of the testing machine but different temperatures, the specimen strain rate is different. To estimate the influence of the thermomechanical parameters on microstructure formation, the experimental values of D, P_L, and S_L must be determined at the same strains and strain rates; therefore, the values of the microstructure parameters obtained for different thermomechanical conditions must be approximated. In this paper, polynomials are used to plot maps of the formation of grains and low-angle boundaries (Figures 9–11) depending on temperature, strain, and strain rate. However, it is not possible to use polynomials to predict nonmonotonically changing data due to a low prediction level. As a result, Figures 9–11 show the maps of the formation of grains and low-angle boundaries only within the experimental range. For a complete analysis of the forming microstructure in the entire strain and strain rate range studied, the constructed maps are insufficient. This problem can be solved by the use of neural networks. They allow us to predict changes in the parameters with acceptable engineering accuracy if the correct architecture and training data are chosen [65,66].

3.2. Constructing the Architecture of Neural Networks and Their Training

To describe changes in the average grain diameter D, the fraction of low-angle boundaries P_L, and the density of low-angle boundaries S_L during deformation at temperatures between 300 and 500 °C, three neural networks with a general scheme (multilayer perceptron) were constructed (Figure 4). The neural networks were trained according to the experimental data from Table 3. Verification was performed according to the experimental data from Table 4, which were obtained at temperatures and strain rates different from those used for training. The images of the microstructures obtained with the thermomechanical deformation parameters from Table 4 are shown in Figure 12.

Table 4. Microstructure parameters of the AlMg6/10 % SiC MMC at temperatures of 350 and 450 °C.

Temp., °C	350			450		
Strain	0.10	0.35	0.54	0.08	0.34	0.54
Strain rate, s^{-1}	0.30	1.14	1.75	0.30	1.20	1.90
D, µm	4.41	3.02	2.19	4.71	3.58	2.67
P_L	0.59	0.80	0.64	0.44	0.64	0.54
S_L, µm^{-1}	0.06	0.18	0.23	0.03	0.15	0.19

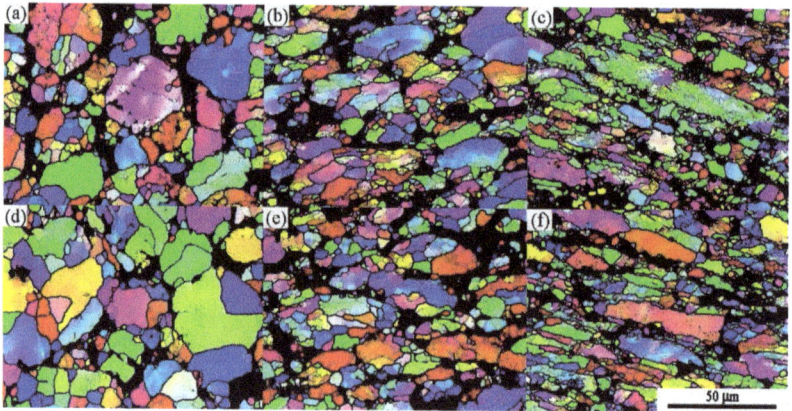

Figure 12. EBSD images of the composite matrix microstructure at deformation temperatures (**a**–**c**) of 350 °C (**d**–**f**) and 450 °C, which correspond to the following strain and strain rate conditions: (**a**) $\varepsilon = 0.10$ and $\dot{\varepsilon} = 0.30$ s^{-1}; (**b**) $\varepsilon = 0.35$ and $\dot{\varepsilon} = 1.14$ s^{-1}; (**c**) $\varepsilon = 0.54$ and $\dot{\varepsilon} = 1.75$ s^{-1}; (**d**) $\varepsilon = 0.08$ and $\dot{\varepsilon} = 0.30$ s^{-1}; (**e**) $\varepsilon = 0.34$ and $\dot{\varepsilon} = 1.20$ s^{-1}; (**f**) $\varepsilon = 0.54$ and $\dot{\varepsilon} = 1.90$ s^{-1}.

As it was shown in Section 3.1, the average grain diameter, and the fraction and density of low-angle boundaries vary nonmonotonically. Therefore, in order to take into account correctly the nonmonotonicity of the relations of the microstructure parameters to the thermomechanical parameters of deformation, which have a form similar to those shown in Figures 9–11, it is necessary to use a large number of neurons (usually more than 20). It should be borne in mind that the number of weight coefficients that need to be determined by training significantly exceeds the number of neurons in the hidden layer. If we choose a neural network structure with one hidden layer and three input neurons (Figure 4), the number of the weight coefficients of the neural network is four times the number of neurons in the hidden layer. In this study, 36 sets of experimental data relating the average grain diameter to temperature, strain, and strain rate were obtained. Similar sets were obtained for the fraction and density of low-angle boundaries. Thus, the maximum number of possible neurons in the hidden layer with such a single-layer neural network is nine. A neural network with this number of neurons makes it possible to describe the evolution of the average grain diameter depending on the thermomechanical parameters of deformation with an average relative deviation of 111%. The average relative deviation for each investigated microstructure parameter is calculated using the formula

$$\delta = \frac{1}{N} \sum_{i=1}^{N} \frac{|\varphi_i - z_i|}{z_i} \cdot 100\% \qquad (2)$$

where φ_i and z_i are the calculated and experimental values of the compared values obtained under the same conditions (in this paper, these values are the average grain diameter D, and the fraction P_L and density S_L of low-angle boundaries); and N is the total number of compared values.

To increase the number of hidden neurons in the neural network, it is necessary to increase the size of the sample by which the neural network is trained. Augmentation is one of the possible ways of increasing the sample size by which the neural network is trained. The essence of this method is experimental data noising. Certainly, this method sometimes allows the problem of neural network uncertainty to be solved. However, as computational experiments have shown, it is impossible to achieve a significant reduction in the average relative deviation for the dependences obtained in this study. As a result, a method of processing experimental data was proposed, which consists of approximating experimental data using a surface, followed by adding data to the training sample. The values of the microstructure parameters from an arbitrary point of the constructed approximating surface are added to the training sample. Let us now consider the algorithm for creating a training sample for the average grain diameter obtained at 500 °C (Figure 13). In this figure, the experimental points are shown in red.

1. First, planes are built on the three nearest points. Then, we select a part of the plane bound by straight intersections with neighboring planes and with a triangle form (Figure 13a).
2. We remove the planes located inside the body. Here, the body means a part of the space bound by the constructed triangles.
3. If the line connecting the experimental point with its projection on the strain–strain-rate plane intersects the triangles, they are removed from the constructed surface of the body (the removed planes are shown in yellow in Figure 13a). This results in a surface approximated by planes in a certain strain and strain rate range (Figure 13b). In this case, a one-to-one correspondence of the strain and strain rate values to the microstructure parameter is achieved.
4. We take an arbitrary point of the constructed approximating surface and add its coordinates to the training sample.

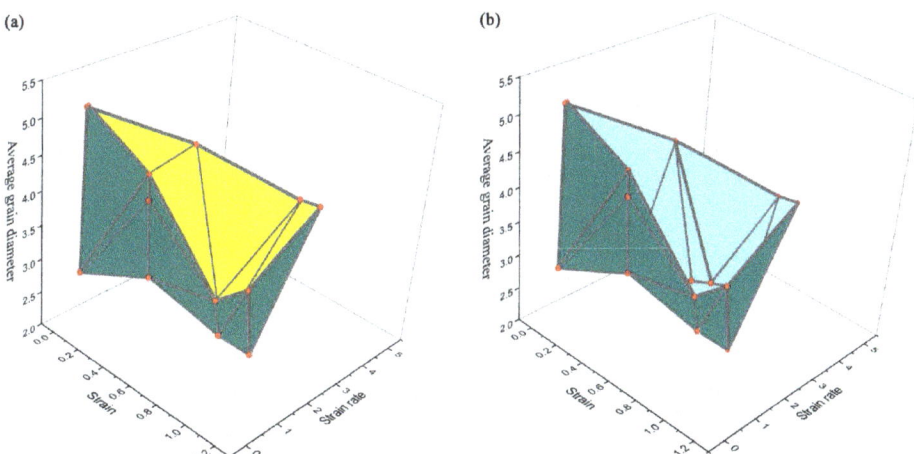

Figure 13. (a) Constructing a surface approximated by planes and (b) the finally constructed surface based on experimental points. The light color in (b) shows the planes that are hidden by the yellow planes in (a).

According to this technique, 1814 values for three test temperatures were included in the training sample. Similar procedures for constructing a training sample were used for experimental data describing the evolution of the fraction and density of low-angle boundaries.

To describe the evolution of the parameters of the composite matrix microstructure depending on the thermomechanical parameters of deformation and single-, two-, and three-layer networks, as well as various activation functions, were considered. The criterion for choosing the composition of the hidden layer and the activation function was the minimum value of the average relative deviation of the experimental data from the calculated ones. Based on this, a single-layer network with 100 neurons in the hidden layer was selected for the dependence of the average grain diameter D and the fraction of low-angle boundaries P_L on the thermomechanical parameters; a two-layer neural network with 75 neurons in the first hidden layer and 11 neurons in the second hidden layer shows the best results for the dependence of the density of the low-angle boundaries S_L on the thermomechanical parameters (Figure 4). A logistic activation function was chosen for all the three neural networks as neural networks using it show the best results in approximating the microstructure parameters depending on the thermomechanical parameters of deformation. The average relative deviation of the experimental data from the calculated ones for the selected neural networks during their training was 5.4, 4.8, and 4.6% for the average grain diameter D, and the fraction P_L and density S_L of low-angle boundaries, respectively.

In order to test the ability of a neural network to correctly predict these microstructure parameters depending on the thermomechanical conditions, neural networks were verified under deformation conditions different from the conditions of training. These microstructure parameters and the corresponding thermomechanical parameters are summarized in Table 4. The average relative deviation of the experimental data from the predicted ones was 4.2, 5.8, and 5.6% for the average grain diameter D, and the fraction P_L and density S_L of low-angle boundaries, respectively. The obtained error values are acceptable for predicting the evolution of the microstructure parameters, and this enables the constructed neural networks to be used to study the mechanisms of softening in the AlMg6/10% SiC metal matrix composite. The constructed neural networks were used to obtain maps of the microstructure parameters, which describe the behavior of the average grain diameter D, and the fraction P_L and density S_L of low-angle boundaries as dependent on temperature, strain, and strain rate. These maps are shown in Figures 14–16.

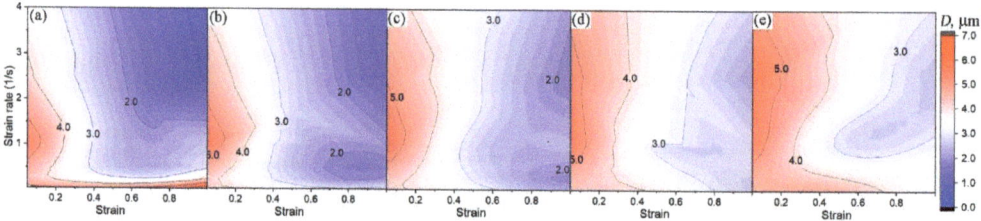

Figure 14. The effect of strain and strain rate on the average grain diameter D depending on temperature, °C: (**a**) 300, (**b**) 350, (**c**) 400, (**d**) 450, and (**e**) 500. The dependences have been obtained from neural network predictions.

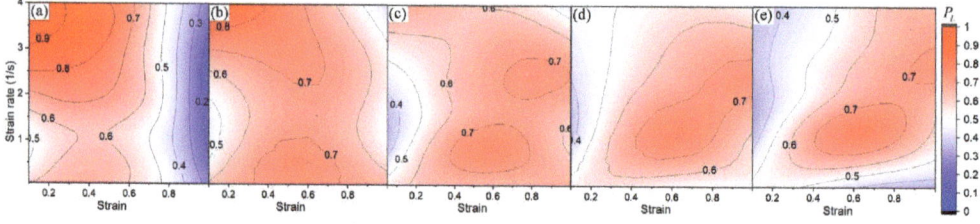

Figure 15. The effect of strain and strain rate on the fraction of low-angle boundaries P_L depending on temperature, °C: (**a**) 300, (**b**) 350, (**c**) 400, (**d**) 450, and (**e**) 500. The dependences have been obtained from neural network predictions.

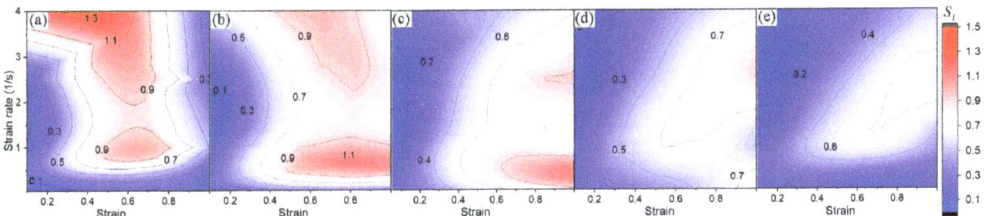

Figure 16. The effect of strain and strain rate on the density of low-angle boundaries S_L depending on temperature, °C: (**a**) 300, (**b**) 350, (**c**) 400, (**d**) 450, and (**e**) 500. The dependences have been obtained from neural network predictions.

3.3. Analysis of Microstructure Formation Based on Neural Network Data

Figure 17 shows the dependences $D - \varepsilon$, $P_L - \varepsilon$, and $S_L - \varepsilon$ at a strain rate of 0.5, 2, and 4 s^{-1} for temperatures between 300 and 500 °C. These dependences are based on the maps shown in Figures 14–16. The analysis of these dependences and maps of the microstructure parameters (Figures 14–16) yields the following generalizing conclusions.

1. The strain dependences of the fraction of low-angle boundaries ($P_L - \varepsilon$) for the entire studied temperature and strain rate range of deformation have a peak (shown by asterisks in Figure 17).
2. For the temperature range between 300 and 350 °C, the strain corresponding to the peak on the dependences decreases with an increase in strain rate. Conversely, for the temperature range between 450 and 500 °C, the strain corresponding to the peak increases with strain rate.
3. For the temperature range between 300 and 500 °C and strain rates above 3 s^{-1}, the strain corresponding to the peak value of the dependence $P_L - \varepsilon$ increases with temperature (Figures 15 and 17).
4. At low strain rates ($\dot{\varepsilon} < 1 s^{-1}$), for the temperature range between 300 and 450 °C, the dependence $D - \varepsilon$ consists of two characteristic portions. In the first portion of the curve $D - \varepsilon$, the average grain diameter decreases, and it remains unchanged with increasing strain in the second portion (steady-state portion). Moreover, as the deformation temperature rises, the strain corresponding to the boundary between these two portions (ε') shifts towards higher strains.
5. For the temperature range between 350 and 500 °C, at strain rates exceeding 1 s^{-1}, the value of the average grain diameter decreases monotonically.
6. An increase in the deformation temperature leads to the formation of a coarser-grained microstructure.
7. The density of low-angle boundaries S_L at the initial stage of deformation increases with strain for the entire range of temperatures and strain rates.
8. At a temperature of 300 °C, after reaching a certain strain value (indicated by squares in Figure 17), the density of low-angle boundaries decreases with further deformation, as is the case with 350 °C, but at strain rates above 3 s^{-1}. For all the other deformation temperatures, the density of low-angle boundaries increases with strain or remains almost unchanged.
9. At a temperature of 300 °C, the strain at which the density of low-angle boundaries S_L decreases with increasing strain shifts towards lower strains with increasing strain rate.

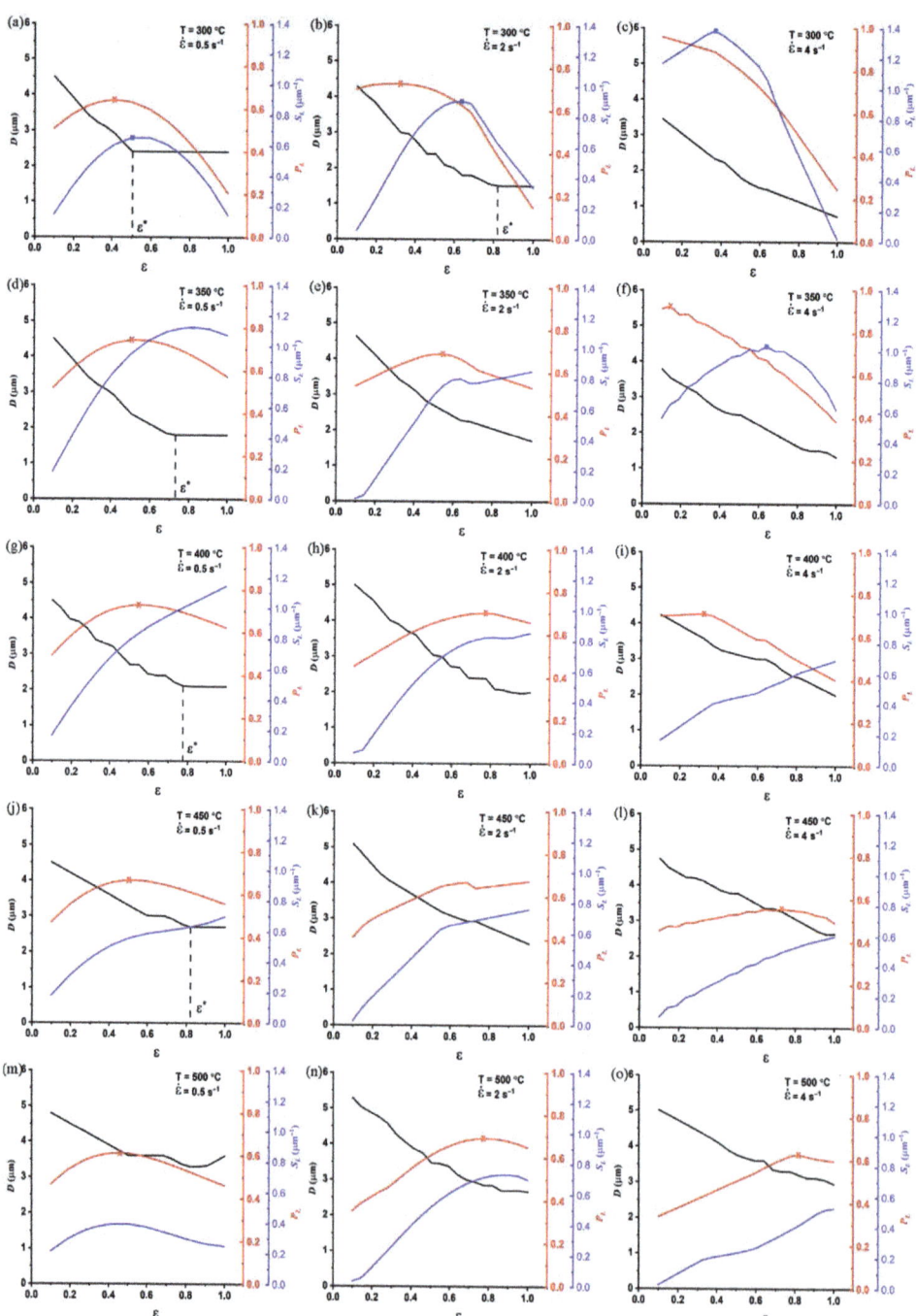

Figure 17. The average grain diameter D, and the fraction P_L and density S_L of low-angle boundaries as dependent on temperature T, strain ε, and strain rate $\dot{\varepsilon}$. The dependences are given for temperature, °C: (**a–c**) 300; (**d–f**) 350; (**g–i**) 400; (**j–l**) 450 and (**m–o**) 500.

The temperature, strain, and strain rate dependences of the fraction of low-angle boundaries P_L for continuous dynamic recrystallization characterizes the influence of these thermomechanical parameters on the formation of new low-angle boundaries and their transformation into high-angle ones. At strains below the strain corresponding to the peak of the dependence $P_L - \varepsilon$, the rate of the formation of new low-angle boundaries prevails over the rate of the transformation of low-angle boundaries into high-angle ones. When the peak is passed, in the region of higher strains, the rate of the transformation of low-angle boundaries into high-angle ones prevails over the rate of the formation of low-angle boundaries. As can be seen from Figure 17, the dependence $P_L - \varepsilon$ of the composite matrix for the temperature and strain rate range under study is described by a convex upwards function. At the same time, under certain strain rate conditions, at temperatures of 300 and 350 °C, the strain dependence of the density of low-angle boundaries ($S_L - \varepsilon$) also has a peak, after which the fraction of low-angle boundaries decreases (Figure 17a–f). This indicates slower polygonization. Judging by the obtained microstructure images for these temperatures, the grains are close in size to the previously formed subgrains, and this indicates the transformation of subgrains into grains. At the same time, at a deformation temperature of 300 °C, the diameter of many subgrains at strains above 0.6 is close to the grain thickness (Figure 5). Together with the data on the evolution of the density and fraction of low-angle boundaries, this testifies that geometric recrystallization occurs in the composite matrix [43,67,68]. The composite matrix has a similar strain dependence of the fraction and density of low-angle boundaries at a temperature of 350 °C and strain rates above 3 s^{-1}. This gives grounds to assert that geometric recrystallization occurs in the composite matrix simultaneously with dynamic continuous recrystallization at a temperature of 350 °C, strains above 0.6 (Figures 15–17), and strain rates above 3 s^{-1}. Apparently, geometric recrystallization is not implemented at temperatures between 400 and 450 °C, strain rates between 0.1 and 4 s^{-1}, and strains below 1.

4. Conclusions

The rheological behavior and softening mechanisms in an AMg6/10% SiC metal matrix composite at temperatures from 300 to 500 °C and strain rates ranging between 0.1 and 4 s^{-1} have been studied. It has been found from mechanical tests that, for the studied range of thermomechanical action, the flow stress curve has a peak value shifted towards higher strains with decreasing temperature and increasing strain rate in the entire range of temperatures and strain rates. This rheological behavior of the composite is due to the interaction of hardening and softening processes. In order to study the mechanisms of softening occurring during deformation, neural networks were built, which required the development of a new technique for processing experimental data, which allows one to form a training sample necessary for a correct description of the initial experimental data by the neural networks. The constructed neural networks have made it possible to describe the evolution of the average grain diameter, and the fraction and density of low-angle boundaries depending on strain, strain rate, and deformation temperature with acceptable accuracy. The use of neural networks together with EBSD images of the formed microstructure during deformation has allowed us to identify the occurrence of the following relaxation mechanisms depending on the thermomechanical conditions of deformation:

1. At a temperature of 300 °C and strain rates ranging between 0.1 and 4 s^{-1}, the composite matrix softens by dynamic recovery and continuous recrystallization, and when a certain value of strain is reached, geometric recrystallization occurs in some grains. At the same time, the strain at which geometric recrystallization starts to intensify in the grains shifts towards lower strains with an increasing strain rate.
2. At a temperature of 350 °C, geometric recrystallization, together with continuous recrystallization, occurs in the composite matrix at strain rates above 3 s^{-1}.
3. At temperatures from 400 to 500 °C and strain rates ranging between 0.1 and 4 s^{-1}, the main softening processes are dynamic recovery and continuous dynamic recrystallization.

Author Contributions: Conceptualization, A.S.; methodology, A.S. and V.K.; validation, A.S. and V.K.; formal analysis, A.S. and V.K.; investigation, A.S. and V.K.; writing—original draft preparation, A.S. and V.K.; writing—review and editing, A.S. and V.K.; project administration, A.S. and A.K.; funding acquisition, A.S. All authors have read and agreed to the published version of the manuscript.

Funding: This research was funded by the RSF (project No. 22-29-00428) in the part of constructing models of metal matrix composites.

Institutional Review Board Statement: Not applicable.

Informed Consent Statement: Not applicable.

Data Availability Statement: The data presented in this study are available on request from the corresponding author.

Acknowledgments: The Plastometriya shared the research facilities of the IES UB RAS, and the shared research facilities of the B. N. Yeltsin Ural Federal University were used. The work on studying the rheological properties of the composites was performed within the research conducted by the Institute of Engineering Science, Ural Branch of the Russian Academy of Sciences, project No. AAAA-A18-118020790140-5.

Conflicts of Interest: The authors declare that they have no conflict of interest. The funders had no role in the design of the study; in the collection, analyses, or interpretation of data; in the writing of the manuscript, or in the decision to publish the results.

References

1. Gopinath, K.; Balasubramaniam, R.; Murthy, V.S.R. Corrosion behavior of cast Al-Al2O3 particulate composites. *J. Mater. Sci. Lett.* **2001**, *20*, 793–794. [CrossRef]
2. Sielski, R. Research needs in aluminum structure. *Ships Offshore Struct.* **2008**, *3*, 57–65. [CrossRef]
3. Davis, J.R. *Aluminum and Aluminum Alloys*; ASM International: Almere, The Netherlands, 1993.
4. Umar, M.; Mohammed Asif, M.; Sathiya, P. Creep and Corrosion Characteristics of Laser Welded AA5083 Al–Mg alloy. *Lasers Manuf. Mater. Process.* **2022**, *9*, 257–276. [CrossRef]
5. Korobov, A.I.; Kokshaiskii, A.I.; Prokhorov, V.M.; Evdokimov, I.A.; Perfilov, S.A.; Volkov, A.D. Mechanical and nonlinear elastic characteristics of polycrystalline AMg6 aluminum alloy and n-AMg6/C60 nanocomposite. *Phys. Solid State* **2016**, *58*, 2472–2480. [CrossRef]
6. Pugacheva, N.B.; Vichuzhanin, D.I.; Kalashnikov, S.T.; Ivanov, A.V.; Smirnov, S.V.; Frolova, N.Y. Studying recovery processes in a strain-hardened Al–Mg–Mn–Fe–Si alloy. *Phys. Met. Metallogr.* **2016**, *117*, 920–926. [CrossRef]
7. Kolachev, B.A.; Elagin, V.I.; Livanov, V.A. *Metal Science and Heat Treatment of Non-Ferrous Metals and Alloys*; MISIS: Moscow, Russia, 2005. (In Russian)
8. Davydov, V.; Rostova, T.; Zakharov, V.; Filatov, Y.; Yelagin, V. Scientific principles of making an alloying addition of scandium to aluminium alloys. *Mater. Sci. Eng. A* **2000**, *280*, 30–36. [CrossRef]
9. Singh, V.; Prasad, K.S.; Gokhale, A.A. Microstructure and age hardening response of cast Al-Mg-Sc-Zr alloys. *J. Mater. Sci.* **2004**, *39*, 2861–2864. [CrossRef]
10. Chhangani, S.; Masa, S.K.; Mathew, R.T.; Prasad, M.J.N.V.; Sujata, M. Microstructural evolution in Al–Mg–Sc alloy (AA5024): Effect of thermal treatment, compression deformation and friction stir welding. *Mater. Sci. Eng. A* **2020**, *772*, 138790. [CrossRef]
11. Smirnov, A.S.; Konovalov, A.V.; Pushin, V.G.; Uksusnikov, A.N.; Zvonkov, A.A.; Zajcev, I.M. Peculiarities of the Rheological Behavior for the Al-Mg-Sc-Zr Alloy Under High-Temperature Deformation. *J. Mater. Eng. Perform.* **2014**, *23*, 4271–4277. [CrossRef]
12. Shvechkov, E.I.; Filatov, Y.A.; Zakharov, V.V. Mechanical and Life Properties of Sheets from Alloys of the Al-Mg-Sc System. *Met. Sci. Heat Treat.* **2017**, *59*, 454–462. [CrossRef]
13. Huang, H.; Jiang, F.; Zhou, J.; Wei, L.; Qu, J.; Liu, L. Effects of Al3(Sc,Zr) and Shear Band Formation on the Tensile Properties and Fracture Behavior of Al-Mg-Sc-Zr Alloy. *J. Mater. Eng. Perform.* **2015**, *24*, 4244–4252. [CrossRef]
14. Baranov, V.; Sidelnikov, S.; Zenkin, E.; Frolov, V.; Voroshilov, D.; Yakivyuk, O.; Konstantinov, I.; Sokolov, R.; Belokonova, I. Study of strength properties of semi-finished products from economically alloyed high-strength aluminium-scandium alloys for application in automobile transport and shipbuilding. *Open Eng.* **2018**, *8*, 69–76. [CrossRef]
15. Hu, Y.; Ou-yang, Q.-B.; Yao, L.; Chen, S.; Zhang, L.-T. A study of interparticulate strain in a hot-extruded SiCp/2014 Al composite. *Int. J. Miner. Metall. Mater.* **2019**, *26*, 523–529. [CrossRef]
16. Chawla, N.; Chawla, K.K. *Metal Matrix Composites*; Springer New York: New York, NY, USA, 2013; ISBN 978-1-4614-9547-5.
17. Yolshina, L.A.; Muradymov, R.V.; Korsun, I.V.; Yakovlev, G.A.; Smirnov, S.V. Novel aluminum-graphene and aluminum-graphite metallic composite materials: Synthesis and properties. *J. Alloys Compd.* **2016**, *663*, 449–459. [CrossRef]
18. Yolshina, L.A.; Kvashnichev, A.G.; Vichuzhanin, D.I.; Smirnova, E.O. Mechanical and Thermal Properties of Aluminum Matrix Composites Reinforced by In Situ Al2O3 Nanoparticles Fabricated via Direct Chemical Reaction in Molten Salts. *Appl. Sci.* **2022**, *12*, 8907. [CrossRef]

19. Gao, M.; Chen, Z.; Li, L.; Guo, E.; Kang, H.; Xu, Y.; Wang, T. Microstructure and enhanced mechanical properties of hybrid-sized B4C particle-reinforced 6061Al matrix composites. *Mater. Sci. Eng. A* **2021**, *802*, 140453. [CrossRef]
20. Gao, M.; Kang, H.; Chen, Z.; Guo, E.; Peng, P.; Wang, T. Effect of reinforcement content and aging treatment on microstructure and mechanical behavior of B4Cp/6061Al composites. *Mater. Sci. Eng. A* **2019**, *744*, 682–690. [CrossRef]
21. Rofman, O.V.; Mikhaylovskaya, A.V.; Kotov, A.D.; Prosviryakov, A.S.; Portnoy, V.K. Effect of thermomechanical treatment on properties of an extruded Al-3.0Cu-1.2Mg/SiCp composite. *Mater. Sci. Eng. A* **2019**, *739*, 235–243. [CrossRef]
22. Liu, Y.; Wang, Q.; Lu, C.; Xue, T.; Hu, B.; Zhang, C. Microscopic residual stress evolution at the SiC/Al interface during nanoindentation via molecular dynamics simulation. *Surf. Interfaces* **2022**, *33*, 102210. [CrossRef]
23. Rollett, A.; Humphreys, F.; Rohrer, G.S.; Hatherly, M. *Recrystallization and Related Annealing Phenomena*; Elsevier Ltd.: Amsterdam, The Netherlands, 2004.
24. Zhang, W.; Li, R.; Yang, Q.; Fu, Y.; Kong, X. Impact Resistance of a Fiber Metal Laminate Skin Bio-Inspired Composite Sandwich Panel with a Rubber and Foam Dual Core. *Materials* **2023**, *16*, 453. [CrossRef] [PubMed]
25. Chaudhuri, A.; Sarkar, A.; Kapoor, R.; Chakravartty, J.K.; Ray, R.K.; Suwas, S. Understanding the Mechanism of Dynamic Recrystallization During High-Temperature Deformation in Nb-1Zr-0.1C Alloy. *J. Mater. Eng. Perform.* **2019**, *28*, 448–462. [CrossRef]
26. Tikhonova, M.; Kaibyshev, R.; Belyakov, A. Microstructure and Mechanical Properties of Austenitic Stainless Steels after Dynamic and Post-Dynamic Recrystallization Treatment. *Adv. Eng. Mater.* **2018**, *20*, 1700960. [CrossRef]
27. Huang, H.; Jiang, F.; Zhou, J.; Wei, L.; Zhong, M.; Liu, X. Hot deformation behavior and microstructural evolution of as-homogenized Al-6Mg-0.4Mn-0.25Sc-0.1Zr alloy during compression at elevated temperature. *J. Alloys Compd.* **2015**, *644*, 862–872. [CrossRef]
28. Chen, B.; Tian, X.L.; Li, X.L.; Lu, C. Hot Deformation Behavior and Processing Maps of 2099 Al-Li Alloy. *J. Mater. Eng. Perform.* **2014**, *23*, 1929–1935. [CrossRef]
29. Gariboldi, E.; Lo Conte, A. Damage mechanisms at room and high temperature in notched specimens of Al6061/Al2O3 particulate composites. *Compos. Sci. Technol.* **2008**, *68*, 260–267. [CrossRef]
30. Luo, X.; Zhao, K.; He, X.; Bai, Y.; De Andrade, V.; Zaiser, M.; An, L.; Liu, J. Evading strength and ductility trade-off in an inverse nacre structured magnesium matrix nanocomposite. *Acta Mater.* **2022**, *228*, 117730. [CrossRef]
31. Smirnov, S.V. Accumulation and healing of damage during plastic metal forming: Simulation and experiment. *Key Eng. Mater.* **2013**, *528*, 61–69. [CrossRef]
32. Smirnov, A.; Smirnova, E.; Konovalov, A.; Kanakin, V. Using the Instrumented Indentation Technique to Determine Damage in Sintered Metal Matrix Composites after High-Temperature Deformation. *Appl. Sci.* **2021**, *11*, 10590. [CrossRef]
33. Li, J.; Wu, X.; Cao, L.; Liao, B.; Wang, Y.; Liu, Q. Hot deformation and dynamic recrystallization in Al-Mg-Si alloy. *Mater. Charact.* **2021**, *173*, 110976. [CrossRef]
34. Lin, Y.C.; Chen, X.-M. A critical review of experimental results and constitutive descriptions for metals and alloys in hot working. *Mater. Des.* **2011**, *32*, 1733–1759. [CrossRef]
35. Gourdet, S.; Montheillet, F. An experimental study of the recrystallization mechanism during hot deformation of aluminium. *Mater. Sci. Eng. A* **2000**, *283*, 274–288. [CrossRef]
36. Wang, Y.; Yang, B.; Gao, M.; Guan, R. Deformation behavior and dynamic recrystallization during hot compression in homogenized Al–6Mg–0.8Mn alloys. *Mater. Sci. Eng. A* **2022**, *840*, 142576. [CrossRef]
37. Shun, T.; Wan, C.M.; Byrne, J.G. A study of work hardening in austenitic FeMnC and FeMnAlC alloys. *Acta Metall. Mater.* **1992**, *40*, 3407–3412. [CrossRef]
38. Tsuzaki, K.; Matsuzaki, Y.; Maki, T.; Tamura, I. Fatigue deformation accompanying dynamic strain aging in a pearlitic eutectoid steel. *Mater. Sci. Eng. A* **1991**, *142*, 63–70. [CrossRef]
39. Smirnov, A.S.; Konovalov, A.V.; Belozerov, G.A.; Shveikin, V.P.; Smirnova, E.O. Peculiarities of the rheological behavior and structure formation of aluminum under deformation at near-solidus temperatures. *Int. J. Miner. Metall. Mater.* **2016**, *23*, 563–571. [CrossRef]
40. Razali, M.K.; Wan, K.S.; Irani, M.; Kim, M.C.; Joun, M.S. Practical quantification of the effects of flow stress, friction, microstructural properties, and the tribological environment on macro- and micro-structure formation during hot forging. *Tribol. Int.* **2021**, *164*, 107226. [CrossRef]
41. Kodzhaspirov, G.E.; Terentyev, M.I. Modeling the dynamically recrystallized grain size evolution of a superalloy. *Mater. Phys. Mech.* **2012**, *13*, 70–76.
42. Cross, A.J.; Ellis, S.; Prior, D.J. A phenomenological numerical approach for investigating grain size evolution in ductiley deforming rocks. *J. Struct. Geol.* **2015**, *76*, 22–34. [CrossRef]
43. Huang, K.; Logé, R.E. A review of dynamic recrystallization phenomena in metallic materials. *Mater. Des.* **2016**, *111*, 548–574. [CrossRef]
44. Sun, Z.C.; Wu, H.L.; Cao, J.; Yin, Z.K. Modeling of continuous dynamic recrystallization of Al-Zn-Cu-Mg alloy during hot deformation based on the internal-state-variable (ISV) method. *Int. J. Plast.* **2018**, *106*, 73–87. [CrossRef]
45. Trusov, P.; Kondratev, N.; Podsedertsev, A. Description of Dynamic Recrystallization by Means of an Advanced Statistical Multilevel Model: Grain Structure Evolution Analysis. *Crystals* **2022**, *12*, 653. [CrossRef]

46. Smirnov, A.S.; Konovalov, A.V.; Muizemnek, O.Y. Modelling and simulation of strain resistance of alloys taking into account barrier effects. *Diagn. Resour. Mech. Mater. Struct.* **2015**, *1*, 61–72. [CrossRef]
47. Su, Z.-X.; Sun, C.-Y.; Fu, M.-W.; Qian, L.-Y. Physical-based constitutive model considering the microstructure evolution during hot working of AZ80 magnesium alloy. *Adv. Manuf.* **2019**, *7*, 30–41. [CrossRef]
48. Karhausen, K.F.; Roters, F. Development and application of constitutive equations for the multiple-stand hot rolling of Al-alloys. *J. Mater. Process. Technol.* **2002**, *123*, 155–166. [CrossRef]
49. Konovalov, A.V.; Smirnov, A.S. Viscoplastic model for the strain resistance of 08Kh18N10T steel at a hot-deformation temperature. *Russ. Metall.* **2008**, *2008*, 138–141. [CrossRef]
50. Wang, Y.; Peng, J.; Zhong, L.; Pan, F. Modeling and application of constitutive model considering the compensation of strain during hot deformation. *J. Alloys Compd.* **2016**, *681*, 455–470. [CrossRef]
51. Svyetlichnyy, D.S.; Muszka, K.; Majta, J. Three-dimensional frontal cellular automata modeling of the grain refinement during severe plastic deformation of microalloyed steel. *Comput. Mater. Sci.* **2015**, *102*, 159–166. [CrossRef]
52. Chen, F.; Zhu, H.; Chen, W.; Ou, H.; Cui, Z. Multiscale modeling of discontinuous dynamic recrystallization during hot working by coupling multilevel cellular automaton and finite element method. *Int. J. Plast.* **2021**, *145*, 103064. [CrossRef]
53. Chen, F.; Zhu, H.; Zhang, H.; Cui, Z. Mesoscale Modeling of Dynamic Recrystallization: Multilevel Cellular Automaton Simulation Framework. *Metall. Mater. Trans. A Phys. Metall. Mater. Sci.* **2020**, *51*, 1286–1303. [CrossRef]
54. Zhu, H.J.; Chen, F.; Zhang, H.M.; Cui, Z.S. Review on modeling and simulation of microstructure evolution during dynamic recrystallization using cellular automaton method. *Sci. China Technol. Sci.* **2020**, *63*, 357–396. [CrossRef]
55. Barrett, C.D.; Imandoust, A.; Oppedal, A.L.; Inal, K.; Tschopp, M.A.; El Kadiri, H. Effect of grain boundaries on texture formation during dynamic recrystallization of magnesium alloys. *Acta Mater.* **2017**, *128*, 270–283. [CrossRef]
56. Sun, X.; Li, H.; Zhan, M.; Zhou, J.; Zhang, J.; Gao, J. Cross-scale prediction from RVE to component. *Int. J. Plast.* **2021**, *140*, 102973. [CrossRef]
57. Babu, K.A.; Prithiv, T.S.; Gupta, A.; Mandal, S. Modeling and simulation of dynamic recrystallization in super austenitic stainless steel employing combined cellular automaton, artificial neural network and finite element method. *Comput. Mater. Sci.* **2021**, *195*, 110482. [CrossRef]
58. Jin, Y.; Zhao, J.; Zhang, C.; Luo, J.; Wang, S. Research on Neural Network Prediction of Multidirectional Forging Microstructure Evolution of GH4169 Superalloy. *J. Mater. Eng. Perform.* **2021**, *30*, 2708–2719. [CrossRef]
59. Fratini, L.; Buffa, G.; Palmeri, D. Using a neural network for predicting the average grain size in friction stir welding processes. *Comput. Struct.* **2009**, *87*, 1166–1174. [CrossRef]
60. Smirnov, A.S.; Shveikin, V.P.; Smirnova, E.O.; Belozerov, G.A.; Konovalov, A.V.; Vichuzhanin, D.I.; Muizemnek, O.Y. Effect of silicon carbide particles on the mechanical and plastic properties of the AlMg6/10% SiC metal matrix composite. *J. Compos. Mater.* **2018**, *52*, 3351–3363. [CrossRef]
61. Konovalov, A.V.; Smirnov, A.S. Identification of the metal flow stress model based on the results of compression tests of specimens. *Zavod. Lab. Diagn. Mater.* **2010**, *76*, 53–56. (In Russians)
62. Gourdet, S.; Montheillet, F. A model of continuous dynamic recrystallization. *Acta Mater.* **2003**, *51*, 2685–2699. [CrossRef]
63. Maizza, G.; Pero, R.; Richetta, M.; Montanari, R. Continuous dynamic recrystallization (CDRX) model for aluminum alloys. *J. Mater. Sci.* **2018**, *53*, 4563–4573. [CrossRef]
64. Sakai, T.; Belyakov, A.; Kaibyshev, R.; Miura, H.; Jonas, J.J. Dynamic and post-dynamic recrystallization under hot, cold and severe plastic deformation conditions. *Prog. Mater. Sci.* **2014**, *60*, 130–207. [CrossRef]
65. Gopi, E.S. *Pattern Recognition and Computational Intelligence Techniques Using Matlab*; Springer Nature Switzerland AG: Basel, Switzerland, 2020; ISBN 978-3-030-22273-4.
66. Kalinin, S.V.; Sumpter, B.G.; Archibald, R.K. Big-deep-smart data in imaging for guiding materials design. *Nat. Mater.* **2015**, *14*, 973–980. [CrossRef] [PubMed]
67. Gholinia, A.; Humphreys, F.; Prangnell, P. Production of ultra-fine grain microstructures in Al-Mg alloys by coventional rolling. *Acta Mater.* **2002**, *50*, 4461–4476. [CrossRef]
68. Nagira, T.; Liu, X.; Ushioda, K.; Fujii, H. Microstructural evolutions of 2n grade pure al and 4n grade high-purity al during friction stir welding. *Materials* **2021**, *14*, 3606. [CrossRef] [PubMed]

Disclaimer/Publisher's Note: The statements, opinions and data contained in all publications are solely those of the individual author(s) and contributor(s) and not of MDPI and/or the editor(s). MDPI and/or the editor(s) disclaim responsibility for any injury to people or property resulting from any ideas, methods, instructions or products referred to in the content.

MDPI
St. Alban-Anlage 66
4052 Basel
Switzerland
www.mdpi.com

Applied Sciences Editorial Office
E-mail: applsci@mdpi.com
www.mdpi.com/journal/applsci

Disclaimer/Publisher's Note: The statements, opinions and data contained in all publications are solely those of the individual author(s) and contributor(s) and not of MDPI and/or the editor(s). MDPI and/or the editor(s) disclaim responsibility for any injury to people or property resulting from any ideas, methods, instructions or products referred to in the content.

www.ingramcontent.com/pod-product-compliance
Lightning Source LLC
LaVergne TN
LVHW070727100526
838202LV00013B/1185